T0318735

RISK MANAGEMENT OF COMPLEX INORGANIC MATERIALS

RISK MANAGEMENT OF COMPLEX INORGANIC MATERIALS

A Practical Guide

Edited by

VIOLAINE VEROUGSTRAETE

ACADEMIC PRESS
An imprint of Elsevier

Academic Press is an imprint of Elsevier
125 London Wall, London EC2Y 5AS, United Kingdom
525 B Street, Suite 1800, San Diego, CA 92101-4495, United States
50 Hampshire Street, 5th Floor, Cambridge, MA 02139, United States
The Boulevard, Langford Lane, Kidlington, Oxford OX5 1GB, United Kingdom

Library of Congress Cataloging-in-Publication Data
A catalog record for this book is available from the Library of Congress

British Library Cataloguing-in-Publication Data
A catalogue record for this book is available from the British Library

ISBN 978-0-12-811063-8

For information on all Academic Press publications visit our
website at https://www.elsevier.com/books-and-journals

Working together
to grow libraries in
developing countries

www.elsevier.com • www.bookaid.org

Publisher: Mica Haley
Acquisition Editor: Rob Sykes
Editorial Project Manager: Tracy Tufaga
Production Project Manager: Priya Kumaraguruparan
Cover Designer: Miles Hitchen

Typeset by SPi Global, India

Contents

13 Risk Assessment of Alloys

EIRIK NORDHEIM, TONY NEWSON

14 Emerging Tools in the Assessment of Metals: Current Applicability

KOEN OORTS, JELLE MERTENS, VINCENT DUNON,
PATRICK VAN SPRANG, FREDERIK VERDONCK

Contributors

William Adams Red Cap Consulting, Lake Point, Utah, United States

José Jaime Arbildua Universidad Adolfo Ibanez, Santiago, Chile

Rüdiger Vincent Battersby EBRC Consulting GmbH, Hannover, Germany

Tina Bodenschatz EBRC Consulting GmbH, Hannover, Germany

Arne Burzlaff EBRC Consulting GmbH, Hannover, Germany

Richard Carbonaro Mutch Associates, Ramsey, NJ, United States

Ruth Danzeisen Cobalt Institute, Umkirch, Germany

Benjamin Davies International Council on Mining & Metals, London, United Kingdom

Karel De Schamphelaere Ghent University, Gent, Belgium

Katrien Delbeke European Copper Institute, Brussels, Belgium

Johannes A. Drielsma European Association of Mining Industries, Metal Ores and Industrial Minerals, Brussels, Belgium

Vincent Dunon ARCHE Consulting, Ghent, Belgium

Kevin Farley Manhattan College, Riverdale, NY, United States

Kate Heim NiPERA Inc., Durham, NC, United States

Rayetta Henderson ToxStrategies, Wilmington, NC, United States

Federica Iaccino ARCHE Consulting, Leuven, Belgium

Katia Lacasse European Copper Institute, Brussels, Belgium

Dominique Lison Université catholique de Louvain, Brussels, Belgium

Carol Mackie Copper Compound Consortium, Loanhead, United Kingdom

Graham Merrington WCA Environment Ltd, Faringdon, United Kingdom

Jelle Mertens European Precious Metals Federation, Brussels, Belgium

Carina Müller EBRC Consulting GmbH, Hannover, Germany

Tony Newson Consultant, Rotherham, United Kingdom

Eirik Nordheim Aluminium REACH Consortium, Brussels, Belgium

Charlotte Nys Ghent University, Gent, Belgium

Adriana Oller NiPERA Inc., Durham, NC, United States

Koen Oorts ARCHE Consulting, Leuven; ARCHE Consulting, Ghent, Belgium

Adam Peters WCA Environment Ltd, Faringdon, United Kingdom

Kevin Rader Mutch Associates, Ramsey, NJ, United States

Patricio H. Rodriguez Universidad Adolfo Ibanez, Santiago, Chile

Jutta Schade EBRC Consulting GmbH, Hannover, Germany

Erik Smolders KU Leuven, Leuven, Belgium

Frank Van Assche International Zinc Association, Brussels, Belgium

Patrick Van Sprang ARCHE Consulting, Ghent, Belgium

Marnix Vangheluwe ARCHE Consulting, Ghent, Belgium

Isabelle Vercaigne ARCHE Consulting, Ghent, Belgium

Frederik Verdonck ARCHE Consulting, Ghent, Belgium

Violaine Verougstraete Eurometaux, Brussels, Belgium

Daniel Vetter ARCHE Consulting, Leuven, Belgium; EBRC Consulting GmbH, Hannover, Germany

Hugo   Waeterschoot Eurometaux,   Brussels, Belgium

About the Editor

Violaine Verougstraete studied medicine and toxicology at the catholic University of Louvain, did a DEA in Public Health and obtained her PhD in Public Health in 2005 from the catholic University of Louvain (Belgium).

For 8 years, she worked as a researcher at the Industrial Toxicology and Occupational Medicine Unit of the Catholic University of Louvain and collaborated in the (European Commission) EU Risk Assessment "Cadmium and Cadmium Oxide."

Between May 2005 and December 2011, she worked for Eurometaux as Health and Alloys Manager. Her main task consisted of coordinating Eurometaux's scientific activities and projects, for example, the HERAG and MERAG projects on risk assessment methodologies for metals, the GHS Joint Project, and human/environmental toxicology-related activities. More specifically with regard to REACH and CLP, she coordinated work on Exposure Scenarios, Exposure Modelling, and Classification Tools, as well as technical projects on metal specificities backing up the EU REACH registration and notification dossiers.

Since 1 January 2012, she has been the EHS Director at Eurometaux and coordinates all EU REACH and CLP activities.

She also attends the European Chemicals Agency Risk Assessment Committee meetings, as a sectorial observer and participates in the OECD Working Party meetings on Exposure and Hazard Assessment.

She is teaching as a visiting professor at the Universiteit Gent and at the Université catholique de Louvain.

1

Introduction

Violaine Verougstraete*, William Adams‡,
Hugo Waeterschoot*, Benjamin Davies§

*Eurometaux, Brussels, Belgium ‡Red Cap Consulting, Lake Point, Utah,
United States §International Council on Mining & Metals, London,
United Kingdom

1.1 FROM METALS…

Metals have been critical to the progress of society for thousands of years—ranging from those at the roots of our civilisation, such as copper, gold, silver, tin, lead, and iron, to those extracted later during industrial developments, such as aluminium, zinc, nickel, tungsten, platinum, and, more recently, rare earths. It is widely recognised that metal released into the environment must be managed. Proper management requires an in-depth understanding of the composition of metal substances and their constituents, which are used in commerce. Commercial products often contain multiple and varied elements and minerals. This book is designed to provide insight into the production, composition, and management of these complex materials.

Metals are used in almost every aspect of modern society, from transport to energy, housing to healthcare, food to technology. Innovation continues apace to develop ever more useful metal-containing materials for wider and more specialised applications. As society changes, its product and infrastructure requirements also alter, along with the demand for a particular metal depending on its use in new, or future products. There is, for example, an accelerating demand for metals and elements used in new energy technologies (e.g., solar cells, batteries, wind generation, etc.) and the production of complex electronic devices and networks enabling sustainability (UNEP, 2013).

Despite sufficient reserves of industrially important metals to supply these anticipated needs, there is an environmental and social imperative to increase the efficiency with which materials are being produced and used. A growing global population consuming metals at the rate seen in western industrialised society is likely to be unsustainable. Society increasingly expects all actors to behave sustainably, finding a fair balance between ensuring appropriate levels of supply and limiting the environmental footprint of production and consumption systems.

1

In addition to efficient production and intelligent use of metals, recycling plays an important role: using secondary resources, locked up in the so-called 'urban mines', ensures that the societal value of extracted metals is optimised and maintained for as long as possible metals if treated responsibly, have an enduring quality: they do not lose their properties through recycling and can be reused again and again as pure metals and alloys in a huge variety of products. Taking into account that most modern technologies also rely on an increasingly wide palette of minor metals that have complex chemistry and often very complex supply chains it is essential to use them wisely and reuse them as much as possible. Indeed, the need for surety of supply of these metals and the complexity of ensuring it has led to them being referred to as 'critical elements'—often attracting special policy considerations.

The abundant and wide variety of uses of metals, as well as their long lifecycle, and sometimes 'dispersive' application means that human beings and the environment come into contact with metals or metal-containing materials on a daily basis.

A growing global awareness of the potential harm to human health and the environment caused by exposure to chemicals (a definition that includes metals) led the World Summit on Sustainable Development (WSSD) in 2002 to make a global political commitment to sound management of chemicals by 2020. International efforts to realise this goal resulted in the adoption of the Strategic Approach to International Chemicals Management (SAICM), by the United Nations Environment Programme in February 2006. The fourth International Conference on Chemicals Management (ICCM4) identified a series of critical 'basic elements' at national and regional levels to the attainment of the overall goal to minimise the adverse effects of chemicals on human health and the environment. Among these is the implementation of systems for the transparent sharing of relevant data and information among all relevant stakeholders using a lifecycle approach such as the implementation of the Globally Harmonised System of Classification and Labelling of Chemicals (UN GHS, 2015). The increasing focus on monitoring and assessing impacts of chemicals on health and the environment and the incorporation of sound chemicals management into corporate policies and practices emphasises the crucial importance of knowledge on chemicals risk assessment and communication for their safe use.

Metals are natural components of the earth and exist in many forms—each with its own specific (geo)chemical characteristics that define its interactions with the environment and living organisms. Their extraction and uses involve a complex and lengthy chain of steps, explained below. The process of mining and metals production in addition to natural releases alters the levels of these natural components in different ecological compartments, meaning that it is important to know what levels can be tolerated, and how the metals are mobilised or sequestered or neutralised. Likewise, potential health risks can be prevented by ensuring the appropriate operational conditions and protection of exposed populations. Ensuring this happens in all part of sound chemicals management. A large part of the management of mining and metal production is therefore the management of chemicals through their lifecycles: their sourcing, production, transport, storage, use, recycling, and—when necessary—their safe disposal, and the management of their occupational health and environmental risks.

1.1.1 …To Complex Inorganic Materials

The metals value chain is summarised in Fig. 1.1.

The very first step in the lifecycle of metals is the extraction and processing of mineral ores to create concentrates. Once the ore is extracted from the earth's crust, the metal-bearing

FIG. 1.1   Metals value chain.

minerals (heads) are separated from waste rock (tails) to form a concentrate using a process known as milling. This is often done by grinding the rock containing the ore to a fine powder (~100 μm) and adding chemicals to the ore materials, then running them through a series of separation processes. The waste from the mill, or tailings, is then pumped to a tailings storage facility. Other techniques include gravity separation, where two or more minerals are separated by movement and the force of gravity and one or more other forces (such as centrifugal, magnetic, or buoyant forces). Once the metal-bearing minerals are separated from waste materials to form a concentrate, the metal content must be removed and refined. This can be done via a number of methods, one of the most common being smelting. The process involves the chemical breakdown of the minerals through heating and melting. Many smelters are uniquely designed to suit a specific concentrate rather than a variety of concentrates, for example, copper smelters are distinctly different from aluminium smelters and are not interchangeable. The outputs of smelting are impure metal shapes that are shipped to refineries for purification. The metal products are then purified into standards set for world metal markets. While there are numerous grades of metal, the bulk of trading is done at about 99.9% purity.

The other main route of processing is hydrometallurgical—where metals are separated from the minerals by dissolving them. This usually entails a chemical process using acids or cyanide. The metals are finally removed from solution-using processes such as solvent extraction and electrowinning—resulting in high-purity metals and avoiding the smelting process.

Refined metals can also be produced from a combination of mined and recycled sources. It has been estimated that over 52% of pure base metals and alloys now come from recycled sources. Metal recyclers can take in a wide variety of valuable input materials, including end-of-life products, scrap, and industrial by-products. Some are relatively 'clean' and already contain high-purity metals, but the more complex raw material streams have variable concentrations of several different metals, and may contain impurities or other substances (e.g., plastics or other organic fractions) requiring extensive separation and purification.

These basic steps have changed little since industrial processing of metals was established centuries ago—but technological development has enabled cleaner and more efficient production at a considerable scale. While the operations described are often thought about as steps in a linear process, metals actually lend themselves well to more 'circular' models of materials use.

A 'circular economy' from mines to end-of-use recovery/recycling can help to manage metals with a minimum of residues and a strong integration of all the steps in the value cycle, making best use of their added value.

However, to make it 'happen' industry needs to possess, develop, and share a deep knowledge of metals' properties and their potential impacts, risks, and functionalities along the whole cycle.

The value chain entails not only pure metals or metal compounds but also a series of more complex materials containing these metals, either before or during their refining (e.g., ores and concentrates or scrap) or during their use (pigments, alloys). The products of the mining and metals industries are classified and regulated as chemicals and are precursors of yet other chemicals.

As mentioned above, the lifecycle of a metal includes a number of steps where it will be present in a 'mixture' with other metal and mineral constituents, embedded in an alloy matrix or the crystalline structure of a pigment, or part of an article going for recovery. In fact there are few applications where metals are used individually or in a pure state. All these metal-containing materials occurring in the lifecycle of the metal are designated, for the purpose of this book, as 'complex inorganic materials'.

Ores and concentrates are typically complex inorganic materials, reflecting the presence of naturally occurring mixtures of metal-bearing minerals in the environment and may contain significant amounts of iron and aluminium silicates, other complex minerals and trace amounts of several elements. Complex inorganic materials also appear in the metals refining and secondary production sector. The refining steps described above aim to remove the so-called non-target constituents and produce (single) enriched metal substances. During these refinement steps, enriched intermediate substances, for example, blister/anode copper, gold doré, lead bullion are produced and further processed at the same site or transported to other facilities. Because of the process conditions and for trading reasons, the enriched metal complex inorganic materials are typically characterised by well-known elemental composition. Additionally, side streams containing both non-target constituents and 'left over' target constituents are removed. Examples of side-product/non-enriched metal complexes are slags, flue dusts, and drosses. These streams are mainly residues grouped with the same main constituents that can potentially come from different furnaces/reactors but that are always 'removed'/'processed' in the same way (e.g., tapping, exhaust, leaching, etc.).

These side-product materials typically contain metal concentrations too low to recover economically. Although when the economy changes they can become a new source of metals.

Both enriched and side products are usually characterised by a high number of constituents, for example, metals and metal compounds that are subsequently removed from the source material to isolate the primary and secondary pure metals. Often there is considerable variability in composition due to the different source materials used during the beneficiation processes (e.g., primary as ore or concentrate, secondary as intermediate side-product substances coming from another metal stream). Complex inorganic materials may take on different forms depending on the origin of the source material (Fig. 1.2 and Table 1.1).

FIG. 1.2   Complex inorganic materials can have different physical forms.

TABLE 1.1   Examples of Complex Inorganic Materials

Slime and sludges	Complex combination of insoluble metal compounds produced by precipitation during electrolytic refining or wining processes
Slags	Produced in the smelting of ores by the reaction of oxygen with sulphur and carbon to produce iron silicates (an inert glass-like dark material)
Matte	A substance resulting from metallurgical processing of primary and secondary sources of base metals to lower the iron content and remove minerals impurities. In the case of Cu, results in a purity of ~60% Cu
Doré	Metallic bars/ingots, grains, or anodes and their residues (spent anodes) resulting from pyro-metallurgy processes applied on primary and secondary feeds with high precious metal content. Doré mainly contains silver and/or gold and copper, lower quantities of platinum group metals (iridium, osmium, palladium, platinum, rhodium, and ruthenium), and other non-ferrous metals in varying concentrations
Flue dust, precious metal refining	Dust generated as a by-product during smelting, refining, and/or use of precious metals and their alloys obtained from primary and secondary sources and recycled plant intermediates. Often recovered from furnace exhaust systems by filtration via cloth bags (bag houses) and cartridge filters
Sinter	Roasting fuses impurities into a brittle product called sinter, which consists of oxides of lead, zinc, iron, and silicon along with lime, metallic elements, and some remaining sulphur

It is important to note that the same complex inorganic material can originate from source materials generated in different processes or process steps. For example, flue dust is generated from all pyro-metallurgical processes producing dust and fumes (smelting, melting, incineration, calcination). Usually flue dusts collected from several pyro-metallurgical processes will be collected, captured, and treated via the same exhaust ventilation or duct, de-dusting, and scrubber system; and hence, it will be practically impossible to re-allocate each flue dust to one or the other step of the metal refining process. The same applies to a number of other complex inorganic materials regularly produced and recovered in metal refining processes and which result in metal-specific materials such as copper slimes and sludges, precious metals slimes and sludges, copper cements, precious metals cements, etc. Slimes and sludges result from a number of electrolytic and hydrometallurgical processes, while cements are formed by precipitating a variety of process residues.

The metals emerging from refineries (finished products for use) usually undergo additional steps to manufacture alloys, pigments, or other metal-bearing materials and finally reach the consumer for use. Alloys are specifically created and designed so that their chemical and physical properties are superior to those of the pure element—metal ingredients. Over 90% of global metal use is in the form of alloys. Some of the properties that can differ between alloys and their ingredients include corrosion resistance, improved wear, electrical or magnetic properties, heat resistance, malleability, etc. (Fig. 1.3).

FIG. 1.3   Copper alloy.

1.1.2 And Their Assessment...

Accurate hazard assessment of these materials in the metals supply chain is essential for ensuring that any potential environmental and health risks associated with their production, transport, and storage are properly managed.

The high volumes of complex inorganic materials used globally and their potential release into the environment has encouraged the development of regulatory systems that can prioritise their assessment and for governments to request industry to share their best practices to limit adverse impacts on exposed populations and the environment.

1.2 WHY THIS BOOK?

The aim of this book is to facilitate the risk assessment and risk management of complex inorganic materials by providing specific and accessible guidance on their assessment. Understanding the potential risks posed by metals along the value chain is a key factor in ensuing that they are produced, handled, and used safely, in keeping with the commitments made at WSSD 2002.

Much work has been done over the last 2 decades to develop robust assessment methods and sound risk management measures that consider the specific qualities of complex metal-bearing inorganic materials. This is an aspect that is often not addressed by generic regulatory guidance documents. While the interest in their management is worldwide, knowledge and tools to support the assessment of complex inorganic materials are dispersed between different interlocutors and locations (authorities in different jurisdictions, industry, academics, and test laboratories, guidance, and support tools, publications). Also, due to the long evolution and history of these industries it is often difficult to identify a single universal nomenclature for the above processes and materials (e.g., a textbook concentrate might be locally referred to as an ore by workers at a particular smelter). A secondary aim of this book is therefore to bring all these pieces of the most advanced methodologies together in a single place, to promote a consistent and coherent approach.

The audience of this book is primarily regulators involved in risk assessment and risk management, industry experts charged with compliance with chemicals management programme requirements, consultants preparing chemicals management files for companies and regulators and academics involved in research on complex inorganic materials.

Ensuring that complex inorganic materials are not only produced, but also handled and used safely is a vital part of ensuring an overall positive contribution of mining and metals towards sustainable development.

KEY MESSAGES

Ensuring that metals are not only produced, but also handled and used safely is a vital part of guaranteeing an overall positive contribution of mining and metals toward sustainable development. A metal's lifecycle includes a number of steps where it will be present in a 'mixture' with other metal and mineral constituents (e.g., ore and concentrate), embedded in an alloy matrix or in the crystalline structure of a pigment, or part of an article going for recovery. These metal containing materials are designed as 'complex inorganic materials' and their accurate assessment is essential to make sure that any potential environmental and health risks associated with their production, transport, and storage are properly managed.

The aim of this book is to facilitate the risk assessment and risk management of such complex inorganic materials in a globalised world, by providing guidance on their assessment.

Significant work has been done to develop robust assessment methods and sound risk management measures, considering the specific qualities of complex inorganic materials. This book also aims at bringing all these types of methodology together to allow a consistent and coherent approach.

References

Reuter MA, Hudson C, van Schaik A, et al.: The Working Group on the Global Metal Flows to the International Resource Panel: Metal Recycling: Opportunities, Limits, Infrastructure: A Report, UNEP, 2013.

UN GHS. United Nations. Globally Harmonized System of Classification and Labelling of Chemicals (GHS), revision 6: 2015. New York and Geneva. ST/SG/AC.10/30/Rev.6. http://www.unece.org/fileadmin/DAM/trans/danger/publi/ghs/ghs_rev04/English/14e_annex10.pdf, 2015.

Further Reading

Eurometaux: https://www.eurometaux.eu/about-our-industry/the-metals-story/.

ICCM4: http://www.saicm.org/Meetings/ICCM4/tabid/5464/language/en-US/Default.aspx.

ICMM (International Council on Mining & Metals): https://www.icmm.com/en-gb/metals-and-minerals.

ICMM (International Council on Mining & Metals): https://www.icmm.com/en-gb/metals-and-minerals/the-role-of-metals-in-society/metals-in-society.

ICMM (International Council on Mining & Metals and Euromines): "Ores and Concentrates: an industry approach to EU Hazard Classification" http://www.icmm.com/en-gb/publications/materials-stewardship/hazard-assessment-of-ores-and-concentrates-for-marine-transport, 2009.

SAICM (Strategic Approach to International Chemicals Management): http://www.saicm.org/StrategicApproach/Towardsnbsp2020/tabid/5499/language/en-US/Default.aspx.

SAICM (Strategic Approach to International Chemicals Management): Overall orientation and guidance for achieving the 2020 goal of sound management of chemicals. The future we want for the sound management of chemicals, http://www.saicm.org/Portals/12/Documents/OOG%20document%20English.pdf, 2015.

US EPA (Environmental Protection Agency): Framework for Metals Risk Assessment. U.S. Environmental Protection Agency Washington, DC 20460. EPA 120/R-07/001. www.epa.gov/osa, 2007.

Sources of Exposure to Inorganic Complex Materials

Violaine Verougstraete*, *Daniel Vetter*[†], *Rüdiger Vincent Battersby*[†], *Jutta Schade*[†], *Frederik Verdonck*[‡], *Marnix Vangheluwe*[‡]

*Eurometaux, Brussels, Belgium [†]EBRC Consulting GmbH, Hannover, Germany [‡]ARCHE Consulting, Ghent, Belgium

2.1 INTRODUCTION

As explained in Chapter 1, the first step in the life cycle of metals is the extraction and processing of mineral ores to create concentrates, both being complex inorganic materials. They then go on to be refined as metals, minerals, and other complex inorganic materials such as pigments and alloys, which are used by a society. An accurate identification and evaluation of exposure/emissions along all these steps is essential for ensuring that any potential environmental and health risks associated with their manufacture, transport, storage, and use are properly managed.

Such an evaluation will cover the workplace (occupational exposure), the different environmental compartments (environmental exposure), the general population exposed to such materials either indirectly via air, food, or drinking water (man via the environment exposure) or as a potential consumer of preparations or goods containing such materials (consumer exposure).

The main principles governing the metals' exposure science remain applicable (see, e.g., Nordberg et al., 2014; Lauwerys and Hoet, 2001; AIHA, 2003; ECHA, 2016a-c; MERAG, 2017; HERAG, 2007; Vetter et al., 2016) when assessing exposure to complex inorganic materials (e.g., importance of physical form/particle size, natural background, and historical releases, speciation, adsorption/desorption behaviour, diffuse sources, differences in bioavailability, etc.).

The key elements of relevance to safe handling of complex inorganic materials are described below, including some general principles aiming at preventing/reducing exposure/emissions. The terminology that will be used in this chapter is summarised in Table 2.1.

9

TABLE 2.1   Terminology used in this chapter of relevance to exposure and fate

Dose	The amount of agent that enters a target after crossing an exposure surface. If the exposure surface is an absorption barrier (i.e., that may retard the rate of penetration of an agent into the target such as skin, respiratory tract lining, and gastrointestinal tract wall), the dose is an absorbed/uptake dose; if the substance does not pass an absorption barrier, the dose is an absorbed/intake dose (WHO, 2004). Specificities of absorption when it comes to complex inorganic materials are further discussed in Chapter 3
Concentration	The quantity of a material or substance contained in unit quantity of a given medium. It can refer 'to the concentration of a chemical in the environment' ('external concentration' or 'external exposure'), or to 'biological tissue concentration' (also called 'internal concentration' or 'internal exposure') to which an organism may be exposed (WHO, 2004)
Natural background concentration	The natural concentration of an element in the environment that reflects the situation before any human activity disturbed the natural equilibrium. As a result of historical and current anthropogenic input from diffuse sources, the direct measurement of natural background concentrations is challenging
Speciation	Chemical form or compound in which an element occurs in both non-living and living systems
Fate	Fate refers to what eventually happens to contaminants released to the environment—some fraction of the contaminants might simply move from one location to the next; other fractions might be physically, biologically, or chemically transformed; and others still might accumulate in one or more media. The different processes affecting fate and potential exposure of organisms (bioavailability), such as inter-compartment transfer, complexation adsorption, and precipitation reactions are important to understand when deriving environmental concentrations of complex inorganic materials in the different environmental media, such as the predicted environmental concentrations (PECs, see Section 2.3)
Bioavailability	The extent to which a substance is taken up by an organism, and distributed to an area within the organism. It is dependent on the physicochemical properties of the substance, anatomy and physiology of the organism, pharmacokinetics, and route of exposure. Hence, metal bioavailability refers to the fraction of the bioaccessible metal pool that is available to elicit a potential effect following the internal distribution: metabolism, elimination, and bioaccumulation
Environmental Exposure Concentration	Exposure benchmark value, which is compared with an environmental threshold value in a risk assessment framework or for compliance checking. The environmental exposure concentration is typically calculated from all individual measured or modelled metal concentrations for a predefined environment taking a high end value (e.g., the P90) of the environmental concentration distribution at a site/region

2.2 OCCUPATIONAL EXPOSURE

Workers may come to contact with complex inorganic materials on various distinct occasions. First, the complex inorganic material may be intentionally manufactured to be sold on the market. Workers may be exposed during its manufacture and uses. For example, if the material represents an inorganic pigment, its uses after manufacture will cover steps such as unpacking of the pigment and its formulation into paints (in industrial or professional settings) and finally its application by professionals.

Second, the complex inorganic material may be a material that needs to be further processed. Such materials include intermediates and any by-products (or even wastes) that undergo chemical modification with the aim of producing something of higher value. Examples may be found in secondary metal production, where sometimes parts per million of a precious metal are recovered from waste materials.

The main difference between both cases is the setting in which the respective complex material is handled. In the first case (i.e., materials placed on the market), workplace exposure considerations will be relevant for both industrial and professional workers, and cover manufacturing and use of the product containing, for example, the pigment. In the second case, further processing of materials will be exclusively handled in industrial settings. Therefore, consumers and non-industrial workers are unlikely to be exposed to this type of complex inorganic material.

On the other hand, recycling and refining processes are usually not restricted to a single complex inorganic material, but involve instead several materials, possibly processed simultaneously, all of which therefore contributing to workers' exposure. In such situations, when protecting workers' health, one needs to account for various constituents of an individual complex material that may affect workers' health and, at the same time, identical constituents contained in various other materials that are handled simultaneously. How to determine workers' exposure in a comprehensive and representative manner thereby becomes the key question.

In general, substances in the workplace may come into contact with the body and possibly enter the body by inhalation (inhalation route), through the skin (dermal route), or potentially by swallowing (e.g., inadvertent ingestion/oral route). Exposure to a particular substance is normally determined by estimating 'external' exposure. For a quantitative exposure assessment and subsequent risk assessment, the exposure estimate will be compared with the effect levels, which are also expressed as 'external' doses. There are some metals for which 'internal exposures' of the organism (i.e., internal dose evaluated by a biological method) will be compared with 'internal effect levels'. Typical examples of metals for which biological monitoring is used are lead or cadmium, complemented by health surveillance or biological monitoring of early effects (Lauwerys and Hoet, 2001).

A quantitative exposure estimation will be needed to consider the following three separate exposure routes:

- Inhalation exposure: the amount of the substance inhaled; usually represented by the average airborne concentration of the substance over a reference period of time in the breathing zone of a worker.
- Dermal exposure: the amount of substance in contact with the exposed skin surface.
- Oral exposure: the amount of substance ingested.

In addition to the exposure routes, the duration and frequency of exposure after which an effect may occur will need to be taken into account when performing the exposure assessment. For comparison with hazards after repeated or continuous exposure (i.e., chronic effects), a reference period of a full shift (normally 8 h) is generally used. Exposures that are typically longer or shorter than the 8-h reference period can be adjusted in magnitude to provide an 8-h time-weighted average estimate so they can be compared with chronic effect levels or occupational exposure limits. If the substance has the potential to cause acute health effects, it may be relevant to identify and evaluate exposure over shorter reference periods (ECHA, 2016c).

Different assessment approaches and tools may be used to assess occupational exposure in general (e.g., ECHA, 2016c) or specifically to metals (e.g., Vetter et al., 2016). Special consideration shall be given to measured exposure data when adequate and representative. If such data are not available, relevant monitoring data from substances with analogous use and exposure patterns and/or analogous properties can also be taken into account, provided that operational conditions and risk management measures when comparing data origin and assessment target are similar to such an extent that would justify such read-across. Finally, modelling tools can be used for the estimation of occupational exposure levels, carefully considering their applicability and boundaries (see also Chapter 9). In some cases, a combination of measured data and modelling approaches may even lead to the most appropriate assessment (ECHA, 2016c).

For inorganic substances, metals in particular, occupational exposure often relies on the measured data. Owing to their hazardous properties, most metals are monitored on a regular basis in the relevant workplaces, both as part of a good occupational hygiene practice in the industry and also following the requirements from legislations on the protection of workers' health. Measurements of the concentration of the substance in the breathing zone of the operator are commonly conducted for inhalation exposure, whereas dermal exposure is commonly assessed by the aid of exposure modelling tools, generating exposure estimates. For inorganics, occupational exposure is commonly assessed by considering the substance as such. Most workplaces do not have knowledge on the precise speciation of air exposures and in most cases it is likely that more than one form of metal may be present. Speciation currently is rather used as a research procedure since it is time consuming and expensive. However, as the importance of speciation has already become more widely recognised by researchers and regulators alike, it may become more commonplace, or even mandatory, to analyse samples for specific species (see also Section 2.2.1.2).

2.2.1 Characteristics of Complex Inorganic Materials Affecting Exposure Assessment

Although important similarities with standard exposure assessments can be found when assessing occupational exposure to complex inorganic materials, some major differences associated with the complexity of the materials (i.e., their variable composition, the importance of the speciation of constituents, and the varying emission potential of individual constituents depending on the conducted processes) shall be addressed to ensure that the exposure assessment remains comprehensive and representative. These can best be handled by small adaptations of the standard exposure assessment approach.

Even though there is no 'golden standard' yet, some sectors have recently gained additional experience in assessing exposure and risks associated with complex inorganic materials such as, for example, the secondary metal production (recycling) sector. In Europe, various input materials of unknown and variable content used in the secondary production of metals had to be registered under the REACH Regulation (REACH, 2006), requiring to carry out a complete risk assessment for human health if the materials were identified as being hazardous to human health. The approach detailed below relies mainly on this experience and its evaluation and aims at tackling the three main factors of complexity: variability, speciation, and varying emission potential due to the physical form along the process.

2.2.1.1 The Issue of Variability

Complex inorganic materials are characterised by the presence of several constituents and a high variability in content (concentrations) of these individual constituents. The constituents as such (or at least their elemental composition) are often known as the composition of the material has generally some market and process implications. For a number of materials such as ores and concentrates or UVCBs produced during manufacturing and recycling, the variability is related to their source. Other complex inorganic materials are less affected by this variability, but may have different compositions related to the expected function or the presence of impurities (e.g., alloys).

For a complex inorganic material containing different constituents, it is proposed that the exposure assessment shall consider all constituents of the material identified as being hazardous for human health. 'Hazardous' should be understood here as any adverse effect reported along the assessment of hazards and that may lead to the derivation of a (no-)effect level or a dose response. These may in some cases exceed the subset of classified endpoints, which are defined by, for example, the UN Globally Harmonised System (UN GHS, 2015) or the EU Classification, Labelling, and Packaging (CLP, 2008). The approach should also consider the fact that several materials can be simultaneously handled at the workplace.

2.2.1.2 The Issue of Speciation

Depending on the composition of the handled complex inorganic materials and operational conditions such as process temperature, different species of the metal constituents may be relevant in a process and at a workplace. It is known that speciation can be a key factor for metals when it comes to expressing hazards. The overall assessment should therefore cover all relevant species, using speciation data, where available, to make the best link with the hazard assessment. Guidance explaining how to perform this kind of refined assessment is offered by authorities and/or the metals associations, for example, for the monitoring of nickel, where also some slightly different effect levels can be derived for the different groups of nickel substances: water-soluble nickel compounds, nickel oxides, sulphidic nickel compounds, and metallic nickel. For companies interested in assessing the speciation of their workplace exposures, several physical and chemical methods can be applied, ideally in combination as all those have strengths and weaknesses (see, e.g., Nickel Consortia, 2012; Oller et al., 2009; Zatka et al., 1992). When there is no knowledge on the specific species present, it is recommended to work on a worst-case basis as further explained in Chapters 7, 8, and 11: where the speciation of the constituent is unknown, the speciation with the worst case classification can be taken forward for the calculation of the classification of the material.

2.2.1.3 The Issue of Varying Emission Potential

It is well known that some of the intrinsic properties of a substance's such as its physical state and form (see, e.g., Fig. 2.1) will influence not only its exposure release pattern but also the subsequent operational conditions and risk management measures to be implemented.

The exposure assessment can therefore be further refined on the basis of granulometry information, as some hazardous effects are not expressed by coarser particles or massive

FIG. 2.1   Variable physical forms of complex inorganic materials associated with different emission potentials.

forms. Plausibility considerations on the substance- and process-intrinsic emission potential (and other operational conditions) and their modification by the prevailing risk management measures should be integrated as well in the assessment. For example, occupational exposure during raw material handling (emptying of received containers into storage silos) shall be assessed. The substance may be handled either as bulk material or in powder form, whereas other operational conditions are similar and the risk management measures implemented vary in a way that powder handling is completely enclosed and the handling of the bulk material may be conducted with or without the local exhaust ventilation (LEV) present. In this example, three exposure situations may be defined when considering inhalation exposure: (i) enclosed handling of powder, (ii) handling of bulk material (with LEV), and (iii) handling of bulk material (without LEV). For more information, see Vetter et al., 2016; ECHA, 2016c. The issue of particle size and its impact on hazard and risk is discussed further in Chapter 3.

2.2.1.4 To Summarise

The proposed approach, which assesses all constituents of the complex inorganic material that are hazardous (and considers factors such as speciation and physical form, where relevant) is expected to reflect variations in composition of the material during its use and cover all hazards of relevance when assessing risks. But it shall also, as further detailed below reflect good industrial hygiene practice in terms of monitoring, represent a 'combined exposure assessment' to individual constituents, and be precautionary to protect workers' health with respect to the requirements of most chemicals management systems.

In such an approach, using monitoring data will provide clear advantages when compared with exposure modelling. For example, such data are intrinsically reflective of the varying content of a constituent in a complex inorganic material and reflect the contribution to exposure of the same constituent contained in materials handled simultaneously.

2.2.2 Inhalation Exposure Assessment

For many constituents of the complex inorganic materials, the existing legislation requires that the employers directly monitor inhalation exposure to those constituents in the workplace atmosphere. Thus, monitoring data are commonly available for the constituents known to be hazardous to human health.

In cases where the measured data are not available or an assessment based on analogous monitoring data could not be established, occupational exposure could be assessed with the aid of a modelling tool. At the first-tier screening level, the MEASE tool (http://www.ebrc.de/mease.html) can be used, such as considering a number of metal specificities (ECHA, 2016c).

2.2.2.1 Assessment Based on Monitoring Data

Monitoring data are often considered as the closed approximation to effective (real) inhalation exposure levels of workers. Exposure levels are monitored by sampling the airborne dust in the workers' breathing zone. The sampled dust is subsequently analysed for individual elements.

The use of inhalation monitoring data from workplaces in which the complex inorganic material is manufactured and/or handled will allow the assessor to address several of the 'complexities' associated with the complex inorganic materials' exposure assessment:

- The variability in composition: the impact of the varying content of any constituent in the material on the exposure levels to that constituent will automatically be reflected in the sample.
- The impact of varying operational conditions and risk management measures will also be directly reflected in any sample.

- The contribution of other substances (including other complex inorganic materials) handled simultaneously at the workplace and having the same constituent(s) as the material that is under investigation will also be automatically reflected in the sample.

The first point is the main reason why monitoring data are commonly the preferred basis for occupational exposure assessments because the introduction of additional uncertainty when assuming the impact of variation in composition, process conditions, and risk management measures on exposure levels can be avoided.

The contribution of other substances to workers' exposure to an individual constituent as listed under the second point is normally not to be addressed in a substance-specific assessment (e.g., as required under EU REACH), but will be of relevance for a workplace assessment. By including such contribution from other substances, the assessment for the substance is intrinsically conservative (i.e., precautionary) with respect to substance-specific legal requirements as in EU REACH. As the exposure assessment is based on these 'aggregated exposure levels', it could already be seen as a combined exposure assessment of all sources of a constituent in a given workplace.

As explained above, the analysis of the sample could include a chemical speciation analysis. This is clearly of relevance if the hazard profile of a constituent varies with its chemical speciation (e.g., the metal has a significantly less hazard potential when compared with its oxide form). Information on speciation of workplace exposure is however sparse. The worst-case chemical species (i.e., the most hazardous) will therefore be assumed in the risk assessment for precautionary reasons unless such worst-case species can be excluded based on plausibility considerations and/or speciation analysis.

Despite the clear benefits offered by the use of monitoring data, it appears that the generation, collation, interpretation, and reporting of monitoring data remain as some of the most challenging tasks in exposure assessment. It is noted that the European non-ferrous metals' sector has recently developed a guidance on these aspects that may be referenced here for the sake of brevity (see Vetter et al., 2016).

2.2.2.2 Inhalation Exposure Assessment Based on Analogous Data

For scenarios for which monitoring data are not available, exposure levels may be extrapolated from similar (monitored) exposure situations. Three types of extrapolation may be distinguished:

1. Data are available for the constituent to be assessed and for the type of activity, but need to be extrapolated to different operational conditions or risk management measures.
2. Data are available for the type of activity, but need to be extrapolated to a different constituent.
3. Data are available for the constituent but need to be extrapolated to a different type of activity.

In addition, any combination of the types listed above may apply. Extrapolation could be conducted by considering the extrapolation factors. On the basis of the above extrapolation types a total extrapolation factor may be systematically described as

$$EF_{total} = EF_{ES} \times EF_{constituent} \times EF_{activity} \tag{2.1}$$

with EF_{total} being the total extrapolation factor, a product of the extrapolation factor for the differences in operational conditions and risk management measures (EF_{ES}), the extrapolation factor for expected differences caused by different constituents ($EF_{constituent}$), and the extrapolation factor for expected differences caused by different activities ($EF_{activity}$).

The exact values for these factors will depend on the observed variability and information on the similarity of the exposure situations to be assessed.

2.2.2.3 *Modelling of Inhalation Exposure*

As explained, the inhalation exposure assessment is preferably based on monitoring data. If such data are not available (neither as direct measurements nor as analogous data) exposure could be assessed by the aid of an exposure estimation tool. The MEASE tool that has been developed for the non-ferrous metals industry in Europe could be used at the first-tier screening level (see also Chapter 9). Higher-tier assessments may be conducted by using the ART tool (https://www.advancedreachtool.com/). The Advanced REACH Tool (ART) version 1.5 incorporates a mechanistic model of inhalation exposure and a statistical facility to update the estimates with measurements selected from an in-built exposure database or the user's own data. This combination of model estimates and data produces more refined estimates of exposure and reduced uncertainty. However, it needs to be noted that ART is not calibrated to estimate exposure from hot, metallurgical processes.

2.2.3 Dermal Exposure Assessment

Dermal exposure data representing actual workplace measurements are commonly not available. Dermal exposure to inorganic substances could be assessed with the exposure assessment tool MEASE (version 1.02.01, available on www.ebrc.de/mease). For dermal exposure, MEASE proposes the exposure bands system of the broadly used Estimation and Assessment of Substance Exposure tool (EASE, EC, 1996), but the resulting exposure estimates are based on the measured data for several metals. For the assessment of dermal exposure with MEASE, different model parameters have to be selected under the pattern of use, pattern of exposure control, and contact level to describe the operational conditions at a workplace that are relevant for the dermal exposure assessment.

2.2.4 Oral Exposure Assessment

For workers, oral exposure is generally assumed to be sufficiently controlled by strict occupational hygiene practices (e.g., not eating and smoking in the workplace, washing hands before eating, etc.) and therefore a quantitative assessment of unintentional ingestion is usually not required under some legislations (e.g., EU REACH) (ECHA, 2016c).

It is however acknowledged that for some constituents of complex inorganic materials, the oral route can become important to consider when particles in the air are large enough to be mainly deposited in the tracheobronchial region, and then cleared to the pharynx primarily by mucociliary clearance, followed by clearance to the gastrointestinal tract (e.g., lead). Where identified as a key contributing route, and no quantitative estimation can be done in routine, oral exposure can be addressed through a qualitative approach aiming to identify the correct risk management measures to be implemented. A useful tool in the context of metals such as lead is the use of biomonitoring as this tool will reflect the contribution from the oral route to the overall exposure and can also be used as a means (or way) to assess the effectiveness of a workplace control strategy.

Biomonitoring has the advantage that contribution to systemic exposure from all routes of exposure could be assessed. For example, as a standard procedure in the EU for workers exposed to lead, biomonitoring in the form of blood measurements is legally required. Thus, such data exist to a large extent in Europe. The uncertainty of the assessment of internal exposure (that is after uptake) is less compared with substances for which exclusively external exposure levels are available. In addition, internal exposure levels already take into account any personal protective equipment as worn by the worker under surveillance.

2.3 ENVIRONMENTAL EXPOSURE

The extent and release rates of metal ions to the environment due to the manufacturing and use of complex inorganic materials should also be evaluated in detail. Fate, behaviour, and bioavailability concepts already developed for individual metals and metal compounds are equally applicable to complex inorganic materials. These concepts have been extensively described in the fact sheets of the Metals Environmental Risk Assessment (MERAG, 2017) and the subsequent sections will only briefly touch on the main principles and assumptions (e.g., natural background, historical contamination and local and diffuse emissions, variation of metal concentration in the natural environment, adsorption/desorption processes, water solubility) that will have to be considered when performing an environmental exposure assessment for complex inorganic materials. For more information the reader is referred to MERAG (2017) and ECHA (2008c).

2.3.1 Environmental Release of Metal Ions From Complex Inorganic Materials

During manufacture and use of complex inorganic materials, metal ions will enter into the different environmental compartments (water, sediment, and soil). Variability in composition, particle size, dissolution kinetics, and speciation characteristics of the complex inorganic material are all elements that will determine the amount of metal that will eventually dissolve in the environment. For the exposure assessment of complex inorganic materials, it is imperative to know which of the constituting metals will be soluble in water, or can be transformed into a soluble form, since the prediction of the environmental concentration (PEC) should be based on the relevant soluble metal ion or species that is bioavailable.

The partitioning behaviour of the released metal ions to sludge/sediment/soil can then be based on the appropriate partition coefficient (K_d) values for soluble ion. In some cases, the complex inorganic material will only be poorly soluble and sufficiently stable to not rapidly transform to a water-soluble form. In these circumstances, the complex inorganic material itself should be assessed taking into account its specific partitioning characteristics. For the aquatic environment, it can be assumed as a first estimate that each of the constituents will dissolve up to its water solubility limit, and that this fraction will be the bioavailable form. Refinement of the assessment may take into account that the kinetics of the dissolution or speciation models may also be used to determine the soluble fraction. For the aquatic environment, transformation/dissolution tests (classification) or simple solubility tests (risk assessment) may help to assess the actual behaviour of complex inorganic materials in the aquatic environment, but more complex set-ups could be needed to determine long-term transformation processes in soils and/or sediments. Once dissolved, abiotic factors of the receiving environment (e.g., organic carbon, hardness, cation exchange capacity, acid volatile sulphides, etc.) will control how much of the released metal ions will actually be bioavailable. The released metals will add up to the metal load that is already available, coming from natural sources or originating from other local and/or diffuse anthropogenic sources. For a proper risk assessment/management of complex inorganic materials, it is necessary that a proper distinction can be made between all these different contributions.

2.3.2 Use of Modelled and Measured Environmental Exposure Data

Metal concentrations in the environment are the result of the natural background, historical contamination, and the local and diffuse emissions associated with the use pattern (anthropogenic) and the complete life cycle of any metal in general (i.e., from mining to waste disposal, including complex inorganic materials). To discern the relative contributions of the different anthropogenic inputs with respect to these natural background/natural sources, diffuse sources analysis is quite often used. A detailed investigation and analysis of diffuse sources allows for understanding their relative contribution to regional/continental emissions, thereby providing crucial and complementary information to environmental compartment monitoring data. Indeed, the comparison of the two often allows a better insight into the relative contribution of the different anthropogenic inputs, also with respect to the local natural background. In addition, diffuse sources analyses enable generic source type allocation (e.g., impurities in inorganic fertilisers, corrosion from metallic structures, releases from brake pads) to be made as well, so as to define their underlying drivers of release. It therefore provides clear insights about the metals emissions originating from products (e.g., corrosion from building materials, tire wearing) and non-product use (e.g., from impurities in oil, from steel/fertiliser manufacturers). A diffuse sources analysis should be conducted per receiving environmental compartment, as the lead emission sources can be different. A combination of modelled exposure data and measured (monitoring data) is, however, needed to fully elucidate the contribution of complex inorganic materials to the environmental exposure on a regional/continental scale.

The tools available to assess environmental exposure and risks associated during the manufacturing and treatment of complex inorganic materials on an industrial local site are being discussed in Chapter 9. Shortly, specific environmental release categories (SPERCs) have been developed to characterise the environmental releases from the manufacture, processing, and downstream uses of the metals and their compounds. These SPERCs can equally be used for modelling local releases from manufacturing sites dealing with complex inorganic materials in their supply chain.

2.3.3 Bioavailability

Bioavailability is the fraction of the metal, released by complex inorganic materials, present in the environment that is available for biological uptake which depends on various biotic and abiotic parameters. For purposes of risk assessment of metals, it is recommended that, besides background and ambient metal concentrations, the distribution of parameters that determine metal bioavailability are also described and integrated in the exposure assessment (Oorts and Schoeters, 2014). To normalise toxicity data towards physicochemical conditions, different datasets for abiotic factors (and environmental concentrations) should be considered depending on the goal of the assessment (i.e., threshold derivation, site-specific assessment, etc.). More specifically, datasets of abiotic factors as well as environmental concentrations should be representative of the area under investigation. The breadth of the datasets will usually be proportional to the scope of the assessment, that is, broader datasets will be necessary for regional assessments with national to continental scales due to spatial variability, compared with local assessments which (only) address site-specific operational scales. It is particularly important to take relevant abiotic factors into account for the metals under investigation. For most metals, dissolved organic carbon (DOC), pH, and hardness are the most important parameters to measure in the water compartment. For the soil compartment, total organic carbon, pH, and effective cation exchange capacity (eCEC, i.e., CEC at prevailing soil pH) are the key parameters controlling metal bioavailability. In sediments, the bioavailability of divalent metals is governed by its binding to acid volatile Sulphides (AVS) and organic carbon.

Assessing environmental exposure is not only important to evaluate the direct toxic effects to ecological receptors but is also necessary to assess the indirect exposure of humans following the man via the environment route (Section 2.5).

2.4 CONSUMER EXPOSURE

The consumer is a member of the general public who may be of any age, either sex, and in any state of health, who may be exposed to a substance by using consumer products, or by being present when others (e.g., professionals) are using the products. A consumer product (substance, mixture, or article) is a product that can be purchased from retail outlets by members of the general public. This also includes chemicals and materials for construction works or car maintenance sold to both professionals and consumers (do it yourself products) (ECHA, 2016b). The assessment of consumer exposure starts with an initial screening to determine if the substance under investigation is actually used as or contained in consumer products or whether the expected exposure is so low that it can be neglected when addressing the risk, as compared with other contributions. If the consumer exposure is not considered to be 'negligible', then a quantitative exposure assessment is desirable, assessing the three exposure routes (inhalation, oral, dermal), using the measured data or modelling tools.

As indicated in the introduction, some complex inorganic materials will not be directly in contact with the consumers (e.g., UVCBs, ores, and concentrates) as they are processed in industrial plants. Other complex inorganic materials such as alloys (in metal objects, jewellery) may however be part of a normal 'consumer environment' and result in exposure, mainly via the oral and dermal routes. In these cases, it is assumed that the risk will be associated with exposure to the metal ion, released by the material. The importance of this release of the metal ion has, for example, been recognised in the EU REACH Annex XVII Restriction on lead in consumer articles (REACH Annex XVII), which stipulates that the articles covered by the restriction will not be allowed on the market if the concentration of lead, expressed as metal, that they or their parts contain, reaches 0.05% by weight. The limit would not apply if it can be demonstrated that the rate of lead release from an article, or any accessible parts of an article, whether coated or uncoated, does not exceed $0.05\,\mu g/cm^2/h$.

The following stepwise approach was proposed in HERAG (2007): for initial screening purposes, default assumptions on release rates as reported for several metals in existing risk assessments can be used. However, for a more refined assessment, the assessor should give preference to measured release rates in physiological media (such as artificial sweat, for example) of metal ions from a particular object or alloy product as explained in Chapters 8 and 13).

2.5 MAN VIA THE ENVIRONMENT

Indirect exposure of humans may occur via the typical exposure pathways: inhalation, ingestion, and dermal contact. Usually the following exposure scenarios are considered: exposure via inhalation of air, drinking water, via food consumption with concentrations in food derived using the concepts of bioconcentration in fish and biotransfer from soil and air to plants, to meat and milk. Direct soil ingestion and dermal contact are often not taken into account in the assessment because exposure through these routes is considered to be very unlikely. Only in cases of extremely polluted soils (e.g., in dump sites or through accidents) or for specific subpopulations (e.g., young children) can these routes provide significant contributions to the total exposure (ECHA, 2016a; HERAG, 2007).

In general, the assessment of indirect exposure is carried out following a stepwise procedure:

- assessment of the concentration in intake media (food, water, air, soil)
- assessment of the intake rate of each medium
- combination of the concentration and the intake rate, and if necessary a factor for the bioavailability of the substance through the respective route of intake
- finally, an internal dose is derived which is taken forward to the risk characterisation.

Indirect exposure is principally assessed on a local and a regional spatial scale. In this context, local and regional environments are not the actual sites or regions, but standardised defined environments. The local scale, in which all food products are derived from the vicinity of one point source, represents a worst-case situation. In contrast, the regional scale assessment depicts a highly averaged exposure situation, which cannot ensure protection of the individual.

The assessment of indirect exposure following the manufacture and use of complex inorganic materials follows these general principles. However, as they are composed of metal and other inorganic constituents, some metal-specific principles apply. These are further detailed for example in the MERAG and HERAG fact sheets (MERAG, 2017; HERAG, 2007).

In a nutshell, aspects to consider can be summarised as follows:

- The assessment can be structured in tiers, depending on the level of available data, with impacts the precision and conservatism of the exposure assessment, starting from basic modelling of concentrations in air, water, and food (Tier 1) to the use of higher-tier exposure models and data (Tier 2), up to biomonitoring data which allow to compare the actual internal dose (biomarkers such as urinary cadmium or blood lead) of individuals with the effect level (NOAEL or LOAEL), preferably in a specific target organ for repeated toxicity. Where such biomarkers of exposure are not available, data of external exposure (modelled or measured concentrations of metals in foodstuff, air, and water, and intake rate assessed via dietary studies, etc.) need to be converted to internal exposure by taking into account appropriate absorption factors for each exposure pathway (inhalation, oral), eventually refined with exposure conditions (e.g., oral absorption factors for fasting vs. non-fasting). The conversion of external to internal exposure then allows to aggregate exposure via various exposure pathways. For endpoints on systemic effects, it is essential to consider aggregated exposure instead of one (major) pathway, otherwise risk may be underestimated.

- It shall be noted that the currently available model approaches, which are based on partition equilibria, are applicable to metals only to a limited degree. For example, the modelling approach of Trapp and Matthies (1995) included in EUSES (Vermeire et al., 1997) to estimate the levels in leaves and roots due to uptake from soil and air proposes the physicochemical properties that are out of range for metals and discards the fact that plants have mechanisms to regulate metal uptake (homeostasis). The alternative is the use of empirical models based on the measured bioconcentration (BCF) factors (root uptake) or the plant deposition model (air route). Research is also ongoing to propose an alternative to/refinement of EUSES that will allow to better address metals and inorganics.
- Where relevant, the assessment of indirect exposure via the environment should be performed separately for susceptible or particularly sensitive subpopulations. This is the case, for example, for children exposed to lead via uptake from soil in contaminated areas. This pathway can be modelled, for example, using the US EPA IEUBK model (https://www.epa.gov/superfund/lead-superfund-sites-software-and-users-manuals) or using a more refined assessment scheme based on toxicokinetic modelling and on comparison with biological monitoring data.
- The general use of 90th percentiles (P90) of quantitative exposure measures is not likely to be applicable to metals, because a multiplication of such values for concentration in environmental media and their intake rates may result in overestimates of internal exposure. However, the use of reasonable worst-case values may, under certain circumstances, allow a consideration to be given to (potentially sensitive) known subpopulations.
- Most key metals are ubiquitous in the environment: reflection must therefore be given to a correct distinction between natural ambient and anthropogenic concentrations.
- Where known, metal–metal interactions need to be addressed, since the nutritional (metal) status or food composition with respect to metal content has been shown to influence the uptake (examples: calcium as a modulator of lead absorption, zinc vs. copper interactions, etc.).

2.6 CAN EXPOSURE BE PREVENTED?

The same principles of reduction of exposure apply for complex inorganic materials as for any other hazardous substance. Once exposure is assessed and documented, the manufacturers and users of complex inorganic materials should consider the measures needed for control risk/exposure, both at the workplace and concerning the environment.

For the workplace, several approaches exist. The EU Chemicals Agent Directive (CAD) provides, for example, the following general workflow:

- Eliminate risks by limiting the use of the substance on the market, or the modification of process by using intrinsically safe equipment or by automatisation.
- Reduce risk by limiting the concentration of a substance, and/or change the form of physical state, and/or apply closed processes, and/or install effective local exhaust ventilation,
- General area ventilation and other workplace-related measures (such as segregation of dirty departments, safe storage, fire/explosion protection, and prevention, eyebaths/ showers).
- Other collective risk management measures aimed at protecting the population of workers, for example, organisational measures limiting the number of exposed workers or the duration of their exposure.
- Personal protective equipment (respiration, skin, eyes) where exposure cannot be prevented by other means.

For the environment different risk mitigation measures can be taken:

- Air pollution control systems (e.g., wet scrubbers, electrostatic precipitators, cyclones, fabric or bag filters, ceramic, and metal mesh filters) can be installed to reduce airborne pollution and avoid as such soil contamination.
- Minimise surface run-off/contamination of rainwater through good practice [following the best available techniques (BAT)] for the storage of raw materials, as well as by good maintenance and cleaning of the whole production plant. Following BAT, the treatment methods are very much dependent on the specific processes and the metals involved. Direct water emissions should be reduced by implementing one or more of the following risk management measures:
- Chemical precipitation: used primarily to remove the metal ions [e.g., $Ca(OH)_2$, pH 11 precipitation: >99% removal efficiency; $Fe(OH)_3$, pH 11: 96% removal efficiency]
- Sedimentation (e.g., Na_2S, pH 11, >99% removal efficiency)
- Filtration: used as the final clarification step (e.g., ultrafiltration, pH 5.1: 93% removal efficiency, nanofiltration: 97% removal efficiency, reverse osmosis, pH 4–11: 99% removal efficiency)
- Electrolysis: for low metal concentration (e.g., electrodialysis: 13% removal efficiency within 2 h at 2 g/L, membrane electrolysis, electrochemical precipitation, pH 4–10, >99% removal efficiency)
- Reverse osmosis: extensively used for the removal of dissolved metals
- Ion exchange: the final cleaning step in the removal of heavy metal from process wastewater (e.g., 90% removal efficiency for clinoptilolite and 100% removal efficiency for synthetic zeolite)

Fugitive emissions should be reduced from material storage and handling, reactors or furnaces and from material transfer points by following the hierarchical measures: process optimisation and minimisation of emissions, sealed reactors and furnaces, targeted fume collection, and avoid storage of dusty materials in open spaces.

KEY MESSAGES

The accurate assessment of exposure/emissions down the production line and the use of complex inorganic materials are essential to ensure that any potential environmental and health risks associated with their manufacture, transport, storage, and use are properly managed. The assessment should cover the workplace, the different environmental compartments and the general population exposed to such materials either indirectly via air, food, or drinking water or as potential consumers of preparations or goods containing such materials. The main scientific principles governing the metals' exposure remain applicable; however, the variable composition of the complex inorganic materials, the speciation of their constituents, and the varying emission potential are factors to consider when evaluating human exposure. For the assessment of environmental exposure, it is imperative to know which of the constituting metals will be soluble in water, or can be transformed into a soluble form, since the PEC should be based on the relevant soluble metal ion or species that is bioavailable. Different tools exist to help the assessor to best apprehend the specificities of the materials and to propose the best risk management measures.

References

AIHA (American Industrial Hygiene Association): In Dinardi S, editor: *The occupational environment: its evaluation, control, and management*, ed 3., Fairfax, VA, 2003, AIHA Press.

EC (European Commission): Technical guidance document in support of Commission Directive 93/67/EEC on risk assessment for new notified substances and commission regulation (EC) No. 1488/94 on risk assessment for existing substances. Brussels: European Commission. ISBN: 92-827-8011-2, 1996.

ECHA: Guidance on information requirements and chemical safety assessment. Appendix R.7.13-2: environmental risk assessment for metals and metal compounds http://echa.europa.eu/documents/10162/13632/information_requirements_r7_13_2_en.pdf, 2008c.

ECHA (European Chemicals Agency): Guidance on information requirements and chemical safety assessment Chapter R.14—occupational exposure estimation, version 3.0, ISBN: 978-92-9495-081-9, 2016c.

ECHA (European Chemicals Agency): Guidance on information requirements and chemical safety assessment Chapter R.15: consumer exposure assessment Version 3.0, ISBN: 978-92-9495-079-6, 2016b.

ECHA (European Chemicals Agency): Guidance on information requirements and chemical safety assessment Chapter R.16: environmental exposure assessment, version 3.0, ISBN: 978-92-9247-775-2, 2016a.

EU CLP: Classification, Labelling and Packaging of Substances and Mixtures Regulation. EC N° 1272/2008, 2008.

HERAG (Health Risk Assessment Guidance for Metals): Indirect exposure via the environment and consumer exposure, http://www.icmm.com/en-gb/publications/health-risk-assessment-guidance-for-metals-herag, 2007.

Lauwerys RR, Hoet P: *Industrial chemical exposure. Guidelines for biological monitoring*, ed 3, Boca Raton FL, 2001, Lewis Publishers/CRC Press.

MERAG (Metals Environmental Risk Assessment Guidance), http://www.icmm.com/en-gb/publications/metals-environmental-risk-assessment-guidance-merag, 2017.

Nickel Consortia: Incorporating data on particle size and chemical composition of aerosols into site-specific exposure scenarios (workers). http://www.nickelconsortia.eu/guidance-documents.html, 2012.

Nordberg G, Fowler B, Nordberg M: *Handbook on the toxicology of metals*, ed 4, Elsevier, 2014, Academic Press.

Oller AR, Cappellini D, Henderson R, et al: Comparison of nickel release in solutions used for the identification of water-soluble nickel exposures and in synthetic lung fluids, *J Environ Monitor* 11(4):823–829, 2009.

Oorts K, Schoeters I: Use of monitoring data for risk assessment of metals in soil under the European REACH regulation. In Reimann C, Birke M, Demetriades A, Filzmoser P, O'Connor P, editors: *Chemistry of Europe's agricultural soils—Part B: general background information and further analysis of the GEMAS data set*, Hannover/Stuttgart, 2014, Geologisches Jahrbuch (Reihe B)/Schweizerbarth, pp 189–202.

REACH EC (European Council) Regulation (EC) No. 1907/2006 of the European Parliament and of the Council of 18 December 2006 Concerning the Registration, Evaluation, Authorisation and Restriction of Chemicals (REACH). Off J Eur Union L396: L136/133–L136/280, 2006.

Trapp S, Matthies M: Generic one-compartment model for uptake of organic chemicals by foliar vegetation, *Environ Sci Technol* 29(9):2333–2338, 1995.

UN Globally Harmonized system of Classification and Labelling of chemicals (GHS), Sixth revised edition ST/SG/AC.10/30/Rev.6. United Nations, New York and Geneva, 2015, https://www.unece.org/fileadmin/DAM/trans/danger/publi/ghs/ghs_rev06/English/ST-SG-AC10-30-Rev6e.pdf.

Vermeire T, Jager T, Bussian B, et al: European Union System for the Evaluation of Substances (EUSES). Principles and structure, *Chemosphere* 34(8):1823–1836, 1997.

Vetter D, Schade J, Lippert K: Guidance on the assessment of occupational exposure to metals based on monitoring data, final report, http://www.reach-metals.eu/index.php?option=com_content&task=view&id=216&Itemid=3242016, 2016.

WHO (World Health Organization Geneva). IPCS Risk Assessment Terminology Harmonization Project: Document No. 1: Part. 1, IPCS/OECD key generic terms used in chemical hazard/risk assessment/International Programme on Chemical Safety Joint Project with OECD on the Harmonization of Hazard/Risk Assessment Terminology—Part. 2, IPCS glossary of key exposure assessment terminology/IPCS project on the Harmonization of Approaches to the Assessment of Risk from Exposure to Chemicals, ISBN: 92-4-156267-6, 2004.

Zatka VJ, Warner JS, Maskery D: Chemical speciation of nickel in airborne dusts: analytical method and results of an interlaboratory test program, *Environ Sci Technol* 26:138–144, 1992.

Further Reading

Chemicals Agents Directive (CAD): Council Directive 98/24/EC of 7 April 1998 on the protection of the health and safety of workers from the risks related to chemical agents at work (fourteenth individual Directive within the meaning of Article 16(1) of Directive 89/391/EEC), http://eur-lex.europa.eu/legal-content/EN/TXT/?uri=celex%3A31998L0024, 1998.

REACH Annex XVII to Regulation (EC) No 1907/2006, column 2 of entry 63, lead restriction in consumer articles, 2014.

Mechanisms Underlying Toxicity of Complex Inorganic Materials

Violaine Verougstraete, Ruth Danzeisen[†], Arne Burzlaff[‡],
Adriana Oller[§], Kate Heim[§], Daniel Vetter[¶], Carina Müller[‡],
Rüdiger Vincent Battersby[‡], Koen Oorts[¶], Dominique Lison[∥]*

[*]Eurometaux, Brussels, Belgium [†]Cobalt Institute, Umkirch, Germany [‡]EBRC Consulting
GmbH, Hannover, Germany [§]NiPERA Inc., Durham, NC, United States [¶]ARCHE Consulting,
Leuven, Belgium [∥]Université catholique de Louvain, Brussels, Belgium

3.1 INTRODUCTION

Complex inorganic materials, as defined in Chapters 1 and 2, have specific properties that set them apart from organics, imparted by the metals they are containing. This chapter focuses on the key mechanistic aspects for an appropriate risk management of complex inorganic materials. Metals are ubiquitous in the environment, and some are essential for humans and/or all forms of life. However, they are usually required only in trace amounts and can be toxic in excess. Metals can neither be created nor destroyed. However, their forms may be changed, converted to different ionic forms (valence states), and transferred to various ligands, thereby changing their biological availability, activity, and, consequently, their toxicity. Metals are also present in various sizes, from small particles to large masses.

Metal toxicity depends on the chemical species formed by the metal and is mainly correlated to free ion concentration and at target organ sites. The main targets for metal toxicity are generally cellular organelles and their components, for example, cell membranes, mitochondria, lysosome, endoplasmic reticulum, nuclei, and some enzymes involved in metabolism, detoxification, and damage repair (Tchounwou et al., 2012).

The main principles governing the mechanisms of toxicity of the metals (e.g. Nordberg et al., 2014; Casarett and Doull's, 2013; Newman, 2014) remain therefore applicable when assessing complex inorganic materials, particularly in relation to the rate and extent to which these materials can produce soluble (bio)available ionic and other metal-bearing species (e.g. metal complexes).

3.2 HUMAN HEALTH

3.2.1 Main Mechanisms of Toxicity of Complex Inorganic Materials

The toxic effects of metals usually involve the initial interaction of the free metal ion with the toxicological target (Hollenberg, 2010). The toxicity of a complex inorganic material containing metals will therefore be related to its capacity to release metal ions that may interact with the biological target. However, it should be acknowledged that for some metals, and thus for some complex inorganic materials, not only the free metal ion will have to be considered, but also some counterions (e.g. chromate in zinc chromate or lead chromate), which may play a role for some adverse effects (e.g. local toxicity effects).

Metals are involved in various metabolic reactions as constituents of metalloenzymes or as cofactors of enzymes involved in the synthesis of nucleic acids, proteins, or carbohydrates. Common mechanisms by which metal ions may act and generate toxicity include inhibition of enzymes, disruption of the structure and/or function of subcellular organelles, covalent modification of proteins, displacement of other critical metals in various metal-dependent proteins, and inhibitory or stimulatory effects on the regulation of expression of various proteins (Hollenberg, 2010). Other mechanisms that have been reported include the generation of free radicals (Jomova and Valko, 2011). These various mechanisms have a variety of targets and may cause damage to various organs including the kidney, respiratory, endocrine, and reproductive systems. In addition, some metal ions (e.g. Cr VI) can directly interact with DNA or indirectly result in genotoxicity or carcinogenesis (e.g. Co II). Importantly, even though some metals share the same mechanisms and targets, each metal is unique in its essentiality and/or toxicity profile.

3.2.2 Key Mechanistic Aspects to Consider When Managing the Human Health Risk of Complex Inorganic Materials

3.2.2.1 Generic Aspects

Humans and animals are generally systemically exposed to toxicants—including metals—through the inhalation route (for metal fumes, metal particles), the dermal route (for those metals that are able to cross the skin barrier), and the oral route (for metals that are ingested via food, drinking water, or soil). The oral route is relevant as well for inhaled metal particles that will be deposited in nasopharyngeal and tracheobronchial compartments and transported through the mucociliary action to the gastrointestinal tract. Finally, particles deposited in the alveolar region can be phagocytosed by macrophages and transported to the respiratory tract lumen or the lymph nodes.

Once metal ions reach the systemic circulation, they will be transported, distributed in the body, and eliminated like any other substance but with a pattern and mechanism that depends on the metal. Transport and distribution models are not always clear-cut: blood and plasma (with binding to a ligand such as albumin) are typically the main transport routes. Some metals may compete with other ionised species such as calcium and zinc to move through membrane channels in the free ionic form. The retention in tissues will occur for some metals and is generally related to the formation of inorganic complexes or metal protein complexes (e.g. lead in bone, cadmium in tissues bound to the low-molecular-weight protein metallo-thionein (Goyer et al., 2004)). Most metals are excreted unchanged via the kidney in urine and to a lesser extent by the gastrointestinal tract but their half-lives may vary widely (e.g. up to decades for cadmium or a couple days for some nickel compounds). Some metalloids such as arsenic are methylated prior to excretion. Interindividual variability in the excretion of some metals and metalloids is well known. Toxicokinetic models are useful in providing a quantitative description of the overall body turnover of metals and can help in establishing dose–response relationships.

Some metals are essential trace elements (ETC) and adverse effects may also occur when concentrations are below the levels required for optimum nutrition. These elements will be subject to homeostatic control mechanisms, which enable adaptation of the organism to varying nutrient intakes to ensure a safe systemic supply (see also 'Essentiality' section).

The toxicological systemic effects of metals will vary according to the metal in question, its physical and chemical form, the total dose absorbed, and the exposure duration (acute or chronic). It shall also be noted that while systemic effects—following the absorption and distribution of the metal—commonly serve as the basis for risk assessment, local effects including irritation or sensitisation may also be associated with exposure to metals contained in inorganic materials. Sensitisation is further discussed in Section 3.2.3.2.

Overall, toxicity will occur when the toxic species reaches a critical target within the body and overwhelms the defence and repair mechanisms of the cells. Aspects like the intracellular context and nature of ligand or protein binding may influence the potential or availability of the metal for interacting at a specific cellular target, such as an enzyme or transport protein, to produce toxicity (Goyer et al., 2004).

The toxicity of metals will be influenced by other factors such as age, with individuals at both ends of the life span generally believed to be more susceptible to metal toxicity, nutritional factors, or immune status (for metals that produce hypersensitivity reactions such as contact dermatitis). Also, the role of genetics in the various types of responses and the threshold levels of exposure leading to various responses is an area that is continuously being investigated (Hollenberg, 2010).

Although each metal is known to have unique features and physicochemical properties that confer its specific essential or toxicological mechanisms of action, some common aspects when assessing the toxicity of the metals, and hence the complex inorganic materials are discussed below.

3.2.2.2 Specificities of Inorganic Materials

BIOAVAILABILITY AND WATER SOLUBILITY

For metal compounds and consequently for complex inorganic materials containing them, it is generally assumed that the potential toxicity is related to the presence of the metal ion in body tissues and its capacity to interact with its target, that is, its bioavailability. Metal bioavailability is specific to a metal compound, and depends on speciation and particle size, the route of entry, its dose, and the exposure matrix. For example, it has been recognised that the bioavailability of lead varies as a function of chemical speciation, age of the exposed individual, level of exposure, the matrix within which lead is contained, and the nutritional status of the exposed individual (e.g. Deshommes et al., 2012; Fowler, 2010; Polak et al., 1996;

US EPA, 1994a,b). Ideally, information on the bioavailability of metal substances is derived from toxicokinetic studies or toxicological tests. In situations where the bioavailability of a substance/material is not known or where it is not feasible to determine it in vivo, bioaccessibility may be used to estimate the potential bioavailability of the metal ion. The bioaccessibility testing involves the determination of released metal ions in synthetic fluids mimicking the different exposure routes and is defined as the "fraction of a substance that dissolves under surrogate physiological conditions, potentially available for absorption into systemic circulation" (Ruby et al., 1999) (see also Chapter 8).

Water solubility of the metal compound is also often used as a surrogate for bioavailability, as it provides a first indication of the potential availability of the metal ion. It has been shown, based on in vivo data, that for metals such as nickel and zinc, large variations exist between the bioavailability of metal ions from soluble salts, the metal itself, oxides, or other very poorly soluble metal substances. For example, for inorganic nickel compounds, a number of subcategories for grouping and read-across of hazard properties have been suggested based on different ranges of aqueous solubility (Hart, 2007). A more recent approach based on bioaccessibility in synthetic fluids relevant to each route of exposure demonstrated that some changes to the conclusions of the hazard assessment based solely on water solubility are warranted (as those appeared to be either over- or underprotective) and that bioaccessibility in relevant fluids is a more appropriate basis for performing grouping and read-across of hazard properties of nickel substances (Henderson et al., 2012). In contrast, for other metals such as lead, it has been shown that differences in solubility do not necessarily impact bioavailability under physiological circumstances. These differences between metals indicate that predictions of bioavailability based on water solubility alone cannot be assumed a priori, but should be demonstrated on a case-by-case basis (HERAG, 2007a).

When evaluating toxicity, it is also important to keep in mind that, very often, toxicity data have been generated for bioavailable forms of a metal: such forms are easier to handle experimentally to achieve high internal doses for the study of toxic effects. However, these forms will most probably not correspond to the ones the assessor of complex inorganic materials will be confronted with. There are several possible approaches to address this: (1) use a default assumption that the metal in the complex inorganic material samples is in its most toxic form (most bioavailable); (2) use additional available data (e.g. toxicokinetics) to derive an adjustment to the effective dose identified in the experimental study; (3) conduct new toxicology studies using the metal form present in the complex inorganic material. The latter scenario will seldom be applicable.

To conclude, bioavailability is the key for determining systemic toxicity and should be assessed as appropriately as possible. This is particularly valid for complex inorganic materials, where the simple presence of the metal in the material will not necessarily grant that material with the biological properties of metal ion. It is generally the actual bioavailability of the metal ion (i.e. its release from the material) that will drive the toxicity. In the absence of bioavailability data from toxicity studies, in vitro bioaccessibility data can be used to estimate it, as will be further explained in Chapter 8. In situations where data or tests are insufficient to address bioavailability rigorously, the assumptions made regarding bioavailability should be clearly detailed in the health assessment, as should the associated estimated impact on results.

SPECIATION AND VALENCE

IUPAC describes chemical species as a specific form of an element defined as to isotopic composition, electronic or oxidation state, complex or molecular structure (Duffus et al., 2009). For the risk assessment of a given metal, the term 'speciation' refers to its different chemical species, including its physicochemical characteristics that are relevant to bioavailability. Different chemical species of an element show differences in solubility, bioavailability, and in the persistence of metals and metal compounds in the environment. For metals such as zinc, for example, some chemical forms of zinc may render the zinc ions less bioavailable, whereas for others such as lead, speciation alone may have less impact (or primarily in combination with factors like particle size). For some metals, different chemical species will exert different effects (e.g. inorganic arsenic and organic compounds, inorganic and organic mercury compounds, inorganic nickel compounds, and elemental nickel). Within a same 'metal family', such as chromium or cobalt, different forms will present different patterns in solubility, bioavailability, and hence toxicity. For example, Co (III) is only stable in an oxidic form like Co_3O_4. When dissolved it is reduced to Co (II), which is the most stable form in aqueous media. The poorly bioavailable Co (III) forms cause toxic effects at oral doses that are much higher than the readily bioavailable Co (II) forms (Kim et al., 2006; FDRL, 1984).

Speciation should initially be considered in any metals risk assessment to identify potential differences among the forms. Factors that contribute to the impact of the 'metal speciation' include: carrier-mediated processes for specific metal species, the valence state, the nature of metal binding ligands, whether the metal is an organic or inorganic species, and biotransformation of metal species. This is further discussed, for example, by Yokel et al. (2006) and some excerpts are as follows:

- Membrane transport proteins (carriers or channel molecules in the plasma cell membrane) have been identified for a number of metals and can be quite selective, transporting one metal species but not another, due to selective binding sites that recognise specific chemical species (e.g. chromium (III) is unable to enter cells but chromium (VI) enters cells through membrane anion transporters). Divalent cations such as cadmium, cobalt, copper, iron, lead, and zinc are substrates for the divalent metal transporter 1 (DMT-1 also known as DCT-1 or nramp-2) that may mediate their intestinal uptake (Gunshin et al., 1997; Picard et al., 2000 cited by Yokel et al., 2006).
- Valence, which is determined by the number of electrons in the outer shell of the atom that it will lose, add, or share when reacting with another atom, is also another contributing factor to 'speciation'. Several metals have more than one biologically relevant valence state (e.g. selenium (II, IV, and VI)). Valence can affect the absorption, distribution, biotransformation, and elimination of a metal and therefore its toxicity. The available evidence does generally not allow to draw a direct correlation between the exact valence and the severity of toxicity, except for elements such as mercury, where Hg(II) produces more toxicity than Hg(I) and Cr(VI) which is more toxic than Cr(III) (Yokel et al., 2006). The severity of toxicity seems to be dependent on how metal valence states interact with other physiological variables that determine the adverse response.
- Ligands are atoms, ions, or functional groups that can bind to one or more central atoms or metal, to form a complex. The ligand will affect the metal's chemical form and therefore its speciation, and influence its absorption, distribution, biotransformation,

and elimination. For example, intestinal aluminium absorption is facilitated by its complexation with citrate (Goyer and Clarkson, 2001). In the absence of a complexing ligand such as citrate, hydrolysis of aluminium complexes results in sparingly soluble species that are less well absorbed (Caruso et al., 2006).

- The comparison of the relative absorption of inorganic with organic metal species generally suggests a greater absorption of organic forms with distribution, biotransformation, and excretion often differing as well.
- Many metals with completely filled inner electronic orbitals will not readily undergo in vivo biotransformation among valence states (e.g. aluminium, lead, tin, zinc). However, others such as chromium (VI) are unstable in vivo and will rapidly be reduced to chromium (V) which is labile and converts to chromium (IV) and ultimately to chromium (III), which is quite stable due to its complexation with DNA and proteins (Kasprzak, 1991).

Given the recognised effects of speciation state on the 'fate' of the metal, it is recommended, when feasible, to incorporate speciation/valence state-specific differences in the assessment of toxicity. The assessment of the effect of elemental species will also depend significantly on the quality of the analytical and toxicological input. The most recent developments come from the field of proteomics and metallomics, dealing with the determination of trace metals in biomolecules (Buscher and Sperling, 2005).

In the meantime, when the speciation of the complex inorganic material is not known, in particular for Unknown or Variable composition, Complex reaction products, or Biological materials (UVCBs), possible ways forward are proposed (see Chapters 8 and 11).

PARTICLE SIZE

When assessing inhalation exposure to metals and complex inorganic materials, not only should the amount of the specific substance of interest present in air be considered, but also its chemical speciation and particle size distribution, if exposure is to dust. The latter parameter critically determines the most likely deposition of the substance in the respiratory tract and its subsequent translocation, with consequences on the overall absorption of inhaled particles (e.g. ECHA, 2016; HERAG, 2007b; Oberdörster et al., 1994; Oberdörster 1993).

Several methods exist for the determination of the particle size distribution of the airborne dust of a substance, in the lab or at the workplace. A review of these is given in a draft guidance document by the ECB (2002) or in a publication by Volkwein et al. (2011), which gives a brief overview of recent materials and methods. A comprehensive overview of the methodologies is also provided by Kenny et al. (1997) and Vincent (2007).

If aerosol particle size data for a particular substance are available, then a mass–median–aerodynamic diameter together with a geometric standard deviation can be calculated commonly under the assumption of an underlying lognormal particle size distribution. Such parameterised distribution can be used to predict the fraction of inhaled dust that deposits in the respiratory tract with the aid of, for example, the Multiple Path Particle Deposition Model, which was developed to calculate the deposition and clearance of aerosols from ultrafine to coarse mode in the respiratory tract of humans and rats (CIIT Centers for Health Research, 2011; Anjilvel and Asgharian, 1995; RIVM, 2002).

The particle size distribution of the aerosol does not, however, need to be known in every situation. The general approach in occupational sites has been to use fractions of the airborne dust that are sampled according to specific sampling conventions (e.g. inhalable, thoracic, or respirable as defined in EN 481, 2008; Nieboer et al., 2005; Oller and Oberdorster, 2010) except in the case of fibres. In the EU, based on the publication of EN 481 2008, different fractions of airborne dust have been defined, of which the following are often referred to:

- Inhalable aerosol fraction: is the fraction (aerodynamic diameter $\leq 100\,\mu$m) of total airborne particles that could enter the body through the nose and/or mouth during breathing.
- Thoracic aerosol fraction: is a subfraction of the inhalable fraction (corresponding to fractions of total aerosol of 50% cut-off point at aerodynamic diameter $= 10\,\mu$m and 1% at aerodynamic diameter $= 28\,\mu$m) composed of particles which could penetrate into the tracheo-alveolar region of the lung (i.e. the whole region below the larynx).
- Respirable aerosol fraction (or alveolar fraction): is the subfraction of the inhaled particles (corresponding to fractions of total aerosol of 50% cut-off point at aerodynamic diameter $= 4\,\mu$m and 1% at aerodynamic diameter $= 10\,\mu$m) that could penetrate into the alveolar region of the lung (i.e. includes the respiratory bronchioles, the alveolar ducts, and sacs).

This means that if measurements of airborne dust take place, it should be mentioned for which of the aerosol fraction(s) (e.g. inhalable, thoracic or respirable), the measurements have been performed. ECHA guidance (2016) stipulates that unless information on the particle size distribution or the aerosol fraction occurring in the workplace is available, it should be assumed that all particles are respirable (i.e. have the potential to penetrate into the alveolar region of the lungs) (ECHA, 2016).

The deposited, and subsequently retained, doses in regions correlate closely with long-term toxic effects. Still, differences in deposited doses between animals and humans due to various factors, including particle size differences of aerosols, have not been consistently taken into account in risk assessment. The US EPA has introduced a concept that allows adjustments for differences between animals and humans to be made. It considers several factors such as particle size distributions that differ between a laboratory and a workplace exposure, as well as biological differences between animals and humans, such as lung morphology. This concept allows, as a first level adjustment, the calculation of a 'regional deposited dose ratio (RDDR)', which represents the ratio between the dose deposited in a given region of the respiratory tract when animals are exposed to a given concentration of the particle in air, and the dose to the same respiratory tract region received by humans exposed to the same air concentration. As a second level adjustment, Oller and Oberdorster (2010) describe an approach to calculate human equivalent air concentrations (HEC) using workplace particle size information instead of the laboratory aerosol particle size distribution. Workers' exposure to the HEC results in the same deposited dose in the respiratory tract (per unit surface area) as achieved in animals exposed to the experimental particle size distribution. The HEC approach has been applied to some metals under the EU REACH Regulation (Oller et al., 2014; Buekers et al., 2015).

ESSENTIALITY

The traditional criteria to define essentiality for human health are that absence or deficiency of the element from the diet produces either functional or structural abnormalities and that the abnormalities are related to, or a consequence of, specific biochemical changes that can be reversed by the presence of the essential element (WHO, 1996, 2002). Metals identified as ETE for humans include copper, chromium, iron, magnesium, cobalt, manganese, molybdenum, selenium, and zinc, and when not present in sufficient concentrations can cause impairment of growth and development, neurological effects, inefficient reproduction, loss of tissue integrity, or defects in physiological and biochemical functions. ETEs are important constituents of several key enzymes and play important roles in various oxidation–reduction reactions. It is not the 'free metal ion' that is essential for humans, but always a complex formed by the metal ion with a biological target. Examples are iron, which is essential as the centre of heme, where it is complexed by four nitrogen atoms and a carbon via a methine bridge, or the copper ion, which is tetrahedral and coordinated by four sulphydryl bridges in superoxide dismutase, an enzyme that also binds zinc for structural stabilisation. Copper is one of the best studied examples of an ETE. It is essential, however, to know that 'free' copper ion is undetectable in human cells (Kaplan and Maryon, 2016). Copper is always bound to chaperone proteins, transporters or to its essential targets, including catalase, superoxide dismutase, peroxidase, cytochrome c oxidases, ferroxidases, monoamine oxidase, and dopamine β-monooxygenase.

ETEs are subject to homeostatic mechanisms that maintain optimum tissue levels over a range of exposures and may involve metal interactions. These mechanisms enable adaptation to varying nutrient intakes or excretions to ensure a safe and optimum systemic supply of ETEs for the performance of essential functions (HERAG, 2007c). The efficiency of these homeostatic mechanisms may vary within populations and with levels of intake of the ETE, related to factors that influence absorption, age-related factors, and dietary and nutritional interactions. The homeostatic mechanisms may also involve an interaction with another essential metal (Goyer et al., 2004), for example, high zinc intake inhibits copper absorption by competing for transport. Deleterious effects occur when ETE exposure is above or below the range which can be accommodated by homeostatic mechanisms. The possibility to have 'deficiency' poses a challenge to basic assumptions of risk assessment where the underlying paradigm aims at minimising exposure as far as possible: for an ETE, an unbalanced concern over high dose effects may lead to recommendations that cause harm from deficiency. The WHO (1996) has therefore proposed to use a U-shaped dose–response curve rather than a 'classical' linear dose–response to consider both deficiency and toxicity (Fig. 3.1). The acceptable range of oral intake (AROI) covers the range in the U-shaped dose–response curve from essentiality to toxic levels.

It is therefore recommended that when assessing the hazard and risk of a complex inorganic material to check whether it includes an ETE, as the way to establish dose–response relationships should in that case consider the possibility of deleterious effects occurring both at (too) low and (too) high doses.

POSSIBLE EFFECT OF COUNTERIONS

The assumption that the free metal ion is responsible for the effect may lead the assessor to conclude that the counterion present in the metal compound is largely irrelevant in

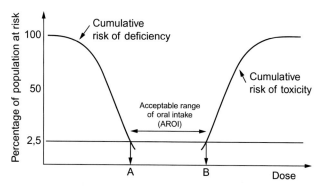

FIG. 3.1 WHO concept on deficiency and toxicity from excess oral intake of ETEs. As ETE intakes drop below A (lower limit of AROI where 2.5% of the population under consideration will be at risk of deficiency), an increasing proportion will be at risk of deficiency. At extreme low intakes all subjects will manifest deficiency. As ETE intakes exceed B (where 2.5% of the population under consideration will be at risk of toxicity), a progressively larger proportion of populations will be at risk of toxicity. *Based on WHO (World Health Organization): Principles and methods for the assessment of risk from essential trace elements, Geneva, 2002, World Health Organization, Series Environmental Health Criteria, No. 228, ISBN 92-4-157228-0.*

producing the effects to be assessed. This assumption could, however, be affected by interactions between the metal ion and the counterion (such as cyanates, oxalates, chromates), which could obscure the role of the metal ion either in acute or in repeated dose studies. The counterion might alter the toxicological properties of the metal ion, leading to a different hazard conclusion compared to other ionic pairs of that metal, and its influence should be checked for each endpoint. If there is reason to believe that the counterion present in the compound can significantly influence its toxicity, it should be taken into account in the evaluation (OECD, 2014). This is particularly important in the comparison of acute toxicity and/or local dermal effects where the counterion may result in the substance being classified as corrosive (NTP, 2016; Appel et al., 2001).

METALLOTHIONEIN

Toxicological effects will occur when the metal ion reaches a critical target within the body and the defence and repair mechanisms of the cells are exceeded. One of the mechanisms of protection the cell may employ against metal ion toxicity is the formation of nontoxic metal–protein complexes with various proteins such as metallothioneins. Metallothioneins are a group of low-molecular-weight proteins, rich in sulphydryl groups that serve as ligands for several metals. Metallothioneins have multiple binding sites with different affinities for metals. In vitro studies have found that silver has the highest affinity, then in descending order mercury, copper, bismuth, cadmium, lead, and zinc ions (Kagi and Kogima, 1987 cited by Goyer, 2004). The types of metal ions bound to metallothioneins differ according to the species, organ, and previous exposures to metals, but most metallothioneins contain at least two different types of metals (e.g. metallothioneins isolated from adult human livers contain mainly zinc and copper, while those from human kidneys contain cadmium, copper, and zinc) (Goyer, 2004).

In most cases, the metallothioneins are inducible and perform a number of functions, including serving as a storage protein for zinc and copper in the liver, kidney, brain, and possibly skin and having an important protective role in cadmium toxicity (Goyer and Clarkson, 2001). There is some evidence on the activity of metallothionein as a scavenger of free radicals (Carpene et al., 2007).

METAL–METAL INTERACTIONS

Goyer et al. (2004) distinguish three classes of interactions between metals: between essential metals, nonessential metals, and between essential and nonessential metals. Antagonism between metals, and much of the uptake and/or sequestration behaviour of metals, occurs as a result of commonalities in uptake mechanisms. For example, the protective effects of zinc against toxicity from excess copper are most likely due to a reduced gastrointestinal uptake of copper. Such interactions are also at play in the consideration of essential and nonessential metals. These interactions are currently under further investigation, in particular in the context of mixture toxicity. One practical consequence of these existing interactions is that testing 'excessively' high doses for essential elements may yield results which require careful interpretation. For example, if very high doses of zinc are administered to a tested animal, zinc will interfere with the copper metabolism, inducing copper deficiency with corresponding deleterious effects.

NEW TOXIC ENTITIES FORMED BY THE ASSOCIATION OF METALS WITH OTHER ELEMENTS (METALLIC OR NOT): SINTERED (WC-CO), DOPED (ITO) MATERIALS

Owing to their unique properties and reactivity, many metals have significant uses in industry as metal compounds, or as ingredients of mixtures. In addition to alloys, where chemical bonds are quite specific, some industrially relevant mixtures combine metallic and nonmetallic elements to form materials with desired properties, such as improved optical properties [doped indium-tin-oxide (ITO) materials], increased hardness (such as cobalt combined with tungsten and carbide), or chemical catalysis (metals combined with organic ligands called 'carboxylates'). In some cases, the association of metal elements with other metallic or nonmetallic elements can generate new properties that cannot be predicted by the addition of the properties of individual ingredient alone. For example, WC-Co (tungsten carbide–cobalt) demonstrates an increased in vitro genotoxicity, as compared to the properties of its individual components, which may be due to a 'combination effect' leading to an increase in the ability to generate reactive oxygen species (ROS) (De Boeck et al., 2003; Lison et al., 2001). The possibility that the association of metals with other elements, metallic or not, can lead to novel properties that will require special attention if they were to occur in complex inorganic materials.

3.2.3 Human Health Hazard Endpoints Requiring Specific Approaches

The endpoints requiring consideration of metal specificities when assessing the underlying mechanism of toxicity are briefly listed and explained below.

3.2.3.1 Toxicokinetics

Toxicokinetics investigate the movement of the metal through the body, from its uptake until its excretion. For some metals such as lead and cadmium, biological measurements may be available, reporting concentrations in blood or urine. Such data represent 'internal exposures', reflecting intakes as well as absorption/toxicokinetics, and allow a direct comparison with levels causing toxicity. For most metals, however, only 'external exposures' data are obtainable and some 'translation' into systemic doses via absorption factors will be needed before comparing with toxicity levels:

- Inhalation absorption factors are not commonly available for most metals, and are also difficult to measure. In the absence of substance-specific inhalation absorption factors, some defaults are proposed but those can be unrealistic and unnecessarily conservative. The fundamental basis for the evaluation of the absorption of metals from inhaled particles is the aerodynamic size-dependent deposition of the particles, and a tiered approach making use of particle size distribution data and the Multiple Path Particle Dosimetry Model (MPPD) model has been proposed in the Health Risk Assessment Guidance (HERAG) project (HERAG, 2007b).

- Absorption after intake via ingestion is known to vary strongly between metals, and can be influenced considerably by chemical speciation, solubility, dietary composition, and nutritional status. Metal–metal interactions are also known to affect absorption: examples are copper–zinc or copper–iron where any one of these essential elements at high intake is thought to compete with transport mechanisms of the other and thereby possibly induce deficiency. An alternative example is lead, which has no known physiological function in the human body, but partly enters the body by utilising calcium transport mechanisms. Nonlinear kinetics usually govern the absorption of metals from the gastrointestinal tract, (with a linear component at low intake levels followed by saturation occurring at higher intake rates, like for lead and copper). Where available, oral absorption data generated in humans should be considered first; animal data can, however, be useful to supplement information on bioavailability of different chemical species and saturation mechanisms. Metal- or metal compound-specific information on oral bioavailability was collected along the HERAG project to derive general conclusions on gastrointestinal uptake as well as information on modifying factors such as speciation, particle size, solubility, etc. In the absence of any metal-specific data, default absorption values (e.g. 100%) as proposed by some guidance documents may be used.

- Guidance documents usually suggest that in the absence of data on dermal absorption, a choice between two default values (10% and 100%) may be made, based on observations with organic molecules where substances with molecular weight (MW) >500 and extreme log Pow [(octanol/water partition coefficient) values (below -1 or above $+4$)] display a limited extent of skin permeation. These considerations do not apply to metals, as inorganic compounds require dissolution involving dissociation to metal cations prior to being able to penetrate the skin by diffusive mechanisms. Deviating from the default values, that is, using an alternative dermal absorption percentage value is generally accepted in regulatory forums, provided data are available and a scientific justification for the use of the alternative percentage is provided. In the context of the HERAG for metals project, the available information on dermal absorption of metals and proposed alternative default absorption factors have been collected in a fact sheet (HERAG, 2007d). In vitro studies of several metals (zinc, nickel, cadmium, antimony, copper, lead) have demonstrated that the penetration of the dermis by dissolved metal cations is generally low, that is, in the range of 0.1%–1%, depending on the resolution of the test system. In the case of lead, a combination of in vitro testing and toxicokinetic modelling has demonstrated that the assumption of a dermal absorption rate in excess of 0.01% would be in conflict with available biological data (blood lead) and is therefore implausible. Under industrial circumstances, many applications involve handling of dry powders, substances, and materials; however, dissolution is a key prerequisite for any percutaneous absorption. Considering the lack of dissolution in 'dry' scenarios, the fact sheet concludes that a lower default absorption factor may be assigned to scenarios where the product is not in aqueous or other liquid media (like in the in vitro experiments). The following default dermal absorption factors (reflective of full-shift exposure) have been proposed for screening risk assessment purposes: from exposure to liquid/wet media: 1.0%, from dry (dust) exposure: 0.1% (HERAG, 2007d).

Following absorption, the transport and the distribution of the metals in the human body will depend on several factors, including the form in which the metal is present in the blood. Among metals there will be large differences between the fraction that is transported to various organs and the fraction that will be excreted, with several factors at play like protein binding, the rate of organ vascular perfusion and biotransformation, the availability, and turnover rate of intracellular ligands for the metal. Various routes also exist by which a metal will be excreted from the body. The main routes are the gastrointestinal and renal but pathways like salivary excretion, perspiration, exhalation, lactation, exfoliation of skin, etc., may be considered in some specific cases. They can both reflect elimination and be used as index of exposure. The biological half-time will also vary greatly among metals and depends on the tissue. Thus, overall, each metal has its specificities when it comes to toxicokinetics and toxicodynamics. Toxicokinetic models that describe the processes of absorption, distribution, biotransformation, and excretion help refine the understanding of complex quantitative dose behaviours by helping to delineate and characterise the relationships between (i) the external/exposure concentration and target tissue dose of the toxic moiety, and (ii) the target tissue dose and observed responses. These are biologically and mechanistically based models and can be used to extrapolate the toxicokinetic behaviour of chemical substances from high to low dose, from route to route, between species, and between subpopulations.

The reader is referred to handbooks dealing in more detail with the toxicokinetics/toxicodynamics of metals for example, Nordberg et al. (2014).

3.2.3.2 Sensitisation

For a number of metals and metal salts, sensitising properties have been reported. These include beryllium, chromium(VI), cobalt, nickel, and chloroplatinates. Metals with sensitising properties are usually considered to be skin sensitisers, some of them also being recognised as respiratory sensitisers, depending on the metal, the route of exposure, and the type of allergic reaction.

Sensitisation to metal substances can be either mediated by humoral antibody-mediated reactions (type I hypersensitivity and to lesser extent type II and III hypersensitivity according to Coombs & Gell) or by T-cell mediated reactions (type IV hypersensitivity according to Coombs & Gell), with mechanisms differing between some metal substances. A more detailed overview can be found in Haschek et al. (2010).

For metals, skin sensitisation requires that an individual becomes immunologically sensitised to a metal by prolonged intimate contact between metal and skin to allow for sufficient time for (1) corrosion of the metal, (2) dissolving the corrosion product into metal ions, and (3) dissolution of sufficient metal ions (above the threshold) through the skin to interact with the immune system. For water-soluble metal substances, the time for steps (1) and (2) are not needed but step (3) is still relevant and takes time. This is termed as the induction or sensitisation phase, which can occur with sufficient exposure (above threshold) to metal substances in the workplace or by contact with consumer products such as clothing buttons or jewellery. Metal substances generally do not penetrate the skin easily, and the risk for sensitisation is elevated in individuals with damaged skin (e.g. body piercing), as well as in conditions of heat or humidity, leading to increased metal corrosion, ion release in sweat, and increasing skin permeability. In a sensitised individual, re-exposure to smaller amounts of the metal ions (less than needed during the sensitisation phase) can cause an allergic reaction, termed elicitation. Elicitation can occur as a skin reaction at the site of the re-exposure, but may also (less commonly) occur in remote skin locations. Until now, no standardised animal model or in vitro model for the evaluation of respiratory sensitisers has been established; information on human exposure is the main source of data for the assessment of respiratory sensitisation of metal substances.

The differentiation between respiratory sensitisation and irritation remains a challenging task, since involved biochemical processes are not yet completely investigated and the development and validation of distinct test systems is therefore difficult. So far, only a limited number of effects allow the differentiation between respiratory irritation and sensitisation, such as the balance of immunologic cells and the effects on lung tissue and other lung functions. The major point of differentiation between sensitisation and irritation is that allergic responses require a preceding sensitising exposure event or an increasing severity of the lesions on successive exposure to smaller quantities of an antigen than the initial sensitising exposure.

One characteristic of differentiation between respiratory sensitisation and chronic alveolar irritation is manifested by proliferation and persistence of type II epithelial cells, fibrosis following interstitial thickening and accumulation of mononuclear cells, whereas (acute) irritation is represented by reversible effects like dyspnoea, rhinitis, redness, hyperaemia, oedema, inflammation, and thickened mucous layer, as well. The problematic issue is that occupational exposure often entails a chronic irritation, which comprises immunological processes, such as inflammation, similar to those seen in respiratory sensitisation. In consequence of chronic irritation, eosinophils and lymphocytes migrate from the blood into the affected tissues and balance of immunologic cells is disrupted in a manner similar to sensitisation reaction. In both sensitisation and chronic irritation cases, persistent inflammation results finally in structural changes (remodelling), like smooth muscle sickening, tissue fibrosis, and mucus cell hyperplasia. In the case of chronic exposure to irritating or sensitising agents, which is the case in occupational settings, a clinical differentiation between irritation and sensitisation remains problematic.

A prerequisite for any potentially skin sensitising substance is its bioavailability, which is its ability to penetrate the stratum corneum of the skin. It is generally assumed that sensitisers enter the skin by diffusion, and this is aided by a small molecular weight (MW < 500) and a high Pow (octanol/water partition coefficient). However, as metals may become bioavailable in sweat mainly as cations, and penetration into intact skin is usually low; the Pow value is therefore not relevant to metals.

Recent evidence suggests that the steps of binding to antigen-presenting cells and intra-cellular processing of antigens (i.e. type I hypersensitivity) is the only one mechanism by which metals can induce hypersensitivity. Another, more prominent, mechanism by which metals induce hypersensitivity is the T-cell specific reaction (type IV hypersensitivity). In addition, there is also evidence that some metals also bind to toll-like receptors. A human toll-like receptor (TLR4) has been identified as a binding target for nickel, cobalt, and pal-ladium, and represents one pathway of how metals may induce a response of the innate immune system (Rachmawati et al., 2013; Raghavan et al., 2012; Schmidt et al., 2010; Schmidt and Goebeler, 2011).

Of the sensitising metals, beryllium needs to be discussed separately, as it has a very spe-cific mode of action. Individuals can develop a beryllium-specific sensitisation reaction, which can be detected with a blood test (beryllium lymphocyte proliferation test). Of the sensitised subjects, some may develop chronic beryllium disease (CBD), a slowly progressive respira-tory disease characterised by granulomas in the lung. The skin is not affected in berylliosis. However, in order to avoid sensitisation, inhalation as well as dermal exposure to beryllium must be strictly avoided. While genetic predisposition may play a role in the sensitisation to other metals as well, the details of the interactions are unclear.

MULTIPLE SENSITISATION

Multiple sensitivity to metals is often described in the literature, but mechanisms of the im-munologic reactions are only partially elucidated. Multiple sensitivity can be roughly divided into co-sensitisation and cross-sensitisation. The term co-sensitisation describes a concomi-tant reaction to substances that occur side by side and induce allergic reactions. Hereby each allergen induces specific reactions. A prominent example is co-sensitivity to nickel and cobalt where animal testing has demonstrated no cross-reactivity but only co-reactivity (Lachapelle and Maibach, 2012; Wahlberg and Lidén, 2000; Lidén and Wahlberg, 1994). Thus, it is import-ant to evaluate concomitant factors such as impurities or confounding factors such as envi-ronmental/occupational conditions for the assessment of sensitisation hazard.

In the literature, the term cross-sensitisation often occurs, which should be explicitly sepa-rated from co-sensitisation, since it describes the case that elicitation is not only induced by the primary allergen, but also from substances similar in structure (secondary allergens). In this form of sensitisation, structural analogues to the primary inducing agent can trigger the allergic cascade (Migueres et al., 2014). Examples of cross-sensitisation are some structural analogue antibiotics such as neomycin, kanamycin, and gentamycin (Lachapelle and Maibach, 2012).

For risk management of sensitisation, the only effective method is the avoidance of known sensitisers above exposure amounts exceeding the thresholds, and in the case of known cross-reactivities additional precautions might be necessary for sensitised individuals com-prising the avoidance of secondary metal allergens.

3.2.3.3 Genotoxicity–Mutagenicity–Carcinogenicity

A mutagenic effect is a permanent and transmissible change in the amount or structure of the genetic material. Alterations may involve a single gene, a block of genes, a whole chromosome, or a set of chromosomes. A mutation in the germ cells of sexually reproducing organisms is of paramount concern since it may be transmitted to the offspring. A mutagen is thus an agent that gives rise to an increased frequency of mutations. The term genotoxicity is inclusive of mutagenicity but further encompasses a wider range of events, which are not necessarily associated with mutagenicity (e.g. causing cytotoxicity).

The mutagenic and/or genotoxic properties of a substance are a key element upon which the hazard classification of substances or the material containing the substance is based (UN GHS, 2015; EU CLP, 2008). Determination of the mutagenic/genotoxic potential is also crucial when evaluating the risk for carcinogenicity associated with exposure to substances, as such or contained in complex inorganic materials. On the basis of the knowledge the assessor has on mutagenicity/genotoxicity, he/she will indeed be able to make a distinction between carcinogens likely causing tumours via interaction with genetic material (genotoxic carcinogens) and carcinogens causing tumours by other mechanisms (nongenotoxic carcinogens). Genotoxic substances are presumed to exert effects without a threshold and exhibit low-dose linearity for the induction of effects (e.g. gene or chromosome mutations). To the contrary, nongenotoxic carcinogens are expected to act via nonstochastic mechanisms and exhibit thresholded dose–response relationships. An example of this nonstochastic mechanism is aneuploidy (defined as a change in chromosome number from the normal diploid or haploid number other than an exact multiple) where the target will not be the DNA but proteins, which are present in multiple copies and thus represent highly redundant targets. Consequently, a minimal level of damage is required before aneugenic effects will be visible. This level is assumed to be thresholded, below which no disturbance of chromosome segregation occurs. Other examples include substances that induce epigenetic changes.

The assessment of the mutagenic/genotoxic properties of substances is typically initially based on in vitro testing, with follow-up work using in vivo testing as appropriate. The determination of the mutagenic/genotoxic potential of metals (and hence, the complex inorganic materials) requires some specificities. More information on the data to collect and on the limitations of current testing strategies can be found in Chapter 6.

The distinction between genotoxic and nongenotoxic carcinogens is often problematic when applied to metals. Almost all metals tested have mixed genotoxic profiles in vitro (e.g. negative Ames test, positive or negative mammalian cell tests) and most will also induce weak positive or conflicting (positive and negative) responses in vivo. Since the process of carcinogenesis can involve a sequence of transition stages that entail genetic alterations and nongenetic events, one cannot per se conclude that in the presence of positive genotoxic data, the metal shall be regarded as carcinogenic. Given this uncertain relationship between genotoxicity and carcinogenicity for metals, alternate rules of procedure are needed whereby the relevance of genotoxicity for metal carcinogenicity can be assessed and applied to risk assessment. Furthermore, the indirect nature of the genotoxic effects of metals (through the generation of ROS) can make consideration of dose–response more difficult. For these reasons, the development of an adverse outcome pathway (AOP) should be considered, when there is evidence for genotoxic events. An AOP is a conceptual construct that portrays existing knowledge concerning the pathway of causal linkages between a molecular initiating event and a final adverse effect at a biological level of organisation that is relevant to a regulatory decision (Ankley et al., 2010). Since genotoxic effects are considered relevant events within an adverse outcome pathway, these should be studied carefully. One needs to differentiate between the primary and the secondary or indirect genotoxicity. Primary genotoxicity might occurs as a direct interaction of the metal with the genetic material and should be detectable in established in vitro and in vivo genotoxicity assays. The term secondary or indirect genotoxicity refers to a substance-induced (non-genotoxic) effect which subsequently leads to genotoxicity. An example is the pathway of DNA damage resulting from reactive oxygen/nitrogen species (ROS/RNS), generated during particle-elicited inflammatory response.

Further information can be obtained from animal studies: a number of metals have been demonstrated to be carcinogenic in animals; however, the mechanism by which carcinogenic responses are induced, and the significance of animal responses for humans vary depending on the metal, and must still be assessed via a careful weight-of-evidence evaluation. Indirect mechanisms appear to be responsible for many carcinogenic responses. Given that the mechanism of action of many metals studied to date appears to be unique, demonstration that a given mechanism of action is, or is not, relevant to humans must be conducted on a case-by-case base. A probable exception to this case-by-case approach would be instances of pulmonary tumours produced by particulate overload mechanisms in the rat. The 'generic' mechanism would be common to noncytotoxic, poorly soluble particles that induce tumours of alveolar origin in the rat following chronic inhalation exposure to excessive (e.g. $250\,mg/m^3$ or above in the case of titanium dioxide) concentrations, exceeding the clearance capacity of rat lungs. Such substances do not generally induce tumours in other experimental animal species (e.g. hamster, mice). The rat's specificity of this effect is supported by the absence of similar responses in the other tested species, coupled with negative human epidemiology, suggesting a lack of relevance of this observation to humans. It has also been shown that the disposition of such inhaled particles is substantially different between nonhuman primates and rats, with the latter being particularly sensitive. The lung cellular responses of rats exposed chronically to particles are hyper-inflammatory and hyperplastic, while primates show normal physiological reactions such as particle accumulation and macrophage responses to inhaled particles. Substances exhibiting such a response profile (e.g. titanium dioxide, tricobalt tetroxide, molybdenum sulphide or dioxide, iron (II, III) oxides, vanadium carbide) should be carefully evaluated and the relevance of these effects should be considered when, for example, classifying the material on the basis of such a profile (ECETOC, 2013).

Epidemiology studies have suggested associations between several metals and human cancer. However, in some cases, initial assumptions regarding causal relationships between specific metals and human cancer have been questioned in view of the significant co-exposures to other metals in the workplace (e.g. simultaneous exposure to arsenic and cadmium where the risk of lung cancer due to cadmium is estimated to be small in comparison with the wide effects of arsenic). While confounding by co-exposures is not a specific property of metals per se, it is reflective of the long industrial history of metal production and the difficulty to fully evaluate the complex co-exposures associated with metallurgical processes. This highlights the need for more comprehensive and refined exposure assessments to be incorporated into future epidemiological studies of metals. Metal ore bodies are, by their intrinsic nature, extremely complex mixtures and the exposure patterns at primary metallurgical facilities are more complex than had been initially realised. While excess cancers may exist at a number of primary metallurgical facilities, the causative agents that are responsible for these excesses can only be defined if more comprehensive and refined exposure assessments are conducted and by considering animal studies with individual substances in conjunction with the results of epidemiological studies with mixed exposures.

Overall all these aspects point to the careful use of weight-of-evidence and expert judgement when evaluating the carcinogenic hazard of complex inorganic materials, where in addition to the number of constituents or ingredients, specificities in their assessment shall be considered.

3.2.3.4 *Reproductive Toxicity*

Metals are diverse in their ability to impact on reproductive organs' fertility and/or to cross the placenta to impact upon the developing foetus (development toxicity), but it is evident that specific mechanisms exist which can regulate the ability of metals to reach these target tissues. This is most evident for metals that are under tight homeostatic control (e.g. copper, zinc) or which bind to common carrier systems and sequestration proteins that modulate the pharmacokinetics and homeostasis of essential metals. Under conditions of excessive exposure, this binding can result in perturbations of ETE homeostasis that can have adverse impacts upon development through indirect mechanisms of action. Thus, excessive amounts of zinc can induce copper deficiency in the developing foetus and it is suspected that cadmium may exert adverse developmental effects via the induction of zinc deficiency. In the instance of both cadmium in excess and zinc deficiency, effects are not mediated by the intrinsic capacity of the material to impair processes such as foetal development but by indirect mechanisms (disruption of homeostatic control mechanisms for trace mineral metabolism). These observations justify paying attention to the design and interpretation of reproductive studies, in particular to consider the toxicokinetic properties of metals and, in certain instances, the unique mechanisms through which metal interactions could influence reproductive system function. In the case of the latter, the interpretation of toxicological studies, and extrapolation from animal studies to humans, is facilitated if the basic dose–response for deficiency and the regulation of metal uptake, distribution, and excretion by homeostatic control mechanisms is fully documented.

3.3 ENVIRONMENT

3.3.1 Main Mechanisms of Toxicity of Inorganic Complex Materials

Many of the aspects and mechanisms described above for metal toxicity to human health are also valid for organisms living in the aquatic, sediment, and soil environments. The first critical driver for the ecotoxicological effects of complex inorganic materials, metals, and metal compounds released into the environment is their fate and behaviour and resulting environmentally available fraction of the constituent metals in the different compartments (water, sediment, soil). These available fractions depend on both the properties of the complex inorganic materials and the released metal compounds (e.g. solubility, adsorption to solid phases) and physicochemical properties of the environmental media (e.g. pH, organic matter). The degree to which metals are available and cause toxicity to aquatic, sediment burying, and terrestrial organisms is indeed determined by site-specific geochemical conditions controlling the speciation/precipitation and/or complexation of metals. This is illustrated in Fig. 3.2.

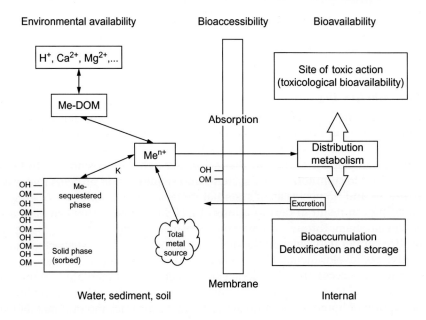

FIG. 3.2   Simplified conceptual outline for metals bioavailability (OECD, 2016). *DOM, dissolved organic matter; OH, binding site occupied by proton; OM, binding site occupied by metal.*

These physicochemical considerations (environmental availability) must consequently be linked to different ecological receptors taking different uptake routes into account. Within the different environmental compartments processes such as adsorption to particles and organic carbon, metal sequestration, and competition with major cations will reduce the free metal ion concentration that is available to bind with a biological membrane and would be bioaccessible. Once inside an organism, the bioaccessible metal pool that is actually available to elicit a potential adverse effect will depend on the way the metal is internally distributed, metabolised, excreted, detoxified, and bioaccumulated.

The following terms are defined in the OECD guidance on the incorporation of bioavailability concepts for assessing the chemical ecological risk and/or environmental threshold values of metals and inorganic metal compounds (OECD, 2016):

- Environmentally available fraction: the total amount of metal in soil, sediment, water, and air that is available for physical, chemical, and biological modifying influences (e.g. fate, transport, bioaccumulation). It represents the total pool of metal at a given time in a system that is potentially bioavailable (McGeer et al., 2004; US EPA, 2007).

- Bioaccessible fraction: the fraction of the environmentally available metal that actually interacts at the organism's contact surface and is potentially available for absorption or adsorption by the organism.
- Bioavailability (or biological availability): the extent to which a substance is taken up by an organism, and distributed to an area within the organism. It depends upon physicochemical properties of the substance, anatomy and physiology of the organism, pharmacokinetics, and route of exposure (UN GHS, 2013).

Hence metal bioavailability refers to the fraction of the bioaccessible metal pool that is available to elicit a potential effect following internal distribution; metabolism, elimination, and bioaccumulation.

3.3.2 Key Mechanistic Aspects to be Considered When Managing the Environmental Risk of Complex Inorganic Materials

The ecotoxicity of complex inorganic materials is related to the bioavailability of the individual metal constituents that they can release into the environment. Therefore, the mechanistic aspects considered for the environmental risk assessment of complex inorganic materials relate to the principles of environmental risk assessment of metals in general.

3.3.2.1 Natural Background and Essentiality

Metals are naturally occurring substances and abundantly present in the natural environment. Naturally occurring background concentrations of metals in water, sediment, and soil, that is, the concentrations that existed before any human activities, vary markedly between geologically disparate areas, and are determined by various factors such as the site- or regional-specific bedrock composition, effects of climate on the degree of weathering, etc. The variation in site-specific conditions has resulted in ranges of naturally occurring background levels that can span several orders of magnitude for one metal. Despite the widespread occurrence of diffuse anthropogenic contribution to metal concentrations in the environment, variations in the natural background concentration at the regional or continental scale is still much larger than the anthropogenic impact at this scale (Reimann and Garrett, 2005; Reimann et al., 2017).

Several metals are also essential elements for living organisms (e.g. Fe, Co, Cu, Mn, Mo, Se, Zn). An element is considered essential when (1) it is present in living matter; (2) it is able to interact with living systems; (3) a deficiency results in a reduction of a biological function, preventable or reversible by physiological amounts of the element (Mertz, 1974). Similar to humans, most environmental organisms are also able to actively control their internal levels of essential elements (homeostasis) within an environmentally relevant range of exposure concentrations. As living organisms have evolved in the presence of the natural metal background concentrations and because some metals are essential elements, living organisms have become conditioned to varying backgrounds. Consequently, the variation in natural background concentrations for metals contributes to an increase in ecosystem differentiation or biodiversity (Crommentuijn et al., 1997). For this reason, exposure of organisms to the natural background level reflects in fact the theoretical lower limit of the concentration, which from an evolutionary perspective, does not present a potential disruption of the genetic pool composition of a species. This theory is applicable to all the metals and is even more crucial for essential metals. As a result, the sensitivity of organisms to metals is determined to a large extent by the environmentally available concentration that the organisms experienced during their lifetime and their developed capability to cope with this concentration. This implies that organisms cultured in an environment with a low (essential) metal concentration may exhibit an overall decreased fitness (deficiency issues) and become more sensitive to stress, including exposure to metals, even essential ones. Conversely, organisms cultured in media with elevated metal concentrations (both essential and nonessential metals, e.g. natural waters or contaminated waters) may become less sensitive to those metals. This phenomenon is related to the 'biogeochemical region' concept (Fairbrother and McLaughlin, 2002). Such phenomena as acclimation/adaptation and essentiality should therefore be taken into account in any risk assessment on metals (MERAG, 2014).

3.3.2.2 Speciation (Environmental Availability)

The form and speciation of metals released from complex inorganic materials in the environment determines the degree to which they are available and cause toxicity to aquatic, sediment burying, and terrestrial organisms. It is generally considered that the dissolved free ionic metal species is far more bioavailable than most complexed metal species (e.g. Zitko et al., 1973). It should, however, be noted that the free ion is not necessarily the best predictor of toxicity for all metals, and other metal species (e.g. AgCl, HgS, SeO^{2-}, AsO_4^{2-}) may contribute to the observed toxicity (Campbell, 1995).

Several site-specific geochemical parameters control the speciation of metals by, for example, adsorption, precipitation, and complexation reactions (Table 3.1).

The distribution of metals over the solid and liquid phase is a key aspect that determines the fate and availability of metals. This solid–liquid distribution is not only controlled by pure adsorption/desorption mechanism, but several other processes like precipitation or encapsulation in the mineral fraction also play a role. Furthermore, unlike for organic substances, the sorption to suspended matter in the water column or solid components in soil and sediment is not driven by sorption to organic constituents only, but also by other solid-phase constituents such as clay minerals and oxides. Environmental conditions (pH, redox conditions, temperature, ionic strength, equilibration time, metal loading, etc.) and the composition of the liquid and solid phase have a strong effect on the solid–liquid distribution of inorganic substances. The relative importance of all these factors varies from metal to metal but pH is generally regarded as a major factor controlling sorption.

In the aquatic environment, speciation of the dissolved metal fraction is further controlled by pH and complexation, especially by dissolved organic carbon (DOC). In addition, several cations (Ca, Mg, Na, K) are known to compete with metal cations for binding to solid phases and affect the concentration of the free metal cations in the solution phase. Similarly, anions such as phosphate, sulphate, and nitrate can act to antagonise the adsorption of metals present as oxyanions (e.g. chromate, molybdate, and selenate).

Metal bioavailability in sediments is governed by processes different from that in water (diffusion, re-mobilisation, burrowing, oxidation/reduction processes, etc.). Multiple sorption phases must be considered for metals (e.g. organic carbon, sulphides, iron, and manganese oxy hydroxides) and their relative importance differs depending on the binding capacity of the metal and general chemical activity. Sulphides are a key partitioning phase-controlling

TABLE 3.1 Relevance of Physicochemical Properties for the Environmental Bioavailability of Metals

Compartment	Parameter	Relevance
Aquatic	pH	The fraction of free metal ion will generally decrease with increasing pH. The pH also determines the amount of protons which can compete with metal ions for binding to the organism's surface.
	DOC	Complexation of metal ions with dissolved organic carbon may affect metal bioavailability (and toxicity).
	Major cations (Ca, Mg, Na K, …)	Presence of cations such as Ca and Mg may compete with the metal cations for binding to sorption sites on solid phase and organism's surface.
Sediment	Sulphides	Sulphides form insoluble metal sulphide complexes with cationic metals, rendering metals unavailable.
	Organic carbon	Increasing organic matter content can result in decreasing bioavailability for both cations and anions.
	pH	Increasing pH causes an increase in sorption for cationic metals, and decreasing sorption for anionic metal ions. However, sediment systems are better buffered than soils.
	Eh	Reduced conditions mitigate metal toxicity (presence of sulphides and change in redox state metals, e.g. Cr^{3+} versus Cr^{6+}).
	Fe/Mn oxides	Increasing oxides content can result in increased sorption of both cations and anions.
Soil	pH	Increasing pH causes an increase in sorption for cationic metals, and decreasing sorption for anionic metal ions.
	Organic carbon	Increasing organic matter content can result in decreasing bioavailability for both cations and anions.
	Clay content	Increasing clay content can result in decreasing bioavailability for both cations and anions.
	Cation exchange capacity	Increasing eCEC (effective cation exchange capacity) indicates increasing sorption capacity and decreasing bioavailability for cationic metals in the soil.
	Fe/Mn oxides	Increasing oxides content can result in increased sorption of both cations and anions.

Based on ECHA (European Chemicals Agency): Guidance on information requirements and chemical safety assessment. Appendix R.7.13-2: Environmental risk assessment for metals and metal compounds, 2008.

cationic metal activity (mercury, silver, copper, lead, cadmium, zinc, nickel) and metal-induced toxicity in the sediment interstitial water system (US EPA, 2005). In addition, organic carbon plays an important role in copper bioavailability and binding to iron/manganese (oxy)hydroxides has been proven to be important for certain metals such as nickel (Costello et al., 2011). Changes in redox potential have a strong impact on speciation, behaviour, availability, and toxicity of metals as demonstrated with chromium where at a negative redox potential chromium is present in its less toxic trivalent form.

Soil properties, such as pH, organic carbon content, and texture, determine the amount and type of metal species available for uptake by plants, invertebrates, and soil micro-organisms, and the resulting toxic response or bioaccumulation of metal. The effective cation exchange capacity (eCEC) is a good predictor of availability for most metal cations, integrating effects of pH, organic matter, and clay content (Oorts et al., 2006; Smolders et al., 2009). For metals occurring as oxyanions, pH and oxide or clay content are better descriptors of metal availability (Song et al., 2006; Oorts et al., 2016). Initial partitioning of metals, taking place within hours after addition of soluble metals to a moist soil, are often followed by much slower reactions, termed fixation, or ageing, that further decrease the bioavailability of added metal with time. Examples of such ageing reactions are diffusion of metals into micro-pores, precipitation of insoluble phases, occlusion into organic matter and inclusion in the crystal structure of soil minerals. Next to the equilibration time, ageing processes also depend on the soil properties such as pH (Ma et al., 2006, 2013; Kirby et al., 2012).

The wide variation of the physicochemical characteristics encountered in the environment is the main reason why no clear relationships have been observed between measured total concentrations of metals in water, sediment, or soil and their potential to cause toxic effects. Therefore, taking speciation and bioavailability into account is a refinement which improves the precision of environmental assessment approaches as it helps to increase the realism of the assessment and can help regulators to better understand the likelihood of the occurrence of adverse effects due to metal contamination.

3.3.2.3 Factors Affecting Bioaccessibility and Bioavailability of Metals

As illustrated in Fig. 3.2, a second critical component of bioavailability, next to the chemical speciation of the metals released from the complex inorganic materials, involves the level that the environmentally available metal species can interact at the organism's surface and is potentially available for absorption or adsorption by the organism. The site of toxic action, or biologically sensitive receptor, is generally referred to as the 'biotic ligand' (BL). For fish this is, for example, hypothesised to be placed on the gill surface. The binding of the metal ion to this biotic ligand results in the manifestation of a toxic effect. However, the free metal ion concentration alone does not sufficiently explain metal toxicity and interactions with other ions for binding on the biotic ligand also need to be considered, resulting in the development of biotic ligand models (BLMs) for aquatic organisms (Pagenkopf, 1983; Di Toro et al., 2001; Paquin et al., 2002). A BLM integrates the chemical speciation calculations for the environment of interest, the binding of the toxic metal species on the biotic ligand, and the relationship between metal binding on the BL and the toxic response in the organism. Such BLMs are now available for several metals and several, mainly aquatic, organisms (OECD, 2016).

The BLM concepts are also illustrated to be valid for terrestrial organisms (plants, invertebrates, and micro-organisms) grown in nutrient solutions (e.g. Lock et al., 2006; Antunes et al., 2007; Mertens et al., 2007) or based on measured composition of the pore water (e.g. Thakali et al., 2006a,b; Lofts et al., 2013). This indicates that the free metal ion in pore water can also be considered as the toxic metal fraction for soil and sediment organisms and that its bioavailability depends on competition with other ions. However, due to the complexity of the solid–liquid distributions of all critical components of the soil solutions (cations, protons, metals, dissolved organic matter, etc.) and the occurrence of ageing reactions (i.e. transition from labile to fixed metal), a fully mechanistic model is not yet available allowing accurate prediction of metal toxicity to soil and sediment organisms starting from total concentrations. Currently, bioavailability corrections for soil are based on empirical regressions between physicochemical properties of soil and metal toxicity on microbial function, plants, and invertebrates, and on ageing factors determined by comparing metal toxicity measured immediately after freshly spiking a soil with a metal salt and soils equilibrated for a long time after spiking (Smolders et al., 2009). Similarly, empirical regressions between sediment properties and toxicity to several sediment organisms have been established (Vangheluwe et al., 2013).

3.3.2.4 *Relative Importance of the Dietary Route*

The BLM concept is focused on exposure via the water phase and as such, dietary exposure is not intrinsically incorporated in the model. Recently, DeForest and Meyer (2015) reviewed the state of science about dietborne-metal toxicity to aquatic biota, with a focus on 13 metals: Ag, Al, As, B, Cd, Co, Cu, Cr, Mo, Ni, Pb, V, and Zn. Of these metals, Ag, As, Cd, Cu, Ni, and Zn have been demonstrated to cause dietborne toxicity to aquatic organisms in laboratory exposures at potential environmentally relevant concentrations. That is, waterborne concentrations at or near existing waterborne criteria and guidelines (e.g. Ambient Water Quality Criteria (AWQC), Environmental Quality Standards (EQS), Predicted No Effects Concentrations (PNECs)) sometimes result in dietborne concentrations which contribute to increased toxicity to the most sensitive species (usually filter-feeding herbivores such as freshwater daphnids and saltwater copepods) beyond the toxicity caused by waterborne exposure alone. However, up till now dietborne exposures have not yet shown adverse ecological impacts, that is, increased tissue concentrations have not been linked to adverse population/community effects for the metals listed above. However, for metals such as selenium the dietary route has been identified as the primary pathway of exposure for both invertebrates and vertebrates (Chapman et al., 2009). Also for mercury, dietary sources have been identified as important where it is primarily the methylated mercury forms that drive biomagnification and toxicity of this element into the food chain (Scheuhammer et al., 2007).

In assessing risks for the sediment compartment, the dietary route could be of a relatively higher importance than for the aquatic compartment. Several sediment organisms (oligochaetes, chironomids, etc.) are dependent on the ingestion and assimilation of sediment particles to survive. Overall, the results support the tenet that sulphide precipitation controls metal toxicity via the pore water in particular with relation to chronic toxicity. Further scientific research is, however, needed to assess the relative importance of the dietary route and its consequences for risk assessment purposes. In soil, the dietary route is only relevant for terrestrial invertebrates that may take up contaminated soil or food via the oral route. Although earthworms ingest large amounts of soil, it is concluded that for metals the dermal route is the uptake route of importance (Vijver et al., 2003) and the aqueous phase of the soil (soil pore water) is considered the main exposure route for most soil invertebrates (Ardestani et al., 2014). As plants and microorganisms are also exposed via the soil pore water, the dietary exposure route may be considered of lower importance in a risk assessment of metal toxicity to soil organisms.

The ecotoxicity of complex inorganic materials is related to the bioavailability of the individual metal constituents they can release into the environment. Therefore, the mechanistic aspects to consider for the environmental risk assessment of complex inorganic materials relate to the principles of environmental risk assessment of metals in general.

KEY MESSAGES

Complex inorganic materials have specific (eco-) toxic properties conveyed by the metals they contain. The main principles governing the mechanisms of the toxicity of metals can therefore be applied when assessing complex inorganic materials, particularly in relation to the rate and extent to which these materials can produce soluble (bio)available ionic and other metal-bearing species. Generally, toxicity will occur when the metal toxic species reaches a critical target within the body and overwhelms the defence and repair mechanisms of the cells. While each metal has unique features and physicochemical properties that confer its specific essential or toxicological mechanisms of action, there are some common aspects such as bioavailability, speciation and valence, essentiality, particle size, and interactions that require to be addressed when assessing the human health toxicity of the metal-bearing complex inorganic materials. These aspects will impact the approaches followed when examining the behaviour of the material in the human body (toxicokinetics) but also for endpoints such as sensitisation or genotoxicity.

The ecotoxicity of complex inorganic materials is related to the bioavailability of the individual metal constituents they can release into the environment. The mechanistic aspects to consider for the environmental risk assessment of complex inorganic materials relate to the principles of environmental risk assessment of metals in general, that is, considering natural background, speciation, essentiality, and bioavailability.

References

Anjilvel S, Asgharian B: A multiple-path model of particle deposition in the rat lung, *Fundam Appl Toxicol* 28:41–50, 1995.

Ankley GT, Bennett RS, Erickson RJ, et al: Adverse outcome pathways: a conceptual framework to support ecotoxicology research and risk assessment, *Environ Toxicol Chem* 29(3):730–741, 2010.

Antunes PMC, Hale BA, Ryan AC: Toxicity versus accumulation for barley plants exposed to copper in the presence of metal buffers: progress towards development of a terrestrial biotic ligand model, *Environ Toxicol Chem* 26(11):2282–2289, 2007.

Appel MJ, Kuper CF, Woutersen RA: Disposition, accumulation and toxicity of iron fed as iron (II) sulfate or as sodium iron EDTA in rats, *Food Chem Toxicol* 39:261–269, 2001.

Ardestani MM, van Straalen NM, van Gestel CAM: Uptake and elimination kinetics of metals in soil invertebrates: a review, *Environ Pollut* 193:277–295, 2014.

Buekers J, De Brouwere K, Lefebvre W, et al: Assessment of human exposure to environmental sources of nickel in Europe: inhalation exposure, *STOTEN* 521-522:359–371, 2015.

Buscher W, Sperling M: Introduction. In Cornelis R, Caruso J, Crews H, Heumann K, editors: *Handbook of elemental speciation II: species in the environment, food, medicine, and occupational health*, Chichester, West Sussex, 2005, John Wiley and Sons Ltd., ISBN 13 978-0-470-85598-0 (chapter 1).

Campbell PGC: Interactions between trace metals and aquatic organisms: a critique of the free-ion activity model. In Tessier A, Turner DR, editors: *Metal speciation in aquatic systems*, New York, 1995, John Wiley.

Carpene E, Andreani G, Isani G: Metallothionein functions and structural characteristics, *J Trace Elem Med Biol* 21(S1):35–39, 2007.

Caruso J, Wuilloud RG, Wuilloud JCA, Harris WR: Modeling and separation-detection methods to evaluate the speciation of metals for toxicity assessment, *J Toxicol Environ Health* 9:41–61, 2006.

Chapman PM, Adams WJ, Brooks MJ, et al: 2009. Ecological Assessment of Selenium in the Aquatic Environment: Summary of a SETAC Pellston Workshop, Selenium 34.

CIIT Centers for Health Research: *Multiple Path Dosimetry, MPPD v.2.11. A model for human and rat airway particle dosimetry*, Research Triangle Park, CA/Albuquerque, NM, 2011, Chemical Industry Institute of Toxicology/ Applied Research Associates (ARA). http://www.ara.com/products/mppd.htm.

Costello DM, Burton GA, Hammerschmidt CR, Rogevich EC, Schlekat CE: Nickel phase partitioning and toxicity in field deployed sediments, *Environ Sci Technol* 45:5798–5805, 2011.

Crommentuijn T, Polder M, Van de Plassche E: *Maximum permissible concentrations and negligible concentrations for metals, taking background concentrations into account*, RIVM, Report 601501001, 1997.

De Boeck M, Lombaert N, De Backer S, Finsy R, Lison D, Kirsch-Volders M: In vitro genotoxic effects of different combinations of cobalt and metallic carbide particles, *Mutagenesis* 18(2):177–186, 2003.

DeForest DK, Meyer SJ: Critical review: toxicity of dietborne metals to aquatic organisms, *Crit Rev Environ Sci Technol* 45(11):1176–1241, 2015.

Deshommes E, Tardif R, Edwards M, Sauvé S, Prévost M: Experimental determination of the oral bioavailability and bioaccessibility of lead particles, *Chem Cent J* 6(138):2012.

Di Toro DM, Allen HE, Bergman HL, Meyer JS, Paquin PR, Santore RC: Biotic ligand model of the acute toxicity of metals. 1. technical basis, *Environ Toxicol Chem* 20:2383, 2001.

Duffus J, Nordberg M, Templeton D: Glossary of terms used in toxicology. ed 2, (IUPAC Recommendations 2007), *Pure Appl Chem* 79(7):1153–1344, 2009.

ECB: *Guidance document on the determination of particle size distribution, fibre length and diameter distribution of chemical substances, draft version of document EUR 20268 EN*, 2002, European Commission Joint Research Centre, Institute for Health and Consumer Protection, Toxicology and Chemical Substances Unit, European Chemicals Bureau.

ECETOC: *Technical report N°122 poorly soluble particles/lung overload*, ISSN-0773-8072-122, 2013.

ECHA (European Chemicals Agency): *Guidance on information requirements and chemical safety assessment Chapter R.14: Occupational exposure assessment, version 3*, 2016, ISBN: 978-92-9495-081-9.

EN 481: *Workplace atmospheres—Size fraction definitions for measurement of airborne particles, 1993EC (European Commission), Regulation (EC) No. 1272/2008 on Classification, Labelling and Packaging (CLP) of Substances and Mixtures*, 2008.

EU CLP. *Classification, Labelling and Packaging of Substances and Mixtures, EC Regulation N° 1272/2008*, 2008.

Fairbrother A, McLaughlin M: *Fact Sheet on Environmental risk assessment: metalloregions*, London, 2002, International Council on Mining and Metals (ICMM) Publ. 12.

FDRL: *Acute oral toxicity study of cobalt oxide tricobalt tetraoxide in Sprague-Dawley rats*, Waverly, NY, 1984, Food and Drug Research Laboratories Inc.

Fowler BA: Molecular and cell biology of lead. In Zalups KJ, editor: *Cellular and molecular biology of metals*, Boca Raton, FL, 2010, CRC Press, ISBN: 978-1-4200-5998-4.

Goyer RA, Clarkson M: Toxic effects of metals. In Klaassen CD, editor: *Cassarett and Doull's Toxicology: The Basic Science of Poisons*, New York, 2001, McGraw-Hill Publisher, pp 811–868.

Goyer R, Golub M, Choudhury H, Hughes M, Kenyon E, Stifelman M: *Issue paper on the human health effects of metals*, Washington, DC, 2004, U.S. Environmental Protection Agency Risk Assessment Forum. 20460 Contract #68-C-02-060.

Gunshin H, Mackenzie B, Berger UV, et al: Cloning and characterization of a mammalian proton-coupled metal-ion transporter, *Nature* 482–488, 1997.

Hart J: *Nickel compounds—a category approach for metals in EU Legislation*, A report to the Danish Environmental Protection Agency, 2007.

Haschek W, Rousseaux C, Wallig M: *Fundamentals of toxicologic pathology*, ed 2, London, UK, 2010, Academic Press, Elsevier Inc.

Henderson RG, Cappellini D, Seilkop SK, Bates HK, Oller AR: Oral bioaccessibility testing and read-across hazard assessment of nickel compounds, *Regul Toxicol Pharmacol* 63:20–28, 2012.

HERAG: *Health risk assessment guidance for metals: gastrointestinal uptake and absorption, and catalogue of toxicokinetic models*, http://hub.icmm.com/document/264, 2007a.

HERAG: *Health risk assessment guidance for metals: occupational inhalation exposure and systemic inhalation absorption*, http://hub.icmm.com/document/262, 2007b.

HERAG: *Health risk assessment guidance for metals: essentiality*, http://hub.icmm.com/document/267, 2007c.

HERAG: *Health risk assessment guidance for metals: occupational dermal exposure and dermal absorption*, 2007d, http://hub.icmm.com/document/261.

Hollenberg PF: Introduction: mechanisms of metal toxicity special issue, *Chem Res Toxicol* 23(2):292–293, 2010.

Jomova K, Valko M: Advances in metal-induced oxidative stress and human disease, *Toxicology* 283(2–3):65–87, 2011.

Kagi JHR, Kogima Y: *Chemistry and biochemistry of metallothionein*, Boston, 1987, Birkhäuser, pp 25–26.

Kaplan JH, Maryon EB: How mammalian cells acquire copper: an essential but potentially toxic metal, *Biophys J* 110(1):7–13, 2016.

Kasprzak KS: The role of oxidative damage in metal carcinogenicity, *Chem Res Toxicol* 4:604–615, 1991.

Kenny LC, Aitken R, Chalmers JC, et al: A collaborative European study of personal inhalable aerosol sampler performance, *Ann Occup Hyg* 41(2):135–153, 1997.

Kim JH, Gibb HJ, Howe PD: In *Cobalt and inorganic cobalt compounds*, International Programme on Chemicals Safety, Geneva, 2006, WHO.

Kirby JK, McLaughlin MJ, Ma YB, Ajiboye B: Aging effects on molybdate lability in soils, *Chemosphere* 89:876–883, 2012.

Klaassen CD, editor: 8e, *Casarett and Doull's Toxicology: The Basic Science of Poisons*, 2013.

Lachapelle JM, Maibach HI: *Patch testing and prick testing: a practical guide official publication of the ICDTG*, ed 3, 2012.

Lidén C, Wahlberg JE: Cross reactivity to metal compounds studied in guinea pigs induced with chromate or cobalt, *Acta Derm Venereol* 74:341–343, 1994.

Lison D, De Boeck M, Verougstraete V, Kirsch-Volders M: Update on the genotoxicity and carcinogenicity of cobalt compounds, *Occup Environ Med* 58(10):619–625, 2001.

Lock K, De Schamphelaere KAC, Criel P, Van Eeckhout H, Janssen CR: Development and validation of an acute biotic ligand model (BLM) predicting cobalt toxicity in soil to the potworm *Enchytraeus albidus*, *Soil Biol Biochem* 38:1924–1932, 2006.

Lofts S, Criel P, Janssen CR, et al: Modelling the effects of copper on soil organisms and processes using the free ion approach: towards a multi-species toxicity model, *Environ Pollut* 178:244–253, 2013.

Ma YB, Lombi E, McLaughlin MJ, et al: Aging of nickel added to soils as predicted by soil pH and time, *Chemosphere* 92:962–968, 2013.

Ma YB, Lombi E, Oliver IW, Nolan AL, McLaughlin MJ: Long-term aging of copper added to soils, *Environ Sci Technol* 40:6310–6317, 2006.

McGeer J, Henningsen G, Lanno R, Fisher N, Sappington K, Drexler J: *Issue paper on the bioavailability and bioaccumulation of metals*, Washington, DC, 2004, U.S. Environmental Protection Agency Risk Assessment Forum. Contract #68-C-02-060.

MERAG: *Metals environmental risk assessment guidance*, https://www.icmm.com/merag, 2014.

Mertens J, Degryse F, Springael D, Smolders E: Zinc toxicity to nitrification in soil and soil-less culture can be predicted with the same biotic ligand model, *Environ Sci Technol* 41:2992–2997, 2007.

Mertz W: The newer trace elements, chromium, tin, vanadium, nickel and silicon, *Proc Nutr Soc* 33:307–315, 1974.

Migueres M, Dávila I, Frati F, et al: Types of sensitization to aeroallergens: definitions, prevalences and impact on the diagnosis and treatment of allergic respiratory disease, *Clin Translat Allergy* 4:16, 2014.

Newman MC: *Fundamentals of ecotoxicology: the science of pollution*, ed 4, Boca Raton, London, New York, 2014, CRC Press.

Nieboer E, Thomassen Y, Chashchin V, Odland JO: Occupational exposure assessment of metals, *JEM* 7:411–415, 2005.

Nordberg G, Fowler B, Nordberg M: *Handbook on the Toxicology of Metals*, ed 4, 2014, Academic Press, Elsevier.

NTP: Report on carcinogens, monograph on cobalt and cobalt compounds that release cobalt ions in vivo, https://ntp.niehs.nih.gov/pubhealth/roc/listings/c/cobalt/summary/cobalt_index.html, 2016.

Oberdörster G: Lung dosimetry: pulmonary clearance of inhaled particles, *Aerosol Sci Tech* 18(3):279–289, 1993.

Oberdörster G, Ferin J, Lehnert BE: Correlation between particle size, in vivo particle persistence, and lung injury, *Environ Health Perspect* 102(5):173–179, 1994.

OECD: *Series on testing and assessment No. 194 guidance on grouping of chemicals*, ed 2, 2014.

OECD: *Guidance on the incorporation of bioavailability concepts for assessing the chemical ecological risk and/or environmental threshold values of metals and inorganic metal compounds*, Series on Testing & Assessment, N°259, ENV/JM/MONO66, 2016.

Oller A, Oberdörster G: Incorporation of particle size differences between animal studies and human workplace aerosols for deriving exposure limit values, *Regul Toxicol Pharmacol* 57(2–3):181–194, 2010.

Oller AR, Oberdörster G, Seilkop SK: Derivation of PM10 size-selected human equivalent concentrations of inhaled nickel based on cancer and non-cancer effects on the respiratory tract, *Inhal Toxicol* 26(9):559–578, 2014.

Oorts K, Ghesquiere U, Swinnen K, Smolders E: Soil properties affecting the toxicity of CuCl2 and NiCl2 for soil microbial processes in freshly spiked soils, *Environ Toxicol Chem* 25:836–844, 2006.

Oorts K, Smolders E, McGrath SP, Van Gestel CAM, McLaughlin MJ, Carey S: Derivation of ecological standards for risk assessment of molybdate in soil, *Environ Chem* 13:168–180, 2016.

Pagenkopf GK: Gill surface interaction model for trace-metal toxicity to fishes: role of complexation, pH and water hardness, *Environ Sci Tech* 17:342–347, 1983.

Paquin PR, Gorsuch JW, Apte S, et al: The biotic ligand model: a historical overview, *Comp Biochem Physiol C* 133(1–2):3–35, 2002.

Picard V, Govoni G, Jabado N, Gros P: Nramp 2 (DCT1/DMT1) expressed at the plasma membrane transports iron and other divalent cations into a calcein-accessible cytoplasmic pool, *J Biol Chem* 275(46):35738–35745, 2000.

Polak J, O'Flaherty EJ, Freeman GB, Johnson JD, Liao SC, Bergstrom PD: Evaluating lead bioavailability data by means of a physiologically based lead kinetic model, *Fundam Appl Toxicol* 29:63–70, 1996.

Rachmawati D, Bontkes HJ, Verstege MI, et al: Transition metal sensing by Toll-like receptor-4: next to nickel, cobalt and palladium are potent human dendritic cell stimulators, *Contact Dermatitis* 68(6):331–338, 2013.

Raghavan B, Martin SF, Esser PR, Goebeler M, Schmidt M: Metal allergens nickel and cobalt facilitate TLR4 homodimerization independently of MD2, *EMBO Rep* 13(12):1109–1115, 2012.

Reimann C, Fabian K, Birke M, Filmoser P, Demetriades A, Négrel P, Oorts K, Matschullat J, de Caritat P, The GEMAS Project Team: GEMAS: Establishing geochemical background and threshold for 53 chemical elements in European agricultural soil, *Appl Geochem* 2017. (in press), https://doi.org/10.1016/j.apgeochem.2017.01.021.

Reimann C, Garrett RG: Geochemical background—concept and reality, *Sci Total Environ* 350:12–27, 2005.

RIVM (National Institute for Public Health and the Environment): *Multiple path particle dosimetry model (MPPD v 1.0): a model for human and rat airway particle dosimetry*, The Netherlands, 2002, Bilthoven. RIVA Report 650010030.

Ruby MV, Schoof R, Brattin W, et al: Advances in evaluating the oral bioavailability of inorganics in soil for use in human health risk assessment, *Environ Sci Technol* 33:3697–3705, 1999.

Scheuhammer AM, Meyer MW, Sandheinrich MB, Murray MW: Effects of environmental methylmercury on the health of wild birds, mammals and fish, *AMBIO J Hum Environ* 36(1):12–19, 2007.

Schmidt M, Goebeler M: Nickel allergies: paying the Toll for innate immunity, *J Mol Med* 89(10):961–970, 2011.

Schmidt M, Raghavan B, Müller V, et al: Crucial role for human Toll-like receptor 4 in the development of contact allergy to nickel, *Nat Immunol* 11(9):814–819, 2010.

Smolders E, Oorts K, Van Sprang P, et al: Toxicity of trace metals in soil as affected by soil type and aging after contamination: using calibrated bioavailability models to set ecological soil standards, *Environ Toxicol Chem* 28:1633–1642, 2009.

Song J, Zhao FJ, McGrath SP, Luo YM: Influence of soil properties and aging on arsenic phytotoxicity, *Environ Toxicol Chem* 25:1663–1670, 2006.

Tchounwou PB, Yedjou CG, Patlolla AK, Sutton DJ: Heavy metals toxicity and the environment. molecular, clinical and environmental toxicology, *Experientia Suppl* 101:133–164, 2012.

Thakali S, Allen HE, Di Toro DM, et al: A terrestrial biotic ligand model. 1. Development and application to Cu and Ni toxicities to barley root elongation in soils, *Environ Sci Technol* 40:7085–7093, 2006a.

Thakali S, Allen HE, Di Toro DM, et al: Terrestrial biotic ligand model. 2. Application to Ni and Cu toxicities to plants, invertebrates, and microbes in soil, *Environ Sci Technol* 40:7094–7100, 2006b.

UN Globally harmonized system of classification and labelling of chemicals (GHS): *Fifth revision, ST/SG/AC.10/30/Rev. 5*, New York and Geneva, 2013, United Nations. https://www.unece.org/fileadmin/DAM/trans/danger/publi/ghs/ghs_rev05/English/ST-SG-AC10-30-Rev5e.pdf.

UN Globally harmonized system of classification and labelling of chemicals (GHS), Sixth revised edition. ST/SG/AC.10/30/Rev. 6, United Nations, 2015, New York and Geneva. https://www.unece.org/fileadmin/DAM/trans/danger/publi/ghs/ghs_rev06/English/ST-SG-AC10-30-Rev6e.pdf.

US EPA: *Technical Support Document: Parameters and Equations Used in Integrated Exposure Uptake Biokinetic Model for Lead in Children (v0.99d)*, EPA /540/R-94/040, PB94- 963505, 1994.

US EPA: *Procedures for the derivation of equilibrium partitioning sediment benchmarks (ESBs) for the protection of benthic organisms: Metal mixtures (cadmium, copper, lead, nickel, silver, and zinc)*, EPA 600/R-02/11. Washington DC, 2005.

US EPA: *Guidance manual for the integrated exposure uptake biokinetic model for lead in children*, Washington, DC, 1994, Office of Emergence and Remedial Response. Publication No. 9285.7-15-1. EPA/540/R-93/081 PB 93-963510.

US EPA: *Framework for metals risk assessment*, EPA 120/R-07:001, 2007.

Vangheluwe M, Verdonck FAM, Besser JM, et al: Improving sediment-quality guidelines for nickel: development and application of predictive bioavailability models to assess chronic toxicity of nickel in freshwater sediments, *Environ Toxicol Chem* 32(11):2507–2519, 2013.

Vijver MG, Vink JPM, Miermans CJH, van Gestel CAM: Oral sealing using glue: a new method to distinguish between intestinal and dermal uptake of metals in earthworms, *Soil Biol Biochem* 35:125–132, 2003.

Vincent JH: *Aerosol sampling, science, standards, instrumentation and applications*, New York, 2007, John Wiley & Sons Ltd.

Volkwein JC, Maynard AD, Harper M: In Kulkarni P, Baron PA, Willeke K, editors: *Workplace aerosol measurement, in aerosol measurement: principles, techniques, and applications*, 3 ed. Hoboken, NJ, 2011, John Wiley & Sons, Inc.

Wahlberg JE, Lidén C: Cross-reactivity patterns of cobalt and nickel studies with repeated open applications (ROATS) to the skin of guinea pigs, *Am J Contact Dermat* 11(1):42–48, 2000.

WHO (World Health Organization): Trace elements in human health and nutrition. In *Trace element bioavailability and interactions*, Geneva, 1996, World Health Organization, pp 23–41 (chapter 3).

WHO (World Health Organization): *Principles and methods for the assessment of risk from essential trace elements*, Geneva, 2002, World Health Organization, ISBN: 92-4-157228-0. Series Environmental Health Criteria, No. 228.

Yokel RA, Lasley SM, Dorman DC: The speciation of metals in mammals influences their toxicokinetics and toxicodynamics and therefore human health risk assessment, *J Toxicol Environ Fealth Crit Rev* 9(1):63–85, 2006.

Zitko V, Carson WV, Carson WG: Prediction of incipient lethal levels of copper to juvenile Atlantic salmon in the presence of humic acid by cupric electrode, *Bull Environ Toxicol* 10:265–271, 1973.

Further Reading

Büdinger L, Hertl M: Immunologic mechanisms in hypersensitivity reactions to metal ions: an overview, *Allergy* 55:108–115, 2000.

ECHA (European Chemicals Agency): *Guidance on information requirements and chemical safety assessment. Appendix R.7.13-2: environmental risk assessment for metals and metal compounds*, 2008.

Meyer JS: The utility of the terms "bioavailability" and "bioavailable fraction" for metals, *Mar Environ Res* 53:417–423, 2002.

4

Principles of Risk Assessment and Management of Complex Inorganic Materials

Jelle Mertens

European Precious Metals Federation, Brussels, Belgium

4.1 INTRODUCTION

Chemicals have been in use for many thousand years, but their numbers have exploded exponentially since the Industrial Revolution, and have continued to increase with the development of science. In 2015, the 100 millionth CAS registry number was assigned to a chemical (http://www.cas.org/news/media-releases/100-millionth-substance). Most of these chemicals have been intentionally isolated or produced to fulfil a certain function in an industrial process or in a product on the market.

As science developed, it became apparent that the use of and the exposure to some of these chemicals were associated with severe toxic effects on living organisms including man. Well-known examples are historical uses of cosmetics containing lead, arsenic, or mercury or exposure to drinking water with a high arsenic concentration. Nevertheless, all organisms on earth are daily exposed to these natural elements without any effect at all in most cases. It is important to highlight that a chemical only poses a risk when two prerequisites are met: the chemical is associated with a hazardous property and there is exposure to the chemical at a concentration that exceeds a safe threshold.

It was not until the second half of the 20th century that the regulatory regimes were installed to (systematically) assess the risks of chemicals in order to ensure their safe use for the environment and humans. Nowadays, authorities worldwide impose legislations that request risk assessments for chemicals, following the decisions taken in international fora like the 2020 goals for chemical management by the Strategic Approach to International Chemicals Management (SAICM, UNEP 2007). These might target the existing chemicals and/or the newly developed ones. Well-known examples are risk assessments for drinking water (e.g., World Health Organisation, 2011), for food contact materials (e.g., European Food Safety Association), and for industrial chemical production (e.g., Occupational Safety and Health Administration or National Institutes of Safety and Health in the United States), or generic legislations covering the entire life cycle of chemicals such as the EU REACH Regulation (EU REACH 2007).

Regardless of the focus of the risk assessment or the authority in charge, there are some commonalities among all:

- the assessment of the hazards and derivation of toxic threshold concentrations;
- the derivation of safe threshold concentrations;
- the determination of exposure concentrations;
- the characterisation of the risk and the associated uncertainty; and
- the (possible) implication and definition of risk management measures to ensure safe use.

Risk assessments of complex inorganic materials will follow the same building blocks, although with some specificities due to the characteristics of these materials (Fig. 4.1). These aspects will be discussed in more detail in the following chapters.

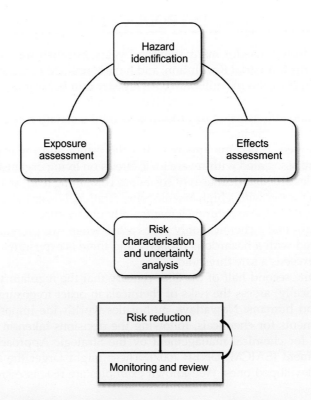

FIG. 4.1   The different building blocks of a risk assessment.

4.2 HAZARD ASSESSMENT

A 'hazard' is defined as *'the inherent capacity of a chemical or mixture to cause adverse effects in man or the environment under the conditions of exposure'*, and 'hazard assessment' as the *'estimation of the relationship between dose or level of exposure to a substance, and the incidence and severity of an effect'* (Van Leeuwen and Vermeire, 2007).

The definition of a hazard inherently implies the following:

- some chemicals are nonhazardous, whereas others are associated with single or multiple hazards;
- some effects appear upon single ('acute') exposure, whereas others require a long-term ('chronic') exposure; and
- the route of exposure (dermal, oral, inhalation) determines if the effect might appear (only relevant for human health hazards).

These principles will be explained in more detail in the following sections.

4.2.1 Environmental Hazard Assessment

4.2.1.1 Environmental Testing

Environmental hazard assessment typically covers organisms living in the aquatic environment, sediment, and soil. When severe concerns arise, higher-tier species like birds are considered as well.

Each of the environmental compartments contains a wide variety of living species, covering multiple taxonomic groups. Even more, compositions differ widely depending on, for example, the geological or climatological conditions. It is obviously impossible to test the effect of the chemical of concern to each species or taxonomic group. A lot of assessment schemes therefore apply a pragmatic approach and select species that represent the different steps in the food chain and that are, for example, easy to keep and test in the laboratory and relevant for the area of concern. For aquatic hazard assessments, typical species are primary producers such as *Pseudokirchneriella subcapitata*, herbivorous species such as *Daphnia magna*, and carnivorous species such as *Pimephales promelas*. Most regulatory regimes request a certain level of standardisation for hazard assessment, and reference is often made to internationally accepted test guidelines like the OECD guidelines 201–203 (OECD, 1992, 2011, 2004) for acute aquatic toxicity. This ensures that the data are relevant and reliable for hazard assessment, and that the test results can be compared. Scientific test data from other species, or following other guidelines might obviously be considered as well but a clear rationale for deviating from the standardised guidelines is often requested by the chemical management systems in place. The requested data can cover acute effects (typically 24–96 h exposure; identification of effects upon single peak exposure) and chronic effects (up to 28 day exposure; identification of effects after long-term exposure at (often) lower concentrations). In ecotoxicity tests, the chemical of interest is homogeneously mixed into the test medium, and the response of the test organism to the chemical (assessed as the effects on motility, survival, growth, reproduction, juvenile growth, etc.) is monitored. Obviously, there is no differentiation in exposure pathway (unless specifically aimed for in the study set-up) as this is a combination of exposure via skin, gill, and/or gut/intestines.

4.2.1.2 Environmental Toxic Threshold Determination

In a next step, the available test data are used to determine a toxic threshold concentration in case the effects are observed during increasing exposure to the chemical of concern. This relationship is called a 'dose–response relationship'. The most frequently derived toxic threshold concentrations for environmental studies are no observed effect concentration (NOEC), lowest observed effect concentrations (LOEC), or ECx values (effective concentrations causing an x% inhibition in response). The NOEC is determined as the highest test concentration at which the response did not differ significantly from the control response, and the LOEC is determined as the lowest test concentration at which the response differs significantly from the control response. The ECx value is calculated from the (modelled) dose–response curve using, for example, a log-logistic model (Doelman and Haanstra 1989).

4.2.1.3 Environmental Classification

The derived toxic threshold concentrations can, in a next step, be used for classification purposes of the chemical. The lowest value for the environmental compartment of concern is selected, compared with the classification threshold for environmental hazards and the appropriate classification can be assigned. Following the international trend to harmonise chemical classification and labelling, the most frequently used classification scheme is UN GHS (UN GHS, 2015). An example of the regional application of the UN GHS is the EU CLP (EU CLP, 2008).

Additional concepts that need to be considered for classification are bioaccumulation, degradation, and M factors. Bioaccumulation is defined as the *'net result of uptake, transformation and elimination of a substance in an organism due to all routes of exposure (i.e., air, water, sediment/soil, and food)'* and degradation as *'the decomposition of organic molecules to smaller molecules and eventually to carbon dioxide, water, and salts.'* M factors ('multiplying factors') are artificial factors assigned to highly toxic chemicals that reduce the classification thresholds accordingly when these chemicals are present in mixtures (UN GHS, 2015).

The classification of metals and metal compounds follows a different approach than nonmetals under UN GHS and linked regulations such as EU CLP (UN GHS 2015, EU CLP, 2008). This is related to some aspects that are specific for metals and metal compounds. First, it is assumed that the ecotoxic effects are caused by the metal ion (with some exceptions such as organometallics or metal cyanides). Second, *bioaccumulation* is of no relevance (K_{ow} for metals far below the threshold $K_{ow} = 4$). Third, *biodegradation* is of no relevance for metals and *'transformation'* (reaction of metal ions with the media to transform to water-soluble forms) needs to be considered instead (UN GHS, 2015).

To classify metals and metal compounds, transformation/dissolution testing (T/D testing; OECD, 2001) is included in the UN GHS classification scheme. The T/D test basically determines the dissolution of metals and metal compounds at different metal loadings (1–100 mg/L) and different timings (24 h for a screening test and 7–28 days for a full test) at environmentally relevant pH (pH 6–8) in standard waters. By comparison and interpretation of the ecotoxicity toxic threshold values with the solubility test data from the T/D testing, metals can be classified (see, e.g., Annex 9.7 and 10 of UN GHS, 2011).

4.2.2 Mammalian Hazard Assessment

The mammalian hazard assessment is based on the comparable principles to those for environmental species: a test species or test system is exposed to increasing concentrations of the chemical of concern, responses of interest are monitored and threshold concentrations are derived. This assessment can be performed for acute toxicity (e.g., skin and eye irritation/corrosion; single exposure to chemical of concern) or chronic toxicity (e.g., repeated-dose toxicity, reproductive toxicity, or carcinogenicity; typically 28 days up to ≥ 2 years or longer daily exposure).

4.2.2.1 Mammalian Testing

For mammalian endpoints, these days there is a strong tendency to move away from vertebrate testing and use nonvertebrate testing instead (same trend also applies to environmental endpoints). The use of alternative methods (in silico, in chemico, in vitro) is encouraged (or even obliged) as a first resort wherever possible (insofar as no test data are already available) before initiating vertebrate testing. This evidently leads to a continuous development and optimisation of alternative test methods. Well-known examples are in vitro assays for skin and eye effects or mutagenicity that have been published by the OECD since the early 2000s (e.g., OECD 476, 2016a, OECD 489, 2016b, OECD 442E, 2016c, OECD 431, 2016d). Besides, a lot of progress has been made in the development of screening tools for chronic endpoints like US EPA's OncoLogic tool (Benigni et al. 2012) and the ToxTracker assay for carcinogenicity (Hendriks et al., 2012, 2016). In vitro assays are not (yet) available for higher tier assays such as repeated-dose toxicity, reproductive toxicity, or carcinogenicity. There is however a strong tendency to optimise the design of these higher tier assays to avoid excessive use of test animals. A good example is the move from the two-generation to the extended one-generation reproductive toxicity study (EOGRTS) under EU-REACH, which saves up to 40% of the test animals per test (Piersma et al. 2011).

A major difference with environmental toxicity is the importance of the exposure route in mammalian toxicity. Exposure mainly takes place via the dermal route, the oral route, or the inhalation route. Dermal exposure is, obviously, most relevant for local effects such as skin irritation or corrosion, but also for other endpoints such as acute toxicity or respiratory sensitisation. However, for most metals and metal compounds, dermal exposure is considered a least relevant exposure route as the absorption of metal ions through the skin is low to negligible (HERAG, 2007). Oral exposure is the most frequently used exposure route for experimental testing (via suspension/dissolution in a vehicle, mixing in food, or exceptionally administered into the stomach directly). The oral route is the major route of exposure to chemicals in daily life. In a working environment, however, hygienic practices and worker's training aim at minimising dermal and oral exposure to chemicals, making inhalation exposure the major route of exposure in occupational settings. For practical reasons, however, inhalation exposure during vertebrate testing is more complex than dermal or oral exposure.

4.2.2.2 Mammalian Toxic Threshold Determination

Once the mammalian toxicity testing is finalised, the data can be interpreted and toxic threshold concentrations can be derived. Here is a major difference with the environmental compartment. For some mammalian toxicity endpoints such as acute or repeated dose toxicity, it is possible to derive a toxic threshold concentration beyond which no effects are observed. For others, however, it is impossible to derive such toxic threshold concentration. This is typically the case for sensitisers, mutagens, or genotoxic carcinogens where you cannot identify a clear cut-off value between effect and no effect. This nonthreshold character has its consequences for classification and risk assessment, as will be explained further.

4.2.2.3 *Mammalian Classification*

Classification of mammalian toxicity endpoints is performed by comparing the experimental toxic threshold concentrations with the response level triggering classification. This includes both threshold effects (e.g., acute toxicity) and some nonthreshold effects (e.g., skin sensitisation). The other nonthreshold effects require 'expert judgement' to decide whether there is a need for classification or not. Again, these are the higher tier endpoints such as reproductive toxicity, mutagenicity, and genotoxic carcinogenicity, and the observed effects need to be interpreted in relation to other relevant data such as the toxicokinetic behaviour of the chemical.

Classification schemes and criteria to be applied are made available and regularly updated via the UN GHS (UN GHS, 2015) or its regional application [e.g., the EU CLP (EU CLP, 2008)].

4.3 EXPOSURE ASSESSMENT

'*Exposure assessment*' is '*the determination of the emissions, pathways and rates of movement of a substance and its transformation or degradation in order to estimate the concentrations/doses to which human populations or environmental compartments are or may be exposed*' (Van Leeuwen and Vermeire, 2007). As explained earlier, a hazardous chemical does not pose a risk if the exposure to this chemical is below a safe threshold. Exposure assessment is therefore essential to decide if the identified hazard effectively implies a risk or not.

Exposure can be determined via modelling or monitoring. Both will be discussed in more detail below. During exposure assessment, it is important to identify all risk management measures (for modelling and monitoring) and input parameters (for modelling) that are relevant. Examples are waste water treatment or ventilation that might significantly impact final exposure values and thus the potential risk that will we determined later.

4.3.1 Environmental Exposure Assessment

4.3.1.1 *Environmental Exposure Modelling*

There are today various models available for modelling environmental exposure to chemicals. A summary of the existing exposure models is available from OECD (OECD, 2012). The exposure models predict the environmental concentration of a chemical based on the physicochemical properties of the chemical (water solubility, vapour pressure, etc.), its behaviour in the relevant environmental compartment(s) (degradation, transport, transformation, etc.), and/or the concentration of the chemical released (amount, periodicity, etc.). The predictions are based on the statistics and correlations between various model parameters. The final exposure estimate depends on the model that is used and its input parameters. It is, as such, critical to select a model that is suited for the chemical of concern (metals vs organics) and the purpose of the assessment (single compartment vs multicompartment models).

Important to keep in mind when modelling environmental metal exposure is that metals are naturally occurring elements and thus always present in the environment, that some metals are essential elements and thus a prerequisite for life, that metals cannot (bio)degrade but only transform, and that parameters such as K_{ow} or vapour 0pressure are not relevant for most metals and metal compounds.

4.3.1.2 Environmental Exposure Monitoring

The alternative (and often preferred) method for environmental exposure assessment is direct monitoring in the environmental compartment of concern. This way, the data you obtain are real exposure values that do not rely on models and model parameters. It is well known that exposure concentrations can vary widely depending on the location: metal exposure levels are mostly higher in industrialised or urban areas compared with rural areas.

As metals are natural elements, their environmental concentration depends on natural factors such as the geology and environmental conditions. Moreover, previous and/or ongoing human activities such as mining or industry might cause local hotspots of peak metal concentrations that exceed the concentration of the surroundings, and might persist for a long time. Next, the sampling and analysis methodology are the critical elements. If not well performed, they might cause severe underestimation or overestimation of the 'real' environmental concentration. Monitoring guidance is available via, for example, MERAG factsheets (MERAG, 2016). In the EU, there have recently been initiatives to adequately determine environmental concentrations of a wide range of metals in soil (Reimann et al. 2014a,b) and natural waters (Salminen et al., 2015).

4.3.2 Human Exposure Assessment

4.3.2.1 Human Exposure Modelling

Similar principles apply for human and environmental exposure modelling.

Human exposure occurs via inhalation, dermal, and/or oral exposure, with the contribution for each route depending on the properties of the chemical of concern and the exposure conditions. Related to the latter factor, exposure assessment is often targeted to specific pathways. As an example, a worker's exposure to metals is often focused on the inhalation and dermal routes, as oral exposure is considered negligible due to good industrial practices. On the other hand, exposure of humans in daily life is often focused on both dermal (via handling of and contact with consumer articles and products) and oral (via food and drinks) exposure.

The total exposure concentration estimated by modelling often severely underestimates or overestimates real exposure. A good example is inhalation exposure, where the fraction effectively reaching the deep lung is often only a fraction of the total fraction in the air: the smallest particles may be deposited or exhaled again, while the largest particles are trapped in the mouth or upper airways and may be translocated to the stomach via the mucociliary action, thus indirectly contributing to oral exposure via air. For dermal exposure, systemic exposure is only relevant for the fraction that passes through the dermal barrier, which is low to negligible for most metals and metal compounds (HERAG, 2007).

Additional factors to consider for human exposure modelling are differences in input parameters that are related to the differences in target population. For instance, body weight or respiration volume is higher for occupational than for daily life exposure (also including sensitive groups such as children or the elderly).

4.3.2.2 *Human Exposure Monitoring*

Human exposure monitoring can mainly be performed by monitoring concentrations that reach the target organs (mainly the lungs, skin, and stomach), or the biomonitoring of effective concentrations in body fluids such as blood or urine. The former gives a more realistic estimate of exposure compared with modelling and allows to differentiate exposure via the different routes. Nevertheless, monitoring gives a worst-case estimate, as factors such as particle size or uptake into the body might significantly reduce effective exposure concentration. A critical factor for monitoring is the technique applied. This is further discussed in Chapters 2 and 3 and in Vetter et al. (2016).

Biomonitoring, on the other hand, gives a realistic estimate of real systemic exposure concentration, taking into account the real bioavailability of the chemical one is exposed to. Nevertheless, the values reflect the exposure via all exposure routes, and no distinction can be made unless the monitoring is explicitly designed for this purpose. Various companies and authorities organise regular biomonitoring campaigns to assess exposure of workers or inhabitants to chemicals such as CDC's National Biomonitoring Programme or the European Human Biomonitoring Initiative (HBM4EU) that will run from 2017 to 2021.

4.4 RISK CHARACTERISATION

'Risk' is the *'probability of an adverse effect on man or the environment occurring as a result of a given exposure to a chemical or mixture'* (Van Leeuwen and Vermeire, 2007). As mentioned earlier, a hazard only becomes a risk if exposure exceeds a safe threshold value. Hazard identification and exposure determination have been discussed in previous sections; the coming ones will focus on the derivation of safe threshold concentrations, risk assessment, and risk management.

4.4.1 Environmental Safe Thresholds and Risk Assessment

Environmental safe thresholds are derived to protect environmental organisms from toxic effects. In reality, and as discussed earlier, it is not possible to test and protect all species, and toxic threshold data are available for a limited amount of species. As a consequence, test data are grouped at the species level, with the lowest reliable value for each species being selected for further consideration. Furthermore, the lowest values for the different species are ranked per environmental compartment, and the lowest value is maintained as the basis for the safe threshold.

Depending on the amount of test data (i.e., number of test species and different endpoints covered per test species), additional safety factors are often applied to anticipate the uncertainty associated with this value. Safety factors typically range between 5 and 1000 or higher. It is assumed that the more species tested and the more extensive the test per species is (e.g., data available for acute and chronic endpoints), the more reliable the toxic threshold is and thus the lower the safety factor that needs to be applied. The safe threshold is then derived as

$$\text{Safe threshold} = \frac{\text{Toxic threshold}}{\text{Safety factor}} \tag{4.1}$$

Some metals such as zinc, nickel, lead, or copper have historically been tested extensively, and have an extensive dataset available with multiple independent tests performed per species and per endpoint. Alternative approaches are developed for these metals, such as the species sensitivity distribution (SSD) approach. In short, if a dataset covers a minimum amount of taxa/species, the toxic thresholds are ranked from low to high on a proportional scale; a sigmoidal curve is plotted through the data; and a safe threshold value is calculated from the curve. A typical value is the hazardous concentration statistically affecting 5% of the species (HC5 value). An additional safety factor (typically 1–5) can be applied to cover the remaining (minor) uncertainty. The SSDs are created on an environmental compartment basis (e.g., freshwater toxicity, soil toxicity).

If no or unreliable data are available for a certain compartment, it is possible to estimate a safe threshold based on the safe threshold for another compartment. For instance, if there is no toxic threshold available for freshwater sediment, it is possible to derive a safe threshold using the value for freshwater aquatic toxicity and the partitioning coefficient as

$$\text{Safe threshold}\,(\text{sediment})\,[\text{mg}\,/\,\text{kg}] = K\,(\text{susp}-\text{water})\,/$$
$$\rho\,(\text{sed})\times \text{safe threshold}\,(\text{freshwater})\,[\text{mg}\,/\,\text{L}] \tag{4.2}$$

with K(susp-water) being the partitioning coefficient of suspended matter and water [unitless] and ρ(sed) being the bulk density of the suspended matter [in kg/m^3].

A similar approach can be taken for, for example, safe thresholds for soils based on freshwater toxicity, K(soil–water) and ρ(soil).

Once the environmental safe threshold is derived, the risk is determined by comparing the exposure concentration to the safe threshold as

$$\text{Risk} = \frac{\text{Exposure concentration}}{\text{Safe threshold}} \tag{4.3}$$

with the risk calculated as a value without unit.

The chemical is assumed to be safe if this ratio is less than 1.

As mentioned before, both exposure and safe threshold are associated with a level of uncertainty. This implies that safety can only be assumed if the risk is 'sufficiently' less than 1. The margin of uncertainty depends on the quality of the exposure assessment and the reliability of the safe threshold.

4.4.2 Mammalian Safe Thresholds and Risk Assessment

Similar principles apply for mammalian and for environmental risk assessment. However, the methodological approaches are different. As mentioned earlier, differences in target population (workers vs consumers), exposure pathways (inhalation vs dermal vs oral), effect levels (local vs systemic effects), and exposure durations (acute vs chronic) are the key for mammalian risk assessment and as such need to be accounted for. This implies that various safe thresholds can be derived for mammalian health by combining these factors. As an example, under EU REACH, the safe thresholds in Table 4.1 need consideration (ECHA, 2012).

The methodology to derive either of these thresholds is similar:

- the applicable toxic threshold concentration(s) is selected based on endpoint relevance and the relevant exposure routes;
- the appropriate dose descriptor is selected for the relevant endpoint(s);
- the dose descriptor is modified for differences between the set-up of the underlying experimental study and the principles behind the threshold concentration, and
- the safe threshold concentration is derived by applying appropriate safety factors to the modified dose descriptor.

As an example, the available dataset of a chemical shows that the safe threshold for long-term, systemic toxicity via inhalation is an effect observed in a 28-day repeated dose toxicity study with rats. In this study, the chemical of concern is administered orally, whereas the threshold concentration is aimed to protect humans via the inhalation exposure. Owing to these differences between the study set-up and the targeted safe threshold, corrections are required to compensate for, for example, morphological differences between rats vs humans and exposure via oral vs inhalation route. This requires a recalculation of the dose descriptor taking into account, for instance, the differences in respiratory volume and body weight between rats and humans, and the differences in absorption of the chemical via oral and inhalation exposure (if applicable). Once recalculated, appropriate safety factors are applied to account for, for example, exposure duration in the test vs lifetime exposure of workers, or interspecies and intraspecies differences. Obviously, this methodology can only be applied if a toxic threshold can be derived.

For nonthreshold effects (e.g., mutagens or carcinogens), an alternative (semi) qualitative risk assessment approach is required. For some, a threshold can be statistically interpolated or extrapolated if data are available, showing a relationship between dose and effect. The technique applied depends on the data and the risk assessor. Once a threshold is determined, the subsequent steps are comparable to the previously discussed quantitative approach, that is, modification of the dose descriptor and application of assessment factors. For chemicals

TABLE 4.1 Safe Thresholds That Need Consideration Under EU REACH Regulation (REACH 2006)

		Inhalation		Dermal		Oral	
		Systemic	Local	Systemic	Local	Systemic	Local
Acute	Workers	✓	✓	NR	✓	NR	NR
	General population	✓	✓	NR	✓	NR	NR
Long term	Workers	✓	✓	✓	✓	NR	NR
	General population	✓	✓	✓	✓	✓	NR

NR = not relevant.

where a threshold can be derived in a quantitative or semiqualitative way, a risk assessment can be performed by comparing the exposure concentration to the safe threshold. The chemical is assumed to be safe for the population and route targeted if this ratio is 'sufficiently less than 1', the latter again depending on the methodologies applied and on the quality of the dataset.

Other nonthreshold chemicals lack any relationship between dose and effect. For these, no threshold can be derived, not even statistically. A simple comparison of exposure to safe threshold can thus not be made and a complete qualitative risk assessment is required. For some chemicals, exposure data are available. These can be compared with a benchmark value derived by, for example, the authorities or applied by the industry, and a cautious prediction on the expected safety can be made. For other chemicals, full qualitative assessments need to be considered by using, for example, exposure and hazard banding approaches where the likelihood of safety vs risk is assessed qualitatively (high/moderate/low) or quantitatively assigning values to expected combinations of exposure vs hazard in a matrix.

4.5 RISK MANAGEMENT MEASURES

Today's regulatory regimes are rightly aimed at ensuring the safe use of chemicals for environmental and human health. The risk assessment described in previous chapters gives the assessor a good impression of the (absence of) risks of a chemical. If safe use can be undoubtedly be shown, then the assessment can stop. If safe use cannot be shown, or if there are remaining doubts, additional risk management measures need to be implemented. These measures are mostly targeted to eliminate or at least reduce exposure, are easy to monitor and report, and have a direct effect on risk reduction if applied correctly. The safe threshold values, on the other hand, are less flexible. They can be refined, for instance, by performing additional testing to reduce the uncertainty associated with a safe threshold value. However, testing and consequently increasing the safe threshold value is generally not the option preferred by authorities and risk assessors, and this approach needs to be well considered for reasons of responsible care.

4.5.1 Environmental Risk Management Measures

The most relevant environmental compartments considered for risk assessment are air, sediment, soil, and water. There are some interlinks between these, as soil pollution is strongly related to deposition via air emissions or application of wastewater treatment sludge as a soil fertiliser. Therefore, risks identified in an environmental compartment might be taken away by implementing measures for another compartment.

There is a wide range of measures that can be taken to reduce environmental exposure (in isolation or combined):

- reduce the amount, duration, and/or frequency of the chemical treated in the facility;
- optimise the industrial process or product characteristics, for example, avoid dust formation or enclose your process;

- optimise emission treatment, for example, installing or refining wastewater or air treatment, or
- treat residues at dedicated facilities, for example, send sludge to a waste treatment plant rather than applying as soil fertiliser.

The environmental exposure can be significantly reduced by applying or optimising one or more of these measures. As said, there are some metal-specific aspects that are critical in this regard, such as the natural background concentrations or historical pollutions that might contribute significantly to the measured exposure concentrations and are often outside the control of industry. In such cases, a slightly adapted strategy from risk assessors or authorities might be advisable. A good example here is the 'added risk approach' that was utilised for the Zinc Risk Assessment under the Existing Substance Regulation (EC, 2010).

4.5.2 Human Risk Management Measures

The reduction of exposure for human health effects follows a tiered approach:

Tier 1: substitute the chemical if possible
Tier 2: apply engineering controls
Tier 3: apply administrative controls
Tier 4: apply personal protective equipment (PPE)

If the chemical can be eliminated and substituted by another, less hazardous alternative, this should be considered as a first option. If not possible, engineering controls should be implemented or optimised so that exposure to the chemical of interest is further reduced (or even completely excluded). Examples are the containment of the process or the installation/optimisation of ventilation. If still at risk, the appropriate administrative controls should be implemented to further reduce exposure. Examples are limitations of exposure duration or reduction of exposure opportunity. These measures complement the engineering controls. If still at risk, a last possibility is the implementation of PPE such as adaptive clothing, gloves, respirators, and goggles. If PPE is required, it is important that it is properly used and maintained to ensure continuous good functioning. The PPE should only be considered as a last resort.

4.6 NOTE

Most chemical management systems have made extensive guidance available, on how to perform risk assessment. However, traditionally, the majority of the guidance materials have focussed on organic chemicals. In many cases, such guidance fails to adequately address the specific characteristics that must be taken into account to perform accurate risk assessments for metals and hence for complex inorganic materials.

Therefore, two projects that aim at addressing the metals and their inorganic compounds more specifically were launched around 2004. The Metals Environmental Risk Assessment Guidance (MERAG) and the Health Risk Assessment Guidance (HERAG) for metals projects have consolidated the most advanced and appropriate scientific concepts for assessing the risk posed by the presence of metals and inorganic metal compounds in peer-reviewed and downloadable fact sheets, released in 2007.

The target audience for the guidance is professionals in the field of environmental science, particularly those working with decisionmakers in the public or private sectors. The goals of MERAG and HERAG are to deliver the basic material to make risk assessments for metals more ecologically relevant—and to provide the readers with material to enable them to adapt to their local, national, or regional assessments.

MERAG and HERAG are intended to be living documents in recognition that the science would evolve as concepts were applied and refined, and new approaches would emerge. MERAG has been significantly updated in 2016 to promote the advances that have been made since the original publication in 2007, and to encourage further integration of metal-specific approaches into risk assessment frameworks.

The background documents and fact sheets can be found here: MERAG: https://www. icmm.com/merag; HERAG: https://www.icmm.com/herag

KEY MESSAGES

In a risk assessment, the acute or long-term effects of chemicals on human health and the environment are compared with exposure/emissions so as to estimate the risks associated with the production and use of these chemicals and hence implement risk management measures where needed. These basic principles of characterising hazard, exposure, and risk are similar for complex inorganic materials: there is no risk if the exposure is (sufficiently) below the threshold concentration. If a (potential) risk is identified, the most appropriate risk management measures need to be implemented. There are however some specificities related to the metals included in the complex inorganic materials that will impact the different steps of the assessment. For example, metals are naturally occurring elements and thus always present in the environment; some are essential elements and exposure cannot be reduced under a minimum level without generating toxicity; and finally some of the commonly used parameters like K_{ow} or vapour pressure are not relevant for most metals and metal compounds. These characteristics of metals, and hence of the complex inorganic material containing them, have been the driver for developing the metals environmental risk assessment guidance (MERAG, HERAG), freely available and updated to include the most recent metals science developments.

References

Benigni R, Bossa C, Alivernini S, Colafranceschi M: Assessment and Validation of US EPA's OncoLogic® Expert System and Analysis of Its Modulating Factors for Structural Alerts, *J Environ Sci Health C* 30(2), 2012.

Doelman P, Haanstra L: Short- and long-term effects of heavy metals on phosphatase activity in soils: an ecological dose-response model approach, *Biol Fertil Soils* 8(3):235–241, 1989.

EC European Commission, JRC Joint Research Centre Institute for Health and Consumer Protection: *European Union Risk Assessment Report Zinc Metal, JRC Scientific and Technical Reports*, 2010, European Union, ISBN: 978-92-79-17540-4.

ECHA (European Chemicals Agency): Guidance on information requirements and chemical safety assessment, Chapter R.8. In *Characterisation of dose [concentration]-response for human health*, 2012, European Chemicals Agency, Helsinki.

EU CLP: *Classification, Labelling and Packaging of Substances and Mixtures Regulation*, EC N° 1272/2008, 2008.

EU REACH: EC (European Council) Regulation (EC) No. 1907/2006 of the European Parliament and of the Council of 18 December 2006 Concerning the Registration, Evaluation, Authorisation and Restriction of Chemicals (REACH), *Off J Eur Union* L396:L136/133–L136/280, 2007.

Hendriks G, Atallah M, Morolli B, et al: The ToxTracker assay: novel GFP reporter systems that provide mechanistic insight into the genotoxic properties of chemicals, *Toxicol Sci* 125(1):285–298, 2012.

Hendriks G, Derr RS, Misovic B, Morolli B, Calléja FM, Vrieling H: The extended toxtracker assay discriminates between induction of dna damage, oxidative stress, and protein misfolding, *Toxicol Sci* 150(1):190–203, 2016.

HERAG: *Health Risk Assessment Guidance for metals: occupational dermal exposure and dermal absorption*, 2007. http://hub.icmm.com/document/261.

MERAG: *Metals Environmental Risk Assessment Guidance: Exposure Assessment*, 2016, ICMM. https://www.icmm.com/website/publications/pdfs/chemicals-management/merag/merag-fs2-2016.pdf.

OECD Organisation for Economic Co-operation and Development: Series on Testing and Assessment N°29: guidance document on transformation/dissolution of metals and metal compounds in aqueous media. ENV/JM/MONO (2001)9, 2001.

OECD Organisation for Economic Co-operation and Development: *Test No. 203: Fish, acute toxicity test*, Paris, 1992, OECD Publishing. http://dx.doi.org/10.1787/9789264069961-en.

OECD Organisation for Economic Cooperation and Development: *Test No. 202: Daphnia sp., acute immobilisation test*, Paris, 2004, OECD Publishing. http://dx.doi.org/10.1787/9789264069947-en.

OECD Organisation for Economic Co-operation and Development: *Test No. 201: freshwater alga and cyanobacteria, growth inhibition test*, Paris, 2011, OECD Publishing, https://doi.org/10.1787/9789264069923-en.

OECD Organisation for Economic Co-operation and Development: Series on Testing and Assessment No. 182: Descriptions of existing models and tools used for exposure assessment. Results of OECD Survey. ENV/JM/MONO (2012)vol. 37, 2012.

OECD Organisation for Economic Co-operation and Development: *Test No. 476: In Vitro Mammalian Cell Gene Mutation Tests using the Hprt and xprt genes*, Paris, 2016a, OECD Publishing, https://doi.org/10.1787/9789264264809-en.

OECD Organisation for Economic Co-operation and Development: *Test No. 489: In Vivo Mammalian Alkaline Comet Assay*, Paris, 2016b, OECD Publishing, https://doi.org/10.1787/9789264264885-en.

OECD Organisation for Economic Co-operation and Development: *Test No. 442E: in vitro skin sensitisation: Human Cell Line Activation Test (h-CLAT)*, Paris, 2016c, OECD Publishing, https://doi.org/10.1787/9789264264359-en.

OECD Organisation for Economic Co-operation and Development: *Test No. 431: In vitro skin corrosion: reconstructed human epidermis (RHE) test method*, Paris, 2016d, OECD Publishing, https://doi.org/10.1787/9789264264618-en.

Piersma AH, Rorije E, Beekhuijzen ME, et al: Combined retrospective analysis of 498 rat multi-generation reproductive toxicity studies: on the impact of parameters related to F1 mating and F2 offspring, *Reprod Toxicol* 31(4):392–401, 2011.

Reimann C, Birke M, Demetriades A, Filzmoser, O'Connor P, editors: *Chemistry of Europe's agricultural soils—Part A: Methodology and interpretation of the GEMAS data set*, Schweizerbarth, Hannover, 2014a, Geologisches Jahrbuch (Reihe B 102). 528 pp. + DVD.

Reimann C, Birke M, Demetriades A, Filzmoser, O'Connor P, editors: *Chemistry of Europe's agricultural soils—Part B: general background information and further analysis of the GEMAS data set*, Schweizerbarth, Hannover, 2014b, Geologisches Jahrbuch (Reihe B 103). 352 pp.

Salminen R, De Vos W, Tarvainen T: *Foregs Geochemical atlas of Europe*, 2015, ISBN: 9516909213 v.1 9516909566 v.2.

UN: *Globally harmonized system of classification and labelling of chemicals (GHS)*, Fourth revised edition ST/SG/AC.10/30/Rev.4, New York and Geneva, 2011, United Nations. https://www.unece.org/fileadmin/DAM/trans/danger/publi/ghs/ghs_rev04/English/ST-SG-AC10-30-Rev4e.pdf.

UN: *Globally harmonized system of classification and labelling of chemicals (GHS)*, Sixth revised edition ST/SG/AC.10/30/rev.6, New York and Geneva, 2015, United Nations. https://www.unece.org/fileadmin/DAM/trans/danger/publi/ghs/ghs_rev06/English/ST-SG-AC10-30-Rev6e.pdf.

UNEP United Nations Environment Programme: a Report of the World Summit on Sustainable Development, Johannesburg, South Africa, 26 August–4 September 2002 United Nations Publication, Sales No. E. 03. II. A. 1 and corrigendum), Chapter I, resolution 2, annex. https://www.saicm.org/Portals/12/Documents/saicmtexts/New%20SAICM%20Text%20with%20ICCM%20resolutions_E.pdf, 2007.

Van Leeuwen CJ: In Vermeire TG, editor: *Risk assessment of chemicals: an introduction*, Dordrecht, 2007, Springer.

Vetter D, Schade J, Lippert K: *Guidance on the assessment of occupational exposure to metals based on monitoring data, final report*, 2016. http://www.reach-metals.eu/index.php?option=com_content&task=view&id=216&Itemid=3242016.

WHO World Health Organisation: *Guidelines for drinking-water quality*, ed 4, Geneva, 2011, WHO, ISBN: 978 92 4 154815 1.

Main Characteristics of Relevance for the Assessment of Complex Inorganic Materials

Marnix Vangheluwe*, William Adams†, Koen Oorts‡

*ARCHE Consulting, Ghent, Belgium †Red Cap Consulting, Lake Point, Utah, United States ‡ARCHE Consulting, Leuven, Belgium

5.1 INTRODUCTION

As mentioned in the previous chapters, ores and concentrates, metal containing—glasses, ceramic, and inorganic pigments, alloys, and UVCBs (Unknown or Variable composition, Complex reaction products or Biological materials) produced during the manufacturing of metals (e.g. slimes, slags, flue dust) are considered as complex inorganic materials, due to their inclusion of several inorganic constituents or ingredients and some specificities complexifying their assessment (e.g. variability in composition). Determining human health and environmental risks from these materials raises unique issues that require special measures in order to make accurate, realistic, and consistent assessments.

In most cases, endpoint specific data are lacking for the complex inorganic materials themselves. Furthermore, the variability in composition may limit extrapolation to other complex inorganic materials by read-across and testing faces some limitations as explained further. In view of these limitations, the assessment will be performed based on the underlying assumption that the fate and hazard properties of the complex inorganic material are driven by the fate and hazard properties of its individual constituents. This, however, does not prevent that a proper understanding of the key characteristics such as typical concentrations and concentration range, speciation of constituents, physical state, and form of the inorganic complex material will be paramount to obtain a robust and realistic risk assessment. For example, often minor classified constituents of a complex inorganic material may result in the most stringent classification if these factors are ignored.

On top of this, traditional aspects of relevance in metal toxicology such as natural occurrence, essentiality, solubility, complexation, and bioavailability should be taken into account when assessing mechanisms of toxicity (see also Chapter 3) and the potential risks of these materials. This chapter explains some of the specific characteristics to be considered, highlighting their importance. Although most of the text in this chapter refers to UVCBs produced during the manufacturing/recycling of metals, the challenges and approaches are equally applicable for alloys, ceramics, and other inorganic complex materials.

5.2 IMPORTANCE OF VARIABILITY AND METAL SPECIATION IN COMPLEX INORGANIC MATERIAL'S RISK ASSESSMENT AND CLASSIFICATION

The accurate characterisation of the complex inorganic material's composition and the collection of existing physicochemical and toxicological data are required to initiate the risk assessment process. However, the risk assessment of some of these complex materials such as UVCBs is challenging due to the observed spatial and temporal heterogeneity in their composition, which is highly dependent on the origin of the source material.

Furthermore, whereas elemental composition is generally known (because of market implications) to have a high degree of variability in composition and mineral forms have been observed and chemical speciation is often unknown or less known. A precise analytical characterisation of the materials is therefore required to ensure a realistic assessment.

Key properties that may drive the risk assessment/classification are the concentration, form (e.g. valence, crystalline versus amorphous, particle size) and solubility of the constituents in the complex inorganic material that may be present as carbonate-, fluoride-, oxide-, or suphfide minerals, or metals (alloys) among others. In the absence of information, a worst-case approach is quite often applied in which all constituents are treated as highly soluble and highly bioavailable. This results in an extremely conservative and unrealistic assessment.

5.2.1 Importance of Variability in Complex Inorganic Materials

The elemental composition of complex inorganic materials is generally precisely known (also because of financial reasons, linked to the market value of the complex inorganic materials), therefore there is no or very low uncertainty on elemental composition. Chemical speciation/mineralogical forms are usually known or estimated based on knowledge (mattes are sulphides for instance), but there may be some unknowns (e.g. which mineral forms are present). Habitually, there is low variability in speciation composition; although for some, typically by-product inorganic UVCBs such as slags, there may be more variability for minor constituents (slags can be oxides/sulphides, etc. depending from which furnace the slag is collected). Variability of physical form and process is generally low but can occur.

For complex inorganic materials with unknown composition, representative samples should be characterised to estimate the typical concentration of each listed constituent. Historical analytical data can quite often be used in establishing typical concentration ranges. For each constituent known to be classified, the maximum of the %-range should be used (for classification or when 'registering' the substance as, e.g. under EU REACH, European

Council Registration, Evaluation, Authorisation and Restriction of Chemicals). However, a quick summation check should be performed to refine the listed constituents so that the total %w/w of all listed constituents equals 100%w/w.

Common practice is to derive a generic, usually worst case, composition based on the overall typical concentrations encountered in the different complex inorganic materials samples by taking a summary statistic (i.e. average or maximum for the purpose of classification). However, a large variability in elemental composition can potentially lead to different hazard profiles. This variability should be considered in the assessment and if data gaps remain, common worst-case classification profile and worst-case effect assessment profiles should be used or additional data on composition and speciation should be obtained to reduce uncertainty in the assessment.

5.2.2 Importance of Solubility/Speciation in Complex Inorganic Materials

Depending on the solubility/speciation of the different mineral forms present in the complex inorganic material, different amounts of free metal ions will be released that may cause toxicity if the concentrations exceeds toxicological threshold values.

As indicated in Chapter 3, it is imperative for the assessment of metals and metal compounds to take into account that metals are naturally occurring and that (eco)toxicity is strongly driven by the amount of metal that is bioavailable.

For the environment, this fraction is a function of the physicochemical characteristics of the test media and the receiving environmental compartment. Extensive guidance on this topic can be found in the OECD guidance document "Guidance on the incorporation of bioavailability concepts for assessing the chemical ecological risk and/or environmental threshold values of metals and inorganic metal compounds" (OECD, 2016). For human health, more information can be found in Chapter 8.

Chemical species are specific forms of an element, defined in relation to their isotopic composition, electronic, or oxidation state, and/or complex mineral or molecular structure. Typically, inorganic UVCBs are chemically characterised by either their elemental composition and/or their speciation and/or mineralogical structure. Metal speciation includes the chemical form of the metal in solution or solid or gaseous phase, either as a free ion or complexed to a ligand. Speciation/mineralogical structure can be described on the basis of their characteristic constituting minerals, crystals, or other chemical species defining a mineralogical (crystallographic) profile. This refers to the speciation information, which is usually known for the major constituents and can be known or unknown (and often variable) for minor constituents. Chemical speciation or mineralogy may refer to the following forms:

- Metallic speciation in the zero oxidation state: for example, Au, Cu, Cd, Ni, Pb, or Zn;
- Soluble and sparingly soluble metal compounds: for example, $CuSO_4$, $AgCl$, or $PbSO_4$;
- Minerals: for example, chalcopyrite, bornite, galena, gibbsite, and/or
- Alloys: metallic speciation in the zero oxidation state.

In addition, each of the above speciation forms may be present as inclusions in other speciation forms (e.g. metallic inclusion in mineral structure) and various crystalline structures or matrices.

Models are able to predict the interactions of metals with the major (Cl^-, SO_4^{2-}, HCO_3^-, CO_3^{2-}, Br, F^-) and minor (OH^-, $H_2PO_4^-$, HPO_4^{2-}, PO_4^{3-}, HS^-) anions as a function of temperature, ionic strength, and pH (MINEQL+, MINTEQA2, PHREEQC, V) (Parkhurst and Appelo, 1999; Allison et al., 1991; Parker et al., 1995).

The importance of bioavailability in the environmental assessment of metals, classification, and development of environmental quality standards (EQS) has been demonstrated scientifically (Ankley et al., 1996; Allen and Hansen, 1996; Di Toro et al., 1991). The concepts and the scientific methods to assess bioavailability of metals have been proposed conceptually over a decade ago (Bergman and Dorward-King, 1996) and continue to gain recognition by regulatory authorities. Although the basic concepts defining divalent metal bioavailability were known widely by the mid-1970s, it was not until 1993 that the US EPA proposed basing water quality criteria for divalent metals on the dissolved phase (Prothro, 1993). This together with the work by Pagenkopf (1983) integrated the foregoing concepts about divalent metal speciation, bioavailability, and bioactivity into a 'gill surface interaction model', the forerunner of the biotic ligand model (BLM) (Di Toro et al., 2001) for aquatic systems. A version of the original BLM copper model now exist for most divalent metals (multiple organisms) and accounts for the interaction of metal species complexation with organic and inorganic ligands as well as organism respiratory membranes (Schecher and McAvoy, 2001; Parkhurst and Appelo, 1999). It is also known that most divalent metals are complexed with organic ligands as described by Tipping and Hurley, (1992). The formation of metal ion-pairs or ion-complexes in natural waters can have a major effect on the rates of redox processes, mineral solubility and biochemical availability (Allen and Hansen, 1996). The form or speciation of a metal in natural waters can change its kinetic and thermodynamic properties. For example, Cu(II) in the free ionic form is toxic to phytoplankton, while copper complexed to organic ligands is non-toxic. The form of a metal in solution can also change its solubility, for example, Fe(II) is quite soluble in aqueous solutions while Fe(III) is nearly insoluble and precipitates rapidly out of solution as ferric hydroxide. Natural organic ligands interactions with Fe(III) can increase the solubility by 20-fold in seawater.

There is evidence to show that neither total nor dissolved aqueous metal concentrations are always good predictors of metal bioavailability and toxicity (Bergman and Dorward-King, 1996; Campbell, 1995).

Knowledge of the speciation/mineral composition offers several advantages as it provides toxicologists with the relevant metal speciation information needed for accurate UN GHS (United Nations Globally Harmonized System) classification as well as information for an informed environmental risk assessment. Most minerals and all elements have direct CAS numbers and for certain endpoints toxicological data are available. But most importantly, metal speciation information may illustrate that the constituents are tightly complexed in a natural mineral matrix, only slightly soluble and essentially nonbioavailable and nontoxic.

Metal speciation can be inferred from Eh–pH diagrams (Pourbaix, 1966), sequential extraction with metal analysis or it can be measured using appropriate analytical techniques. Mineralogical analyses provide information on the form in which each element is present in the UVCB (e.g. oxide, sulphide, silicate, etc.). Recent advances in analytical chemistry (e.g. microprobe assays, advanced X-ray crystallography, diffraction analysis, scanning electron microscopy/electron probe microanalysis, and synchrotron analyses) are available to define more precisely the structure and composition of the mineralogical form present in a UVCB. Frequently ion selective electrodes are used to measure free metal ions (i.e. Me^{++}).

5.3 CASE STUDY: ENVIRONMENTAL HAZARD ASSESSMENT OF COPPER CONCENTRATE

To illustrate the complexity of some complex inorganic materials and an approach for assessing their environmental significance, information on copper concentrate and speciation is presented as an example.

The mineral content of copper concentrate is quite complex with at least 25 different minerals present (Table 5.1). The minerals themselves can contain multiple elements. While copper is the desired end product other elements may be commercially recovered such as molybdenum, silver and gold, and other precious metals (Table 5.2). The determination of these elements in the copper concentrate and their bioavailability to plants and animals, for example, in the event of a spill, on soil or in water can form the basis of a risk assessment and a classification of the hazard of the substance in the market place.

TABLE 5.1   Mineral Content of a Typical Copper Concentrate

Mineral	Weight Percent	Formula
Chalcopyrite	52.46	$CuFeS_2$
Bornite	13.51	Cu_5FeS_4
Chalcocite	0.55	Cu_2S
Covellite	0.26	$CuSe$
Tenn/Tetrah	0.20	$(Cu,Fe)_{12}As_4S_{13}$
Enargite	0.06	Cu_3AsS_4
Cu_Mo	0.01	$CuMO_4$
Molybdenite	3.03	MoS_2
Realgar	0.00	As_2S_2
Pyrite	19.64	FeS_2
Other_Sulphides	0.59	FeS,Fe_2S_3
Quartz	2.23	SiO_2
K_Feldspar	2.77	$K.Al.Si_3O_8$
Plagioclase	0.31	$(Na,Ca)(Si,Al)_4O_8$
Illite	0.39	$(K,H_3O)(Al,Mg,Fe)_2(Si,Al)_4$
Montmorillonite	0.24	$(Na,Ca)O_{,3}(Al,Mg)_2Si_4O_{10}(OH)_2xn(H_2O)$
Biotite/Phlogopite	0.88	$k(Mg,Fe^{++}_2)_3AlSi_3O_{10}$

Continued

TABLE 5.1   Mineral Content of a Typical Copper Concentrate—cont'd

Mineral	Weight Percent	Formula
Talc	0.40	$Mg_3Si_4O_{10}$
Chlorite	0.17	$NaO_{,5}(Al,Mg)_6(Si,Al)_8O_{18}(OH)_{12}x5H_2O$
Olivine/Serpentine (asbestos)	0.01	$(Mg,Fe)_2SiO_4$
Amphibole	0.79	$(Mg,Fe)_7Si_8O_{22}(OH)_2$
Andradite	0.59	$Ca_3Fe^{+++}{}_2(SiO_4)_3$
Other_Silicates	0.08	Si complexes
Fe-Oxide	0.13	FeO, Fe_2O_3
Carbonates	0.42	$(Ca,Mg)CO_3$
Other	<u>0.23</u>	–
	100.00	

TABLE 5.2   Content of Elements in Typical Copper Concentrate

Elemental	Percent in Product
Copper	24.0–36.0
Iron	23.0–30.0
Silica	5.0–11.0
Aluminium	2.0–5.0
Molybdenum	0.1–1.5
Arsenic	0.1–0.5
Cadmium	0.1–0.5
Lead	0.1–0.5
Silver/Gold	0.01–0.1

For purposes of a risk assessment, most often the freely dissociated metal element is measured and compared with known toxicity information for the given metal (ions). Most metal toxicity studies are performed with soluble salts of the element such as copper or nickel sulphate. The salts and other chelates of a metal ion can give rise to quite different toxicities, as exemplified by a range of carcinogenic potential from various nickel species. It is important to note that while metal salts are often readily soluble in moderate concentrations (ppm levels) the corresponding minerals are generally much less soluble and considerably less bioavailable. Hence, it is necessary to estimate the amount of the metal ions that will be released from the material in the media of interest.

There are several approaches to assessing the amount of the metal ion that will be released when the complex inorganic material is placed in water, soil, or ingested. Common approaches utilise standard tests such as transformation/dissolution studies (OECD, 2001)

for natural surface waters, soil leaching studies (e.g. lysimeters), and bioelution studies with gastric fluid (ASTM, 2007). Each study is performed to determine the soluble fraction of the metal that is released from the material. The remaining gangue is considered nonbioavailable. For most metals the soluble, freely dissociated metal (free metal ion) frequently correlates with toxicity responses with some important exceptions. For aluminium and iron, total extractable metal correlates best to toxicity in aquatic systems (Gensemer et al., 2017; Arbildua et al., 2016). Additionally, for example, Erickson et al. (1996) has shown that several copper species in addition to cupric ions contribute to toxicity with fathead minnows.

The importance of speciation comes into play for risk assessments and classification when assessing the toxicity of dissolved metal species derived or extracted from complex inorganic materials. For copper concentrate the importance of speciation, as an example, is demonstrated for aluminium, copper, and iron in the following sections. The common chemical species of these three elements that may occur in aqueous solution across a range of Eh and pH values are shown below in Figs. 5.1–5.3.

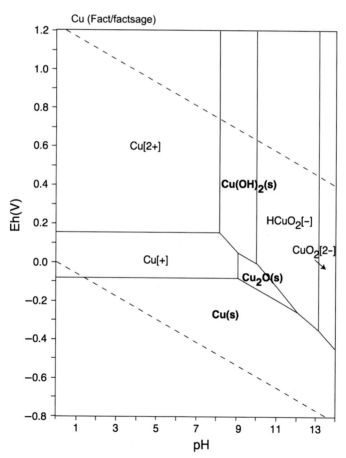

FIG. 5.1   Pourbaix diagram for copper in dissolved water.

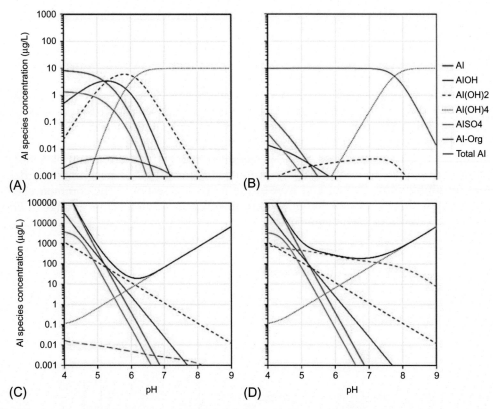

FIG. 5.2 Pourbaix diagram—BLM predicted aluminium speciation at 20°C in soft water over a range of pH values from 4 to 9. In panels A and B the total aluminium concentration is fixed at $10\,\mu g/L$ (since total Al is fixed in these panels it is not shown). In panels C and D the aluminium concentration corresponds to the solubility limit associated with amorphous gibbsite. In panels A and C the DOC concentration is very low ($0.0001\,mg\,C/L$), while in panels B and D the DOC concentration is $5\,mg/L$ (Santore et al. 2017). *From Santore R, Kroglund F, Teien H-C, Ryan A, Rodriguez R, Stubblefield W, Adams W, Nordheim E, submitted to Environ Toxicol Chem 2017.*

In panels A and B the total aluminium concentration is fixed at $10\,\mu g/L$ (since total Al is fixed in these panels it is not shown). In panels C and D the aluminium concentration corresponds to the solubility limit associated with amorphous gibbsite. In panels A and C the dissolved organic carbon (DOC) concentration is very low ($0.0001\,mg\,C/L$), while in panels B and D the DOC concentration is $5\,mg/L$ (Santore et al., 2017).

The following serves as a more complete list of dominant inorganic species and solid phases regulating aqueous concentration and toxicity for selected metals (aluminium, copper, and iron) in aquatic systems (modified and after Morel, 1983).

- Aluminium: Al^{+++} $Al(OH)^{++}$, $Al(OH)^{2+}$, AlF^{2+}, $Al(OH)^{2+}$ Al_2O_3, $Al(OH)_4^-$, $Al_2(SO_4)^3$, Al-silicates, Al-organic;
- Copper: Cu^+ Cu^{2++}, Cu_2O, $CuCO_3$, $CuOH^+$ $Cu_2(OH)_2CO_3$, CuS, Cu_2S, $CuFeS_2$, $CU_2CO_3(OH)_2$, $Cu_3(CO_3)_2(OH)_2$, Cu-organic;
- Iron: Fe^{++}, Fe^{+++}, $FeCl^+$, $FeSO_4$, $Fe(OH)^+$, $Fe(OH)_2$, $Fe(OH)_3$, $Fe(OH)_4^-$, FeS, FeS_2^-, $FeCO_3$, $FePO_4$, $Fe_3(PO_4)_2$, Fe-silicate.

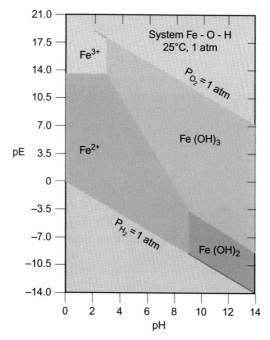

FIG. 5.3 Pourbaix diagram for iron. *Pourbaix M, 1966, Reproduced with permission from NACE International, Houston TX.*

These three elements were chosen for this speciation example because the toxicity of the major species has been investigated for aquatic species and the data have been modelled. A general principle can be applied to all three metals; in that the toxicity becomes less as the pH of the aquatic test solution increases across a range of pH 5.5–8.0, which covers the pH range of most natural waters. Careful pH control is critical when performing toxicity studies with these metals. Work by Erickson et al. (1996) demonstrated forms other than Cu^{++} contribute to the toxicity of copper to fathead minnows. As the pH of the solution increases copper transitions from cupric ions to copper hydroxide, copper oxide, and copper carbonate. The copper BLM accounts for changes in toxicity as a function of the changes in speciation and water chemistry [pH, hardness, DOC, and complexing ions and ligands (Santore et al., 2001)].

Aluminium (Al) speciation is complex and is largely determined by pH. In the absence of organic matter, Al^{3+} is the predominant Al species at low pH (<5.5; Fig. 5.2A). As pH increases above 5.5, Al hydroxide complexes become increasingly important, and can dominate aqueous, Al speciation in waters with low amounts of organic matter (Santore et al., 2017). As pH increases, hydrolyses leads to the formation of $Al(OH)^{2+}$ and $Al(OH)_2^+$ and lower solubility until the anionic aluminate ion $Al(OH)_4^-$ predominates at pHs > 7. The distribution of Al species changes dramatically in the presence of even moderate amounts of organic carbon. At 5 mg/L of DOC, organic Al complexes are by far the most important of the dissolved Al species at pH values below 8. In the presence of organic matter, the organically complexed Al is the most important form of the metal at pHs between 5 and 8 (Fig. 5.2D). The formation of organic Al complexes results in considerable increases in Al solubility in circumneutral pH conditions. The importance of organic complexes to the overall speciation of Al explains why organic carbon has such a large effect on Al solubility and toxicity. An interesting consideration is that Al solubility decreases from pH 5.5 to 6.0 and then begins to increase as pH increases due to changes in the species present. When Al concentrations exceed solubility, polymeric Al hydroxides can form amorphous colloids and crystalline solid phases which precipitate out of solution. The kinetics of this transformation to polymeric species, including aqueous colloids and amorphous precipitates, depends on many factors but typically occurs over a time scale of minutes to several hours. These polymeric species are known to contribute to the toxicity of Al to aquatic species (Santore et al., 2017). The importance of pH, DOC, and hardness on the toxicity of Al to a range of aquatic species as a function of speciation has been recently reported by Gensemer et al. (2017) and Santore et al. (2017) for a number of rivers and lakes are representative for range of conditions typically found in Europe (Tables 5.3 and 5.4).

Iron speciation follows a similar pattern as that of aluminium in that as the pH increases across a range of 5–8, ferrous ions transform to ferrous hydroxide with subsequent transition to ferric ions that complex with hydroxyl, sulphate, and carbonate ions. Due to the insolubility of ferric hydroxide this species often controls concentrations of iron in natural surface waters and laboratory test waters at pH 7 or higher. The formation of ferric hydroxide (precipitates) rapidly increases as pH, dissolved oxygen and temperature increase and can range from hours to minutes. Both iron and aluminium sulphate are used in water-treatment processes due to the fact that the hydroxide precipitates readily form and can be used to remove metals and metalloids from solution. Iron readily complexes with humic acids and in many natural waters controls the amount of dissolved iron remaining is solution. DOC readily provides protection in natural systems against toxicity, thus playing a dual role of increasing soluble forms but decreasing overall availability for most organisms.

Iron toxicity dramatically decreases as a result of pH, DOC, and hardness increasing as demonstrated by Arbildua et al. (2016) with green algae. While toxicity in natural environments is very rare due to iron complexation/precipitation, in very soft waters (hardness < 25 mg/L) with low DOC (<1 mg/L) and pH 6, chronic toxicity values can be in the range of 200–500 µg/L. In water high in hardness and DOC and pH 7–8 the aquatic chronic toxicity often exceeds 1–3 mg/L. (Arbildua et al., 2016; Rodriguez 2017).

TABLE 5.3 Overview of the Physicochemical Characteristics of the Selected European Surface Water Scenarios

Scenario (Country)	pH	Ca (mmol/L)	Mg (mmol/L)	Na (mmol/L)	K (mmol/L)	SO_4 (mmol/L)	Cl (mmol/L)	DIC (mmol/L)	DOC (mg/L)
Ditches (NL)	6.90	1.50	1.10	2.60	0.22	0.79	3.20	6.90	12.00
River Otter (UK)	8.10	1.17	0.48	0.62	0.14	0.35	0.66	2.35	3.20
River Teme (UK)	7.60	1.25	0.34	0.56	0.09	0.24	0.63	2.50	8.00
River Rhine (NL)	7.80	1.72	0.45	1.60	0.15	0.54	2.30	2.46	2.80
River Ebro (ES)	8.20	1.82	0.91	0.23	0.03	0.07	0.16	0.72	3.70
Lake Monate (IT)	7.70	0.34	0.14	0.10	0.02	0.20	0.09	1.06	2.50
Neutral acidic lake (SE)	6.70	0.22	0.06	0.34	0.02	0.11	0.33	0.40	3.80

DIC, dissolved inorganic carbon; *DOC,* dissolved organic carbon; *IT,* Italy; *UK,* United Kingdom; *NL,* The Netherlands; *ES,* Spain; *SE:* Sweden.

TABLE 5.4   Site-Specific Species EC_{10} and $HC_{5, 50}$ Values (μg total Al/L) for the Selected EU Waters

Species	Ditches—Nl	Otter—UK	Theme—UK	Rhine—Nl	Ebro—Sp	Monate-It	Lake-Sw
Danio rerio	636	632	448	483	835	266	182
Lampsilis siliquoidea	852	821	553	654	1193	288	182
Hyalella azteca	1150	440	570	654	663	166	161
Salvelinus fontinalis	1411	1072	876	1022	1475	445	297
Brachionus calyciflorus	1806	1080	840	1071	1599	397	206
Pseudokirchneriella subcapitata	2044	958	1176	1249	1385	369	327
Daphnia magna	1814	541	614	678	829	203	174
Lymnaea stagnalis	4033	2014	1658	2298	3347	602	293
Lemna minor	4126	5717	3968	4583	6848	3267	2154
Pimephales promelas	1380	1099	867	1013	1499	463	299
Ceriodaphnia dubia	4997	750	1137	1310	1287	251	237
Chironomus riparius	6963	2568	2542	3583	4584	688	368
Aeolosoma sp.	12654	5174	4724	6878	9455	1241	610
HC5, 50 (μg total Al/L)	534	317	320	343	456	117	94

The classification and assessment of toxicity of copper concentrate is often determined by the availability of minor mineral species in the concentrate, such as lead or arsenic. Both elements are classified as hazardous for the aquatic environment under UN GHS. Therefore it is important to know the amount and the mineral species that is present. Galena (lead suphfide) is the most common form of lead present in copper concentrate. This species is moderately soluble in water, however lead oxide, if present is very water soluble whereas lead carbonate and lead phosphate are quite insoluble. Hence, determination of the mineral species present can assist with the assessment.

KEY MESSAGES

Endpoint specific data are usually lacking of complex inorganic materials, in particular when it comes to UVCBs produced during manufacturing and recycling of metals. The variability in their composition may limit extrapolation by read-across to other complex inorganic materials and make testing more challenging. The assessment will therefore be performed using a 'constituents-based approach', that is, making use of the information available on the individual constituents. The use of this approach should be combined with the proper understanding of the material's key characteristics, such as typical concentrations and the constituents' concentration range, their speciation, physical state, and the form of the inorganic complex material. Other aspects of relevance in metal toxicology such as natural occurrence, essentiality, solubility, complexation, and bioavailability should be taken into account while assessing mechanisms of toxicity so as to achieve realistic and robust risk estimates, and consequently management of exposures/emission where needed.

References

Allen H, Hansen D: The importance of trace metal speciation to water quality criteria, *Water Environ Res* 68:42–54, 1996.

Allison J, Brown D, Novo-Gradac K: *MINTEQA2/PRODEFA2. A geochemical assessment model for environmental systems: version 3.0 user's manual*, Athens, GA, 1991, Environmental Research Laboratory, Office of Research and Development, US. Enviromental Protection Agency. EPA/600/3-91/021.

Ankley G, Di Toro D, Hansen D, Berry W: Technical basis and proposal for deriving sediment quality criteria for metals, *Environ Toxicol Chem* 15:2056–2066, 1996.

Arbildua J, Villavicencio G, Urrestarazu U, et al: Effect of Fe (III) on *Pseudokirchneriella subcapitata* at circumneutral pH in standard laboratory tests is explained by nutrient sequestration, *Aquat Toxicol*, 2016.

ASTM (American Society for Testing and Materials): Standard test method for determining extractability of metals from art materials. ASTM standard test method D5517-07, Philadelphia, 2007.

Bergman H, Dorward-King E: *Reassessment of metals criteria for aquatic life protection*, Pensacola, FL, 1996, SETAC Press.

Campbell P: Interactions between trace metals and aquatic organisms: a critique of the Free-ion Activity Model. In Tessier A, Turner DR, editors: *Metal Speciation in Aquatic Systems*, New York, NY, 1995, John Wiley, pp 45–102.

Di Toro DM, Allen HE, Bergman HL, Meyer JS, Paquin PR, Santore RC: Biotic ligand model of the acute toxicity of metals. 1. Technical basis, *Environ Toxicol Chem* 20:2383, 2001.

Di Toro D, CZarba C, Hansen D, et al: Technical basis for establishing sediment quality criteria for nonionic organic chemicals using equilibrium partitioning, *Environ Toxicol Chem* 10:1541–1583, 1991.

Erickson R, Benoit D, Mattson V, Nelson Jr. H, Leonard E: The effects of water chemistry on the toxicity of copper to fathead minnows, *Environ Toxicol Chem* 15:181–193, 1996.

Gensemer R, Gondek J, Rodriquez P, et al: Evaluating the effects of pH, hardness, and dissolved organic carbon on the toxicity of aluminum to aquatic organisms under circumneutral conditions, Submitted for Publication, *Environ Toxicol Chem*, 2017.

Morel EM: *Principles of Aquatic Chemistry*, Sommerset, NJ, 1983, John Wiley & Sons.

OECD: *Guidance on the incorporation of bioavailability concepts for assessing the chemical ecological risk and/or environmental threshold values of metals and inorganic metal compounds*, Series on Testing & Assessment, N°259, ENV/JM/MONO(2016)66, 2016.

OECD: *Guidance document on transformation/dissolution of metals and metal compounds in aqueous media*, Series on Testing & Assessment, N°29, ENV/JM/MONO(2001)9, 2001.

Pagenkopf G: Gill surface interaction model for trace-metal toxicity to fishes: role of complexation, pH, and water hardness, *Environ Sci Technol* 17:342, 1983.

Parker D, Norvell W, Chaney R: GEOCHEM-PC—a chemical speciation program for IBM and compatible personal computers. In *Chemical equilibrium and reaction models*, Madison, 1995, Soil Science Society of America, pp 253–269.

Parkhurst D, Appelo C: *User's guide to PHREEQC (version 2), A computer program for speciation, batch-reaction, one-dimensional transport and inverse geochemical calculations*, U.S. Geological Survey, Water-Resources Investigations Report 99-4295, Denver, Colorado, 1999, p 312.

Prothro M: *Office of Water policy and technical guidance on interpretation and implementation of aquatic life metals criteria*, Washington, DC, 1993, U.S. Environmental Protection Agency, Office of Water. 7 pp + appendices.

Pourbaix M: *Atlas of Electrochemical Equilibria in Aqueous Solutions*, (English edition), Oxford, 1966, Pergamon Press. 644p.

Rodriguez P: Personal communication, manuscript in preparation; Toxicity of iron to algae and daphnids as a function of water chemistry, 2017.

Schecher W, McAvoy D: *MINEQL+: A Chemical Equilibrium Modeling System (4.6)*, Hallowell, ME, 2001, Environmental Research Software.

Santore R, Kroglund F, Teien H-C, et al: Development and application of a biotic ligand model for predicting the toxicity of dissolved and precipitated aluminum, Submitted for Publication, *Environ Toxicol Chem*, 2017.

Santore R, Di Toro D, Paquin P: A Biotic Ligand Model of the acute toxicity of metals II Application to acute copper toxicity in freshwater fish and Daphnia magna, *Environ Toxicol Chem* 20:2397–2402, 2001.

Tipping E, Hurley M: A unifying model of cation binding by humic substances, *Geochem Cosmochem Acta* 56:3627–3641, 1992.

Further Reading

UN GHS: *United Nations Globally Harmonized System of Classification and Labelling of Chemicals*, ed 4, 2011. United Nations, New York and Geneva, https://www.unece.org/fileadmin/DAM/trans/danger/publi/ghs/ghs_rev06/English/ST-SG-AC10-30-Rev6e.pdf.

6

Data Needs, Availability, Sources, and Reliability

Federica Iaccino*, *Arne Burzlaff*†, *Jelle Mertens*‡

*ARCHE Consulting, Leuven, Belgium †EBRC Consulting GmbH, Hannover, Germany
‡European Precious Metals Federation, Brussels, Belgium

6.1 INTRODUCTION

The assessment of a complex inorganic material is procedurally comparable to the assessment of any other material. The first step is the compilation and evaluation of all available and relevant information, including (but not limited to) the intrinsic properties of the material (physicochemical, environmental, and toxicological properties), the uses of the material (manufacture, use, and subsequent life-cycle stages), and the associated emissions and exposure.

Most chemical management systems stipulate the requirements for hazard and exposure data collection. When assessing complex inorganic materials, the assessor will be confronted with some specificities and difficulties, and alternative approaches need to be considered and supported with the appropriate data.

This chapter will discuss the data-gathering process, the assessment of data reliability/relevance/adequacy and how to consider and address the specificities for the risk assessment of complex inorganic materials. Limitations of the current (eco)toxicity test protocols for complex inorganic materials will be touched upon.

6.2 DATA REQUIREMENTS TO PERFORM A RISK ASSESSMENT OF A COMPLEX INORGANIC MATERIAL

The first and essential step of the hazard and risk assessment of complex inorganic materials, is defining the identity of the material. The identity includes (but is not limited to):

- the material's name and related identifiers (IUPAC name, CAS number, EINECS number, etc.),
- molecular and structural formula, and
- (analytical) information on the composition.

Material identity is critical to decide on the required information for the subsequent hazard and risk assessments.

The next step is the determination of information requirements (hazard data, exposure information, etc.) and their gathering, via either existing data or newly generated data. Most chemicals management systems outline minimum information requirements. Overall information needs to cover the behaviour and the potential effect(s) of the material under the relevant conditions of use. Two types of data are required: data on the intrinsic material properties (such as physicochemical properties, environmental fate and behaviour, eco- and toxicological hazards) and data on the use(s) and associated exposure/emission (such as manufacturing or further handling) of the material. In addition, some information on metal specific aspects, such as their natural occurrence (i.e. background concentrations) or bioavailability should be collected, as those aspects will impact the assessment. It is recommended to map all available data and to consider the development of an Intelligent Testing Strategy (ITS) to fill remaining gaps.

Complex inorganic materials consist of multiple constituents, meaning that the data identification and gathering will be a more comprehensive and resource-intensive exercise compared to simple chemicals, as explained below.

6.2.1 Data on the Identity of Complex Inorganic Material (Analytical Data)

The identity of a complex inorganic material goes beyond the traditional identifiers of material naming (IUPAC name, CAS number, EINECS number, etc.), and additionally includes information on the molecular and structural formula, (analytical) information on the elemental composition and structural data. A clear and unequivocal identification of the material is crucial for the next steps of the assessment, in particular if the assessment is based on the profile of the complex inorganic materials' individual constituents.

Most complex inorganic materials are associated with an intrinsic variability in material composition. This is mainly related to the variability of the source materials for the metal refining industry (rather than to differences in industrial manufacturing processes). Nowadays, typical source materials are complex raw materials and recyclable goods (e.g. electronic scrap, spent automotive/industrial catalysts or other precious metals containing materials) rather than (more pure) primary resources. Despite the good control of and knowledge about the industrial refining processes, this variability in source materials inevitably leads to a compositional variability of the complex inorganic materials produced during the refining process (see also Chapters 10 and 11). Obviously, when determining the material identity, this potential variability needs to be accounted for.

The material characterisation can be particularly challenging for complex inorganic materials such as UVCBs (chemical substances of Unknown or Variable composition, Complex reaction products and Biological materials) which are associated with this temporal and spatial variability in composition. The analysis of such materials is ideally performed at specialised and accredited laboratories to combine quantitative (e.g. elemental composition) with qualitative analysis (e.g. mineralogical analyses). The proficiency of the analytical lab should be documented in continuous quality control (QC) and quality assurance (QA) activities. Quality control involves the analysis of QC samples, such as calibration standards, certified reference materials, spiked samples, duplicate sample analysis and blanks and ideally also the participation in proficiency testing programs. Quality assurance includes an internal and external audit programs, and implementation of international quality management standards, such as ISO 9001 and ISO 17025.

Quantitative and qualitative analysis can be performed using different techniques depending amongst others on the type and composition of the material. An overview of analytical techniques is given in Tables 6.1 and 6.2. The quantitative information is usually gathered by comparing the signal obtained for the material [(or one/few of its constituent(s)) to the signal of a standard (series) of the constituent(s) of interest]. A qualitative analysis cannot be quantitatively assessed, and requires a careful interpretation of the obtained signal/output of the analyst and expert judgment. A combination of different analyses (i.e. quantitative and qualitative) is recommended for an unequivocal identification and characterisation of the material. Some of the techniques mentioned in Tables 6.1 and 6.2 are often combined. For example: analytics for total C and S, for sulphates or for silicates are usually reported together with total metal analysis, and mineral analysis of materials are usually based on information from various sources such as X-ray diffraction, optical microscopy, and scanning electron microscopy.

6.2.1.1 Physicochemical Properties

Information on physicochemical properties of a complex inorganic material provides fundamental information on the material's properties and its behaviour (such as melting- and boiling point, relative density, and water solubility). An overview of the most relevant physicochemical endpoints is provided in Table 6.3. The information for some of these endpoints is required for the assessment of the physical hazards (such as flammability or oxidising properties) or for the (eco)toxicological hazard assessment (such as solubility or granulometry). Related to the latter, physicochemical properties may directly influence the scope of the environmental or human health assessment.

For most complex inorganic materials, information on physicochemical hazards needs to be generated on a representative sample of the material. The experimental data and corresponding classification(s) resulting from the testing of this representative sample are assumed to apply to all complex inorganic materials with a similar material identity. For example, water solubility of metals and metal compounds is specifically tested following the Transformation/Dissolution (TD) Protocol (OECD, 2001). When complex inorganic materials are tested using this assay, the relative solubility for the individual elemental constituents should be determined in order to use the test data for other materials with a minor constituent variability.

In specific cases, physicochemical information from constituents might be read-across to the complex inorganic material as a viable alternative to testing. When this approach is taken, a read-across justification needs to be provided. This should contain the supporting information balanced with expert judgment on substance handling, and a minimum of experimental validation. As an example, flammable properties of metal powders may correlate with the particle size, or more specifically the relative surface area. Smaller particles exhibit a higher relative surface area and therefore have a larger surface for the oxidation reaction required for flammability. In cases where a physicochemical hazard assessment needs to be performed, one might consider testing selective metal powders for their flammable properties to determine a threshold for flammability.

6.2.1.2 Environmental Fate and Behaviour Data

The physicochemical characterisation often allows assessors to predict the behaviour of complex inorganic materials in the environment. For instance, water solubility or TD

TABLE 6.1   Quantitative Methods Typically Used for Complex Inorganic Materials Identification

Physical Form	Analytical Method	Information Obtained
Solid	ICP (Inductively Coupled Plasma) equipped with Optical Emission Spectrometer (OES), Spark Emission Spectrometer (SES) or Mass Spectrometer (MS): Destructive method—total or sequential dissolution required	Determination of elemental composition for all metals: *Note on quantitative elemental information: in most systems Na, Ca, Mg, etc. are immediately reported as CaO, MgO,TiO$_2$, etc. however the oxygen as such is not measured*
	XRF (X-ray fluorescence) Nondestructive method	Routine metal elemental analysis, typically used for substances from foundry, screening of incoming raw materials, analysis of internal recycled flue dust, analysis of emission filter.
	Titrimetry/volumetric analysis	Routine analysis to determine the concentration of an identified analyte (mostly constituents)
	AAS (atomic absorption spectroscopy)	Analysis using absorption of optical radiation (light) by free atoms in the gaseous state determine ~70 elements in solution or as solid sample
	Inert gas fusion infrared and thermal conductivity detection	Measurement of oxygen, nitrogen, and hydrogen content of inorganic materials, ferrous and nonferrous alloys, and refractory materials
	Electrogravimetry (on totally dissolved solid or raw solid)	Determination of specific elements such as Cu in raw materials and anodes
	TG thermogravimetry (induction furnace + solid-state infrared) (on totally dissolved solid)	Rapid simultaneous determination of Total sulphur and carbon content (e.g. in steel, cast iron, copper, alloys, ores, cement, ceramics, carbides, minerals, coal, coke, ashes, lime, gypsum, sand, and glass) *Note: total C and S are generally reported together with elemental (metal) information*
	SatMagan (SATuration MAGnetization Analyser), that is, comparative weighing in gravitational and magnetic fields (on totally dissolved solid)	Determines amount of magnetic material *Note: total magnetic content is generally reported together with Elemental (metal) information*
	Colourimetry (on totally dissolved solid)	Used for a variety of analysis such as the content of silicate *Note: in most systems, the silicate content is reported as SiO$_2$ in the (quantitative) elemental (metal) information*
	Separation technique: Ion exchange chromatography (of totally dissolved solid)	Determination of the amount of sulphates and other anions *Note: in most systems, the sulphate content is reported as SO$_4$ (CHEM) together with the (quantitative) elemental (metal) information*
	Spectrolaser, an elemental analysis instrument based on a technique called LIBS (Laser-Induced Breakdown Spectroscopy) for analysis of raw solid samples	Elemental metal analysis

Continued

TABLE 6.1   Quantitative Methods Typically Used for Complex Inorganic Materials Identification—cont'd

Physical Form	Analytical Method	Information Obtained
Liquid	Same chemical assays as for dissolved solid (see above)	Quantitative chemical composition
	Liquid (and/or gas) chromatography followed by UV and/or MS detection	Separation techniques to determine hydrophilic (or volatile) chemical components of a liquid (for UV–vis detection: with UV active functionalities only; and for MS detection: comparative to data base spectra)

TABLE 6.2   Qualitative Analytical Methods Typically Used for Complex Inorganic Materials

Physical Form	Analytical Method	Information Obtained
Solid	XRD (X-ray diffraction)	Main mineralogy of crystalline phases
	Optical microscopy (OM) and Scanning Electron microscopy (SEM) (on polished solid) with EDS (energy dispersive spectrometry) and WDS (wave-length dispersive spectrometry) analysers	Mineralogical characterisation
Liquid	Same chemical assays as for dissolved solid	Qualitative chemical composition
	Liquid (and/or gas) chromatography followed by UV–vis and/or MS detection	Separation techniques to determine hydrophilic (or volatile) chemical components of a liquid (for UV–vis detection: with UV active functionalities only; and for MS detection: comparative to data base spectra)

TABLE 6.3   Overview of Physicochemical Endpoints Required Under Most Regulatory Regimes

Endpoint	Further Use
Physical form	–
Melting/freezing point	–
Boiling point	–
Relative density	Inhalation assessment
Surface tension	Assess the potential to react with mucous membranes
Water solubility	Environmental and human health hazard assessment
Partition coefficient n-octanol/water	Environmental fate assessment
Flammability (solids)	Physical hazards
Explosive properties	Physical hazards
Self-ignition temperature	Physical hazards
Oxidising properties	Physical hazards
Granulometry	Inhalation assessment
Viscosity	Aspiration assessment

testing explains the expected dissociation in the water compartment. However, further data are required to cover, for example, the soil and sediment compartments, or to determine whether the substance is associated with bioconcentration or bioaccumulative properties.

When complex inorganic materials need to be tested for this endpoint, a pragmatic approach is taken based on the assessments of the constituents ('constituents-based approach'). This is related to their inherent variability in composition, and the consequent difficulties to select a representative sample for testing. This approach is explained in more detail in the following sections and chapters (see Chapters 7 and 8).

6.2.1.3 *Environmental and Human Health Toxicological Effects*

Preference is given to test data with the complex inorganic material of interest. For some (commercial) materials such as alloys or final slags, data may be available. These can then be used on their own or in a weight of evidence approach. However, for most complex inorganic materials, no experimental data exist. Similar to the environmental fate and behaviour endpoint, testing for environmental and human health effects will also be based on the constituents-based approach.

6.2.2 Data on Uses and Exposure

Information on uses, risk management measures, exposure, and emissions related to complex inorganic materials (or their constituents) are prerequisites for risk assessment. In practice, this usually entails the relevant details on the manufacturing, the various uses of the material and the disposal of the material.

The exposure assessment (mostly) covers the determination or prediction of the constituent metal concentrations at the workplace, for the human population via the environment (e.g. inhalation exposure via air emissions or ingestion via emissions to water) and where relevant via direct use of certain materials (e.g. use of articles made of alloys or stainless steel), and for the different environmental compartments (water, air, soil, sediment).

Metals are naturally occurring elements, and have a long history of (industrial) use. As consequence, measured environmental concentrations are an inherent combination of the natural background levels, potential historical contaminations and the (various) anthropogenic emissions (see Chapter 5). Although often difficult to achieve, gathering exposure and emission data associated with the individual uses or life-cycle stages for the various targets such as workers, general population, or the environment is crucial for chemicals management purposes. This allows an assessment of potential risks associated with each of these and allows identifying the most efficient risk management option(s).

The exposure assessment can be performed via measured/monitored data or modelled data. In the metal sector, monitoring data of hazardous metals are generally available, as measurements in workplaces or the environment are often requested by (national or regional) legislations. These data can then be used as database for modelling or as validation for modelled concentrations. It should be noted that both monitoring and modelling is usually focused on the individual (hazardous) constituents of a complex inorganic material. A constituent-based approach is used to do the exposure assessment of a complex inorganic material and a detailed description of this approach is provided in Chapters 2, 10, and 11.

6.3 QUALITY ASSESSMENT OF AVAILABLE DATA

Once the hazard and exposure data gathering is completed, all retrieved information need to be evaluated for data quality. Data quality includes as basic elements (in order of execution):

- Relevance: describes the suitability of the data to identify a particular hazard endpoint
- Reliability: describes the study's intrinsic quality compared against study-specific criteria, such as compliance with international guidelines or completeness and quality of the reporting
- Adequacy: describes the usefulness of the data for hazard and/or risk assessment purposes.

6.3.1 Relevance of Information

The relevance of information is essentially evaluated by answering the following questions:

- *Is the tested material representative for the material that is evaluated in the risk assessment, that is, are we evaluating the right thing?*
- *Does the study address the endpoint, that is, is the study answering the question 'How hazardous is the material?'*

The representativeness of the tested material emphasises the need for a complete and comprehensive evaluation of its identity in comparison with the material that is evaluated in the hazard and risk assessment. The assessor would otherwise not be in the position to separate the relevant from the nonrelevant information. This is particularly significant in poorly soluble materials being produced and marketed as powder/granules, since the (surface) properties of the particles (e.g. morphology, reactivity, functionalisation) might have an influence on the hazard properties.

6.3.2 Reliability of Information

Several criteria are available to ensure a harmonised assessment of reliability of experimental studies. The most commonly used criteria are the ones established by Klimisch et al. (1997). This system is referred to by a lot of national and international authorities (e.g. ECHA, 2011; US EPA, 2004; OECD, 2005). Although the criteria of Klimisch et al. were initially focused on (eco)toxicological laboratory studies conducted in accordance with (inter)national testing guidelines and under GLP, they are nowadays also commonly used for physicochemical and environmental fate and behaviour studies.

Some studies (Küster et al., 2009; Ågerstrand et al., 2011) clearly showed the need for an updated evaluation system. Recently, the Criteria for Reporting and Evaluating ecotoxicity Data (CRED-criteria) have been developed for assessing reliability and relevance (Kase et al., 2016; Moermond et al., 2016). The CRED-criteria aim at improving reproducibility, transparency, and consistency of reliability and relevance evaluations of aquatic ecotoxicity studies. Also, the CRED-criteria evaluation method provides more detailed guidance on how to evaluate study reliability and it should be noted that it has been questioned whether the Klimisch criteria can be fully applied when assessing the potential adverse consequences of essential trace elements, for which physiological, structural, and/or functional impairment due to either excessive or deficient exposures can be a concern (Plunkett, 2004).

The reliability assessment of human studies requires a different approach compared to studies with animals. Epidemiology studies provide evidence from human populations under real-world conditions and in this differ from animal studies performed under standardised conditions. The evaluation of evidence about a causal relationship between a presumed cause and an observed effect can be performed using the Bradford Hill criteria (Hill, 1965). A guidance document for the evaluation and use of epidemiological studies for risk assessment purposes has been published by the WHO (2000).

Further general data-quality screening recommendations and reading is available in several documents (ECHA, 2011; OECD 2005; EC, 2003).

6.3.3 Adequacy of Information

Adequacy defines the usability of information to allow a clear decision by the hazard/risk assessor whether:

(a) the substance has hazardous properties leading to a hazard classification,
(b) the substance has to be considered as persistent bioaccumulative toxic (PBT), or
(c) the information permits the derivation of effect levels for risk assessment.

The evaluation of adequacy is particularly important in case multiple test results are available for a single endpoint. Opposed to a 'key study approach' where a single highly reliable study is used to fill a particular endpoint, the 'weight of evidence approach' considers data with different reliability ratings developed via different test setups [such as quantitative structure–activity relationships (QSAR) or in vitro/in silico systems]. In case multiple studies are available for a single endpoint, the studies with the highest reliability should be given preference to develop the endpoint conclusion. The adequacy assessment of the various studies requires expert judgement and needs to be well documented, so that the reasoning/assessment is clear to other assessors or regulatory bodies. Note that metal specific aspects such as natural background concentrations, metal essentiality, or metal ions as driver for metals' ecotoxicity have to be taken into account when evaluating hazard or data exposure of complex inorganic materials.

6.4 DATA SEARCH: DATABASES AND WEB SOURCES

6.4.1 Hazard and Exposure Data

Risk assessors initially run a literature search to identify existing information for the complex inorganic material and its constituents. Well-known information sources on fate (eco) toxicity and/or risk data are the typical generic databases such as PubMed, but also other dedicated sources such as:

- eChemPortal: OECD's Global Portal to Information on Chemical Substances. eChemPortal allows simultaneous searching of reports and data sets by chemical name and number, by chemical property, and by globally harmonized system (GHS) classification. Direct links are obtained to collections of chemical hazard and risk information prepared for government chemical review programmes at national, regional, and international levels. Classification results according to national/regional hazard classification schemes or to the GHS of Classification and Labelling of Chemicals are provided when available. In addition, eChemPortal also provides exposure and use information on chemicals (https://www.echemportal.org/echemportal/index.action);

- HSDB ('Hazardous Substances Data Bank'; toxicology database that focuses on potentially hazardous chemicals) (https://toxnet.nlm.nih.gov/cgi-bin/sis/htmlgen?HSDB);
- RTECS (Registry of Toxic Effects of Chemical Substances; compendium of data extracted from the open scientific literature) (https://www.cdc.gov/niosh/rtecs/default.html);
- IRIS (US EPA's Integrated Risk Information System; electronic data base containing information on human health effects that may result from exposure to various chemicals in the environment) (https://www.epa.gov/iris); or
- ECOTOX (US EPA's Ecotoxicology knowledgebase, source for locating single chemical toxicity data for aquatic life, terrestrial plants, and wildlife) (https://cfpub.epa.gov/ecotox/).

For some of the complex inorganic materials or their constituents, a complete hazard and/or risk assessment is already performed and publically available under existing regulatory programmes such as in the EU (REACH available on the ECHA dissemination website, https://echa.europa.eu/nl/information-on-chemicals/registered-substances), by US EPA or by the OECD programme on HPV chemicals.

Focusing on metals and inorganic complex materials, consortia and associations representing the metals' industry have joined forces to collate and share relevant hazard and risk assessment information. The biggest driver hereto was the need for (voluntary) risk assessments under the EU Existing Substance Regulation, and which has now been replaced by the EU REACH, 2007 Regulation. A lot of data have been gathered under these programmes, and these have to be shared amongst the various players. One of the platforms to share data is the multimetallic database (MMD), developed under the supervision of Eurometaux, where several metal consortia agreed to collect and share relevant EU REACH information in a unique database. This ensures consistency in data used for assessments (e.g. derived threshold values and underlying scientific reasoning), in particular for complex inorganic materials.

In the absence of experimental data, alternatives to animal testing should be considered as a first option to fill data gaps.

The possibility for read-across from another material and the use of secondary sources (such as handbook data) should be initially considered. Secondly, in silico tools should be envisaged to fill data gaps. In silico tools are computational methods for estimating the behaviour or toxicity of a chemical using mathematical models. Starting from typical characteristics such as molecular structure and/or physicochemical characteristics, in silico tools can complement (or even substitute) in vitro and in vivo testing, thus minimising the need for animal testing. For organics, QSARs are well developed and are available on dedicated websites (e.g. OECD QSAR Toolbox or EPA EPISuite). For metals, however, the QSAR approach is not directly applicable, and an alternative approach has been developed since the mid-1950s: the 'quantitative cation–activity relationship' (QCAR) or 'quantitative ion–character activity relationship' (QICAR) concept as summarised by, for example, Walker et al. (2003). The basis for the development of QICARs is the available metal fate/(eco)toxicity data and the relevant physicochemical properties of both target and source metals such as molecular weight, electronegativity, or standard electrode potential (Chapter 14). It is important to accurately document the assumptions, applicability ranges and any other relevant information that is considered in case of in silico methods.

A number of databases are available for environmental exposure data on a continental scale. An overview is available in Table 6.4. Harmonised monitoring data for metal concentrations and general physicochemical properties of water, sediment, or soil are critical for a realistic assessment of exposure (and thus risk) for metals in the various environmental compartments. Important factors for harmonisation are land use, sampling and analytical methodologies. The data cited below provides a reliable basis for taking into account the spatial variability for both exposure and effect concentrations via, for example, considering bioavailability through variation in physicochemical characteristics (such as pH or dissolved

TABLE 6.4   Sources for Continental Monitoring Data for Metals

Database	Environmental Compartment	Geographical Area	Parameters Analysed	Reference
FOREGS	Water sediment soil	Europe	• Total + aqua regia extractable element content • pH • Total carbon content (sediment, soil) • Dissolved carbon content (water) • Sulphate, nitrate (water) • Inorganic elements in water	http://weppi.gtk.fi/publ/foregsatlas/
NAWQA	Groundwater, streams, sediment, fish, and clam tissue	United States		http://water.usgs.gov/nawqa/ Mahler et al. (2006); Ayotte et al. (2011); DeWeese et al. (2007)
GEMAS	Soil (arable land 0–20 cm) and grazing land (0–10 cm)	Europe	• Total + aqua regia extractable element content • General soil properties (pH, organic carbon, clay, CEC)	Reimann et al. (2014) http://gemas.geolba.ac.at/
USGS	Soil, ambient background (0–5 cm, A-horizon and C-horizon)	United States	(near) total element content	Smith et al. (2013) http://pubs.usgs.gov/ds/801
National Geochemical Survey of Australia	Soil, ambient background (0–10 and 60–80 cm depth)	Australia	• Total + aqua regia extractable element content pH	De Caritat and Cooper (2011)

organic carbon). The availability of these data avoids worst-case assumptions/predictions in exposure and effects assessments and thereby increases the transparency and reliability of the regional risk assessment.

6.4.2 Data Storage

It is essential to have an accessible and widespread tool for collection, storage, maintenance, and sharing of available information on the complex inorganic material and its constituents. A widely used application for this purpose is the International Uniform Chemical Information Database (IUCLID) tool; available at https://iuclid6.echa.europa.eu/). It is a key software application for international regulatory bodies and the chemical industry, and is used among others in the OECD Cooperative Chemicals Assessment Programme (CoCAP), EU REACH, the US EPA HPV Chemical Challenge Programme, and the Japan HPV Challenge Programme.

IUCLID version 5 was implemented for the first time in the OECD Harmonised Templates. These are standard data formats for reporting study results in order to determine chemicals' properties and their effects on human health and environment. IUCLID version 6 was released in 2016, (https://iuclid6.echa.europa.eu/) and includes a number of additional features such as the 'assessment entity', which will allow a more transparent and comprehensive data reporting for complex inorganic materials.

6.5 WHAT IF THERE ARE REMAINING DATA GAPS?

If all possible data sources have been explored without allowing to fulfil a gap in the information required to perform the assessment and if data on comparable materials do not allow to read-across, the possibility of testing should be envisaged.

However, this requires to address both the issue of variability of some of the complex inorganic materials and to carefully design the test setting to consider some metal-specific aspects.

6.5.1 Variability: What to Test?

Most complex inorganic materials are characterised by a variable composition, and this variability will impact on the representativeness of the sample of material that can be selected for testing. A single sample will usually reveal only the composition of the material the day it is taken and will thus imperfectly reflect the actual variability of the material to be assessed.

An alternative approach is to use and/or generate the information on the constituents of the material and to predict the material's expected behaviour based on the combination of the constituents' specific properties in a worst-case scenario, considering all relevant moieties present at 100%. This is further explained in the following chapters. This constituent-based approach does not, however, consider potential synergistic or antagonistic effects of the constituents, or any matrix effects on the material, which can only be assessed via testing. Also using a worst-case scenario may result in a worst-case assessment, not reproducing the actual exposures and hazards at the plant.

Thus, both approaches entail some uncertainty, which in turns affects the conservatism of the obtained (eco)toxicological test results. Fig. 6.1 compares the two approaches—testing the complex inorganic material or using data on its constituents—in terms of conservatism and uncertainty:

In view of the important variability that can affect some materials such as UVCBs during the refining of metals, but also of the availability of data and practicality of testing, a constituents-based approach is generally preferred by industry or regulators. Fig. 6.1 summarises the key elements of the rationale to be followed and data on constituents can be retrieved as described earlier.

It may however happen that data is missing on some of the constituents and the assessor will need to establish on a constituent-specific basis whether additional testing or exposure information is required to carry out a risk assessment.

6.5.2 Metal Specificities in Data-Generating Tools/Tests/Strategies

Chapters 7 and 8 will explore in detail the specificities of the assessment and in particular the consideration of bioavailability. When generating test data and/or assessing literature data it will be important to assess whether the appropriate protocols have been used, to most appropriately reflect the toxicity of the metals contained in the complex inorganic material. Some specific and standardised protocols exist such as the TD testing (OECD 2001) while others are under development (e.g. bioelution). Difficulties experienced with available test

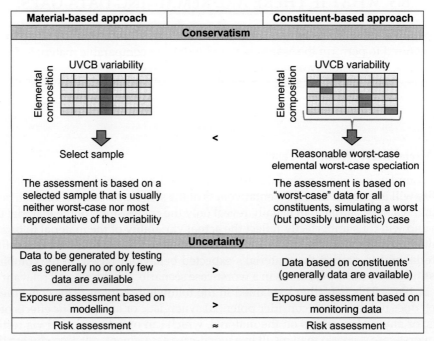

FIG. 6.1   Comparison of standard approach (testing the UVCB) and constituent-based approach related to conservatism and uncertainty.

designs or test strategies are also collected, so as to be able to propose, when possible, some metal-specific refinements.

For environmental testing, the natural metal background concentration, for example, is key information to consider when designing experimental testing or evaluating data. Only data from ecotoxicity tests conducted at metal background concentrations (in the culture media) similar to the region under investigation and within the ecological niche of the organism should be deemed relevant. Species tested at very low background metal concentrations outside their ecological boundaries may be overly sensitive (due to deficiency issues or induced stress levels that would exert toxicity) and conversely, organisms cultured in media with elevated metal concentrations (both essential and nonessential metals) may become less sensitive. In particular, it is recommended that the essential metal concentration in the culture medium should be at least equal to the minimal concentration that is not causing deficiency for the test species used. Concentrations of nonessential metals should fall within the natural background variation of these metals. Defining minimal levels of metal background for the selection of relevant culture media should only be performed in case there is scientific evidence that acclimation/adaptation phenomena are relevant for the metal under investigation. Ideally, the culture and test media that are used should be within the physicochemical boundaries of the environment under study; this relates to pH, hardness, DOC, and natural metal background, etc. If they are reported and test organisms have been cultured in conditions that are outside the natural background concentration ranges, such data should be discarded. It is, however, recognised that this may lead to a reduction in the number of useful ecotoxicity data which may even sometimes limit the possibility of using a species sensitivity distribution. Another complicating factor is that quite often culture conditions are not reported and in that case expert judgment should be used to decide if the study can still be used or not.

For human health testing, metal specificities are also important. A good example is mutagenicity/genotoxicity testing. Usually the assessment of mutagenic properties of chemicals follows a tiered approach, starting with in vitro testing, followed up by in vivo testing as appropriate. Before initiating experimental testing, some key metal characteristics need to be well considered. Some examples:

- the solubility of the test compound might directly affect the study design (sparingly soluble compounds vs. readily soluble compounds);
- although metal uptake in bacteria appears to be limited and reverse mutation tests in bacteria therefore return negative, the conduct of a bacterial reverse mutation test is nevertheless recommended to demonstrate (a lack of) mutagenic activity;
- testing in mammalian cell culture assays for forward mutations is proposed as part of the base testing data set. Preference is given to a system sensitive to the large DNA deletions, as this is believed to be the predominant form of damage induced by the indirect mechanisms of action that characterise many metals;
- although observations of in vitro mutagenicity are common, responses often appear to be elicited in the absence of direct metal interaction with DNA. Some hypotheses are indirect oxidative mechanisms in vitro (Ariza et al., 1998; Yang et al., 1999; Vaglenov et al., 1998; Roy and Rossman, 1992) and in vivo (Huang et al., 1988; Fracasso et al., 2002; Valverde et al., 2001), interference with DNA repair (Hartwig, 1994), interference with function of the mitotic apparatus by metal specific binding to mitotic spindle proteins (Their et al., 2003);

- while control for sources of artefactual responses is advised for any substance, it is particularly important in the testing of essential metals that are involved in processes regulating cell division, differentiation, and apoptosis;
- nonphysiological routes of exposure can increase the precision of dosing, but bypass physiological carrier systems, fail to reproduce patterns of tissue specificity that would characterise oral or inhalation exposures, and may result in levels of internal exposure that could not be achieved via oral or inhalation exposure.

The challenge that is presented for follow-up testing is a determination of whether or not positive in vitro responses are induced via mechanisms that are relevant for intact organisms. Follow-up in vivo studies would ideally be conducted to evaluate the genetic endpoint of concern from in vitro testing (gene mutations, chromosome, or genome effects), coupled with an understanding of the toxicokinetic properties of the substance under study. Tissues known to accumulate high concentrations of metal, or to be the targets of metal toxicity, would be assigned priority for evaluation. If toxicokinetic information is lacking, indicator tests such as the Comet assay could be applied to identify potential target tissues for in-depth study with mutagenicity tests. The route, intensity and duration of exposure would be selected with a view to both maximising mutagenic responses and avoiding high dose effects (such as tissue necrosis or apoptosis) that may produce artifactual false positives.

6.5.3 How to Address Combined Toxicity If a Constituents-Based Approach Is Followed?

Further investigation is still required to understand (potential) interactions between the various constituents in the complex inorganic material and account for such 'mixture effects'. Both for human health and environment, the combined toxicity effects of several metals acting simultaneously are in fact not totally clear. The available mixture toxicity data mainly focus on a limited amount of metal combinations (typically binary mixtures). Moreover, these data are mostly obtained at concentrations resulting in significant effects (e.g. EC50 level or higher), while for risk assessment the interactions at no or low effect concentrations are relevant (e.g. NOEC or EC10 level).

Scientific research on mixture toxicity has been ongoing across sectors for some years. Examples are the 'Toxicity and Assessment of Chemical Mixtures' (SCHER, SCCS, SCENIHR, 2012) and the 'Transitional Guidance on mixture toxicity assessment for biocidal products for the environment' (ECHA, 2014). Despite the need to refine and test the methodology, further developments are under investigation to identify a valid tiered approach for environmental and human health assessment. Improvements and state of the art in the combined toxicity assessment are reported in the other chapters (Chapters 7–9).

Alongside the scientific development, the metal sector considers investing in the creation of an ad hoc database, populated with high-quality data, to ensure a correct dissemination of (scientific) findings and an appropriate implementation of the scientific approaches. Available data such as toxic reference values, fate information, environmental parameters, toxicokinetics information, target organ indications, and mode of action for constituent metals (and metal mixtures if available) will be included and ideally be linked to the metals mixture assessment, to ensure a transparent reporting of the data used to assess mixture effects. Such a database will allow for a comparison between available data (as with the existing MMD, developed under Eurometaux's umbrella) and could possibly push for a higher degree of harmonisation in data screening and assessment of data relevance.

KEY MESSAGES

The assessment of the complex inorganic material starts by compiling and evaluating all available and relevant information, including the intrinsic properties of the material, its uses (manufacture, use and subsequent life-cycle stages), and the associated emissions and exposure.

Usually only limited data are available for complex inorganic materials; it is therefore proposed to use the information on their constituents to predict their behaviour (constituents-based approach). Using data on constituents reduces the need for testing and addresses the difficulty of selecting a representative sample.

Several databases where data can be accessed to retrieve the existing information on constituents (and sometimes on the material as such) are available, and an overview is provided in this chapter.

While some metal constituents are data rich and have been previously assessed, others may be data-poorer and associated with critical data gaps. It is expected that further regulatory developments will continue to boost data development for such metals, and especially for those of highest concern.

Alternatively, in silico tools such as QICARs might deliver useful weight in support of evidence. In certain cases, some testing may be envisaged to complete data gaps on constituents. Metal specificities should be carefully considered when designing such tests.

IUCLID 6 provides a reliable format to store and report the data on complex inorganic materials

References

Ågerstrand M, Küster A, Bachmann J, et al: Reporting and evaluation criteria as means towards a transparent use of ecotoxicity data for environmental risk assessment of pharmaceuticals, *Environ Pollut* 159:2487–2492, 2011.

Ariza ME, Bijur GN, Williams MV: Lead and mercury mutagenesis: role of H_2O_2, superoxide dismutase, and xanthine oxidase, *Environ Mol Mutagen* 31:352–361, 1998.

Ayotte JD, Gronberg JM, Apodaca LE: *Trace elements and radon in groundwater across the United States*, U.S. Geological Survey Scientific Investigations Report 2011-5059, 115 p. 2011.

De Caritat P, Cooper M: *National Geochemical Survey of Australia: The Geochemical Atlas of Australia*, Geoscience Australia Record, 2011/20 (2 Volumes), 557 pp 2011.

DeWeese LR, Stephens VC, Short TM, Dubrovsky NM: Trace-element concentrations in tissues of aquatic organisms from rivers and streams of the United States, 1992-1999: U.S, *Geological Survey Data Series* 309, 2007.

EC (European Communities): *Technical guidance document (TGD) on risk assessment in support of Commission Directive 93/67/EEC on Risk Assessment for new notified substances Commission Regulation (EC) No 1488/94 on Risk Assessment for existing substances Directive 98/8/EC of the European Parliament and of the Council concerning the placing of biocidal products on the market*, 2003. https://echa.europa.eu/documents/10162/16960216/tgdpart2_2ed_en.pdf.

ECHA (European Chemicals Agency): Guidance on information requirements and chemical safety assessment, Chapter R.4. In *Evaluation of available information, version 1.1*, 2011. ECHA-2011-G-13-EN.

ECHA (European Chemicals Agency): *Transitional Guidance on the Biocidal Products Regulation: transitional guidance on mixture toxicity assessment for biocidal products for the environment*, 2014, European Chemicals Agency, Helsinki.

EU REACH: *EC Regulation (EC) No 1907/2006 of the European Parliament and of the Council of 18 December 2006 concerning the Registration, Evaluation, Authorisation and Restriction of Chemicals (REACH), establishing a European Chemicals Agency*, 2007.

Fracasso ME, Perbellini L, Solda S, Talamini G, Franceschetti P: Lead induced DNA strand breaks in lymphocytes of exposed workers: role of reactive oxygen species and protein kinase C, *Mutat Res* 515:159–169, 2002.

Hartwig A: Role of DNA repair inhibition in lead-and cadmium-induced genotoxicity: a review, *Environ Health Perspect* 102:45–50, 1994.

Hill AB: The Environment and Disease: Association or Causation? *Proc R Soc Med* 58(5):295–300, 1965.

Huang X, Feng Z, Zhai W, Xu J: Chromosomal aberrations and sister chromatid exchanges in workers exposed to lead, *Biomed Environ Sci* 1:382–387, 1988.

Kase R, Korkaric M, Werner I, Ågerstrand M: Criteria for Reporting and Evaluating ecotoxicity Data (CRED): comparison and perception of the Klimisch and CRED methods for evaluating reliability and relevance of ecotoxicity studies, *Environ Sci Eur* 28:7, 2016.

Klimisch H, Andreae M, Tillmann U: A systematic approach for evaluating the quality of experimental toxicological and ecotoxicological data, *Regul Toxicol Pharm* 25:1–5, 1997.

Küster A, Bachmann J, Brandt U, et al: Regulatory demands on data quality for the environmental risk assessment of pharmaceuticals, *Regul Toxicol Pharmacol* 55:276–280, 2009.

Mahler BJ, Van Metre PC, Callender E: Trends in metals in urban and reference lake sediments across the United States, 1970 to 2001, *Environ Toxicol Chem* 25(7):1698–1709, 2006.

Moermond CT, Kase R, Korkaric M, Ågerstrand M: CRED: criteria for reporting and evaluating ecotoxicity data, *Environ Toxicol Chem* 35:1297–1309, 2016.

OECD (Organisation for Economic Co-operation and Development): *Series on Testing and Assessment N°29: Guidance Document on Transformation/Dissolution of metals and metal compounds in aqueous media*, ENV/JM/MONO (2001)9, 2001.

OECD (Organisation for Economic Co-operation and Development): Manual for the assessment of chemicals, Chapter 3. In *Data Evaluation, 3.1 Guidance for Determining the Quality of Data for the SIDS Dossiers: (Reliability, relevance and adequacy)*, 2005. http://www.oecd.org/chemicalsafety/risk-assessment/manualfortheassessmentofchemicals.htm. Paris.

Plunkett V: In *Criteria developed for rating the utility and quality of studies of excess or deficiency of essential*, Presented at the US Society of Toxicology Meeting, Baltimore. 2004. cited in HERAG 2007 (Personal communication) Fact sheet on Essentiality, http://www.ebrc.de/downloads/HERAG_FS_07_August_07.pdf.

Reimann C, Birke M, Demetriades A, Filzmoser P, O'Connor P, editors: Chemistry of Europe's agricultural soils—Part A: methodology and interpretation of the GEMAS data set. In Geologisches Jahrbuch (Reihe B 102), Hannover, 2014, Schweizerbarth. 528 pp. + DVD.

Roy NK, Rossman TG: Mutagenesis and comutagenesis by lead compounds, *Mutat Res* 298:97–103, 1992.

SCHER, SCCS, SCENIHR: *Toxicity and assessment of chemical mixtures*, 2012. http://ec.europa.eu/health/scientific_committees/environmental_risks/docs/scher_o_155.pdf.

Smith DB, Cannon WF, Woodruff LG, Solano F, Kilburn JE, Fey DL: Geochemical and mineralogical data for soils of the conterminous United States. In *U.S. Geological Survey Data Series 801*, 2013. http://pubs.usgs.gov/ds/801/. 19 pp.

Their R, Bonacker D, Stoiber T, et al: Interaction of metal salts with cytoskeletal motor protein systems, *Toxicol Lett* 140–141:75–81, 2003.

US EPA: *Interim Guidance on the Evaluation Criteria for Ecological Toxicity Data in the Open Literature: Phases I and II: Procedures for Identifying, Selecting, and Acquiring Toxicity Data Published in the Open Literature For Use in Ecological Risk Assessments*, Washington, DC, 2004, Office of Pesticide Programs.

Vaglenov A, Carbonell E, Marcos R: Biomonitoring of workers exposed to lead. Genotoxic effects, its modulation by polyvitamin treatment and evaluation of the induced radioresistance, *Mutat Res* 418:79–92, 1998.

Valverde M, Trejo C, Rojas E: Is the capacity of lead acetate and cadmium chloride to induce genotoxic damage due to direct DNA-metal interaction? *Mutagen* 16:265–270, 2001.

Walker JP, Enache M, Dearden JC: Quantitative cationic-activity relationships for predicting toxicity of metals, *Environ Toxicol Chem* 22:1916–1935, 2003.

WHO (World Health Organisation): Working Group report on Evaluation and use of epidemiological evidence for environmental health risk assessment: WHO Guideline Document, *Environ Health Perspect* 108:997–1002, 2000.

Yang JL, Wang LC, Chang CY, Liu TY: Singlet oxygen is the major species participating in the induction of DNA strand breakage and 8-hydroxydeoxyguanosine adduct by lead acetate, *Environ Mol Mutagen* 33:194–201, 1999.

Environmental Toxicity Assessment of Complex Inorganic Materials

Frank Van Assche, *Katrien Delbeke*[†], *Karel De Schamphelaere*[‡], *Charlotte Nys*[‡], *Koen Oorts*[§], *Erik Smolders*[¶]

[*]International Zinc Association, Brussels, Belgium [†]European Copper Institute, Brussels, Belgium [‡]Ghent University, Gent, Belgium [§]ARCHE Consulting, Leuven, Belgium [¶]KU Leuven, Leuven, Belgium

7.1 INTRODUCTION

Metals may be used in their elemental form or as a compound, or in the form of more complex inorganic materials such as alloys, where elements are "so combined that they cannot be readily separated by mechanical means" (UN GHS, 2011). Metals are also constituents of series of other types of complex materials that are used and generated during primary and secondary production of metals, ceramics, and inorganic pigments. Examples are ores and concentrates, slimes, sludges, flue dust, and other residues. These materials usually qualify as Unknown or Variable composition, Complex reaction products or Biological materials (UVCBs) (Rasmussen et al., 1999; US EPA, 2005a,b), due to their variability in composition and physical form. The manufacture and use of complex inorganic materials may thus be associated with exposure of the environmental compartments and a thorough assessment of the environmental hazards and risks is required to ensure their appropriate management.

Ecotoxicity information on complex inorganic materials is usually scarce. When data on some complex inorganic materials are available, grouping and read-across approaches as proposed, for example, by OECD or EU REACH may be tried out to help fulfilling information requirements for a 'similar' material under investigation, while minimising testing. These approaches use the available data to predict the effects of the assessed material, providing the materials are sufficiently comparable (e.g. OECD, 2014; ECHA, 2008a, 2011).

While data gaps could also be filled by launching new tests, it will generally be preferred to avoid doing so, for reasons of complexity and variability of the materials, as discussed further under Section 7.2. Considering the overall paucity of data on complex inorganic materials and the complexity associated with direct testing, the tendency is to base the environmental assessment of the complex inorganic material on its constituting elements, and possibly to refine the assessment by applying established methodologies that have been developed for single metals, for example, the quantification of metal solubility and bioavailability.

It is indeed generally recognised that the ecotoxicity of most metals and metal compounds is related to the toxicity of the soluble metal ions (Allen et al., 1980; UN GHS, 2011 Annexes 9.7 and 10; OECD, 2012). The toxicity of metals and metal compounds contained in the complex inorganic material is thus dependent on the capacity of the substance to release the ionic form of respective metals contained in the solution. This means that any assessment of hazard or risk has to address the solubility of the complex inorganic material and the effects of the mixture on soluble ions that arise after dissolution in the environmental compartment.

This 'constituents'-based approach, which makes the best use of the data on the metals and metal compounds included in the material, is however confronted with the challenge of having to predict the combined exposure and effects on the constituents. Section 7.3 proposes some ways forward to address the combined hazard and risk.

This chapter summarises the main principles driving the hazard and risk assessment of complex inorganic materials and discusses refinements that have recently been proposed in this respect. Focus will be put on practical approaches that quantify the solubility and bioavailability (and thus the ecotoxicity potential) of the metals contained, and limit the need for ecotoxicity testing. A way forward on the assessment of combined hazard and risk, using tiered approaches with different levels of refinement is proposed and illustrated with practical examples.

7.2 MEETING ENVIRONMENTAL HAZARD INFORMATION REQUIREMENTS

A number of data should be available to perform the appropriate hazard and risk assessment of the complex inorganic materials. The data to be collected is usually specified in chemicals management legislation (e.g. in EU REACH, UN GHS, and US TSCA).

As indicated above, existing data on the materials themselves is scarce and testing is not preferred. On the other hand, due to the number and the wide variability of existing complex inorganic materials, it is practically impossible to assess each and every material specifically. Some pragmatic solutions therefore need to be found to address data requirements and the hazard/risk assessment.

7.2.1 Grouping of Complex Inorganic Materials

It is therefore proposed, as a first step, to structure the environmental assessments by grouping materials of similar composition, and expected similar physicochemical and ecotoxicological properties. In practice, grouping will be done based on different criteria, starting from general information, for example, within the metal concentrates, groups will be distinguished based on the main metal in the ore (copper concentrates, zinc concentrates, etc.). Alloys can be grouped based on composition and/or existing technical standards and intermediate materials produced during manufacturing processes can be grouped if they have similar chemical compositions (e.g. the lead leach residues from zinc-related metallurgy). Broader groups can be further split into more specific groups, for example, within the zinc concentrates group, the materials originating from sulphidic ore will be taken as a group (see Section 7.5.1). Ultimately, grouping should result in a 'functional complex inorganic materials-group' of materials with expected similar physicochemical and ecotoxicological properties. Within a group, a *generic* assessment can be made on 'representative materials' of the group instead of analysing every single specific material of the group. This requires the representative materials to be relevant for all the materials of the group.

The type of grouping that can be performed has to be in line with the requirements of the chemical management systems. Some legislation, such as the EU REACH regulation, imposes a requirement for 'joint submission of data by multiple registrants'. These provisions require that when the same substance is intended to be manufactured in the EU Community by one or more manufacturer and/or imported by one or more importer, the information relating to the properties of the substance and its classification shall be collected and submitted jointly. Registrants of the same substance have thus to comply with important data-sharing obligations (see also Chapter 6). EU REACH registrants of an inorganic UVCB intermediate substance (e.g. doré) will collect the information reported by the companies involved in its production (e.g. composition, processes) and from the gathered information describe and submit the identity of the material in a *generic* way. If testing (e.g. physicochemical testing or solubility testing) has to be performed, representative materials will be selected.

The information to be collected on these representative materials is outlined in Section 7.2.3.

7.2.2 Selection of Representative Materials

After grouping, representative materials on which the assessment is to be carried out and/or on which testing could be performed, can be selected. Key is to define substances within the 'functional complex inorganic materials-group' that are typical for its characteristics as a whole ('representative') and to keep at hand the reasoning backing up the selection of the representative material. This is further explained under Section 7.4.1.

7.2.3 Information Requirements for the (Representative) Materials

As indicated above, chemicals management legislations usually specify the data to be collected. The stepwise data-gathering approach proposed in Fig. 7.1 has been used to fulfil information requirements under the EU REACH system.

As indicated previously, a key aspect for the metals, and hence for the complex inorganic material toxicity, is the solubility of the metal ion. The data gathering should therefore include information on solubility, on the representative material if available, and on its constituents. As a default approach in hazard assessment, in the absence of metal ion release data, constituents of complex inorganic materials are usually considered as being entirely soluble. However, in many cases, the metals contained in these substances are only partially (e.g. intermediates of metal production) or barely soluble (e.g. alloys, slags). Determination of the solubility of the metals and metal compounds contained in the complex inorganic material is thus key for a realistic hazard assessment.

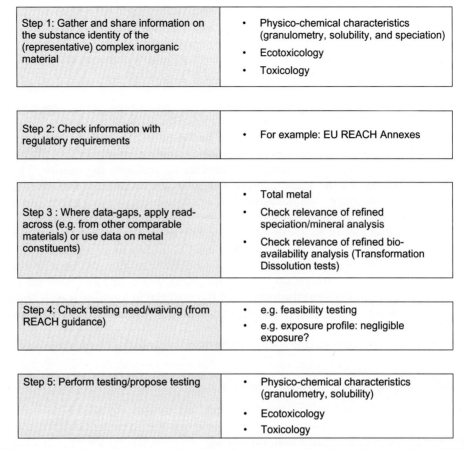

FIG. 7.1   Example of stepwise approach for data-gathering required for the registration of the UVCB under REACH.

Metal solubility also depends on the speciation of the metals. Different metal species (metal, oxides, sulphides, minerals, and alloys) with different solubility, with known and/or unknown hazards may be present within one complex inorganic material. In addition, for metallic substances and inorganic complex materials that result in particulate materials, the solubility of the metals contained may be related to the surface that is exposed to the solution.

Accordingly, the particle size (granulometry) of such substances may be an important factor in toxicity. For some metals, for example, for nickel, different aquatic effect classifications have been derived for the powder and the massive form, respectively (EU CLP, 2008). Selection of a representative particle size may thus be crucial in the assessment of complex inorganic materials in particulate form. Knowledge on the speciation and on the other physicochemical characteristics of the complex inorganic material that define the solubility is an inherent part of their hazard identification. More information on possible data sources can be found in Chapter 6.

7.3 HAZARD AND RISK ASSESSMENT OF THE COMPLEX INORGANIC MATERIAL

Different approaches can be followed to assess the hazard and risk of complex inorganic materials, depending on the data that are available.

For hazard assessment, where available, data obtained on the representative material of the group can used. It could also be generated by testing. This is discussed under Section 7.3.1. However, in practice, the assessment will usually be done using data on the constituents. Some refinement approaches can be applied. The general approach shown in Fig. 7.2 is further detailed in Sections 7.3.1 and 7.3.2.

For risk assessment, the hazard assessment values (e.g. predicted no effect values) and exposure values will be based on the constituents. The risk will be based on the combined risk from the (solubilised) constituents. The environmental exposure resulting from the manufacturing and/or use of the complex inorganic material will be estimated based either on modelling of environmental metal concentrations from emissions, or from direct measurements

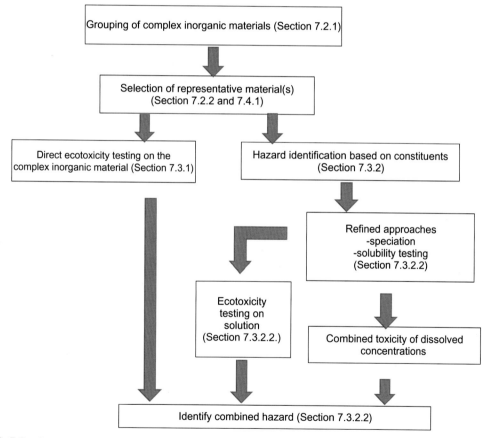

FIG. 7.2   General scheme of hazard identification of the complex inorganic material with the selection of reference material(s) indicated as a first, critical step.

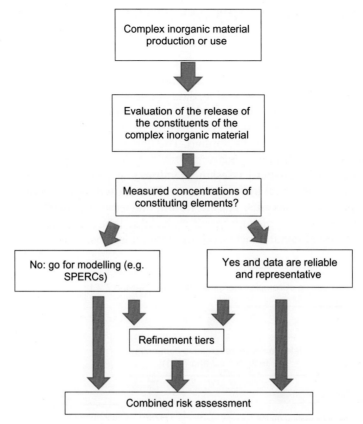

FIG. 7.3   General scheme of risk assessment of the complex inorganic material based on modelled or measured metal releases.

in the receiving environments. Some refinement approaches can be applied. Exposure can be considered in a generic way, where several data sets are combined to reflect the performance of the complex inorganic materials group as a whole, or can be more site-specific. The general approach for the complex inorganic material's risk assessment is outlined in Fig. 7.3 and further detailed under Section 7.3.2.3.

7.3.1 Use of Complex Inorganic Material-Specific Data for Hazard Assessment

Ecotoxicity data on complex inorganic materials is scarce, and one could envisage direct testing of such materials with a standard testing set-up. However, testing encounters have several limitations: the material may be insoluble or only sparingly soluble, it can have varying particulate size and form and, in particular, its composition may be highly variable (UVCBs). These limitations make the selection of one single test material to cover the characteristics of the whole group problematic. For these reasons, direct ecotoxicity testing on the complex inorganic material is only applied in specific cases, for example, for 'reality checks' on hazard assessments using the combined toxicity approach (see Section 7.3.2). Instead of testing the ecotoxicity of material samples directly using classical ecotoxicity tests, preference is given to perform the hazard assessment based on the toxicity of the constituents of the complex inorganic material.

7.3.2 Use of Data on the Constituents of the Complex Inorganic Material for Hazard and Risk Assessment

This approach makes use of the knowledge and data on the constituents of the material. When complex inorganic materials enter the environment, they will release a combination of metal ions in solution depending on the solubility of their constituents. The hazard and risk assessment for the environmental compartment will be based on the consideration of the combined toxicity of this resulting mixture of metal ions, according to the principles outlined in Section 7.3.2.1.

7.3.2.1 Principles of Combined Metal Toxicity

It is known that the effect of a mixture may differ extensively from the effects of the individual constituents in the mixture (Altenburger et al., 2013). This represents a challenge while having to assess the combined toxicity using the ecotoxicity data of the individual metal constituents. However, it is also acknowledged that testing every possible mixture combination occurring in the environmental compartment is unrealistic. Therefore significant efforts have been directed to develop approaches that would best predict mixture toxicity based on the individual toxicity of the constituents of the complex inorganic material. The main principles are explained hereunder.

CONCENTRATION ADDITION AND/OR INDEPENDENT ACTION?

Currently, two mixture reference models are generally accepted in ecotoxicology: the concentration addition (CA) and independent action (IA) models (Jonker et al., 2005, 2011). Both models assess mixture effects based on the effects of the individual mixture constituents. One of the most important differences between these mixture reference models is the assumptions about the underlying mechanisms of how the mixture constituents affect organisms.

In the CA model, originally described by Loewe and Muischnek (1926), the toxic effect is related to the weighted sum of concentrations of mixture constituents, the weighing factor being larger for more toxic constituents. The concept can be mathematically expressed as

$$\sum_{i=1}^{n} TU_i = \sum_{i=1}^{n} \frac{c_i}{ECx_i} = 1 \tag{7.1}$$

where n is the number of mixture constituents, TU_i is the toxic unit (TU) of the ith mixture component. The TU_i is defined as the ratio between c_i, the concentration of the ith mixture component, and ECx_i, the $x\%$ effective concentration of the ith mixture component (when applied singly).

If the CA model holds, the sum of TUs ($\sum TU$) = 1 in a mixture causing $x\%$ effect. The CA model assumes that substances have the same mode of action. Additionally, it assumes that a substance in a mixture can be exchanged for other substances without changing the overall mixture toxicity, as long as the sum of TUs of the mixture does not change.

The IA model assumes that substances have a different mode of action. The IA model is based on the concept of independent random events (Bliss, 1939). It assumes that the joint response to a mixture (y_{mix}) can be calculated as the product of the expected responses to each of the individual components in the mixture (y_i) with y being a value between 0 (full inhibition) to 1 (no inhibition), that is,

$$y_{mix} = \prod_{i=1}^{n} y_i \qquad (7.2)$$

The model is often represented by

$$E_{mix} = 1 - \left(\prod_{i=1}^{n} (1 - E_i) \right) \qquad (7.3)$$

In Eq. (3), E_{mix} is the proportional joint effect of the mixture, and E_i is the proportional individual effect of the ith mixture component if applied singly.

The CA and IA models fundamentally differ in how they assess effects of mixtures wherein the constituents are present at low concentrations. Based on Eq. (3), the IA model assumes that for a mixture where each of the mixture constituents is present at a concentration that does not cause effect, that there would not be a mixture effect either. Only components that are present at a concentration that causes an effect (i.e. $E_i > 0$) will actually contribute to the joint effect. According to the CA model, all mixture components contribute to the overall mixture toxicity, proportional to their TU. This implies that the joint effect of a mixture wherein the mixture components are present at low concentrations, for example, all at their EC1 (i.e. the 1% effective concentration) is dependent on the number of mixture components, and that many of such small effects may add up to a risk.

Because of this higher conservatism of the CA model, there is a regulatory preference for this approach over the IA approach as a default, at least in the absence of adequate mode of action information (e.g. SCHER, SCCS, SCENIHR, 2012; ECHA, 2014). Moreover, for pragmatic reasons, CA is more applicable since it can be used with single data points or single substance data, such as EC50-, NOEC-, or PNEC-values whereas IA requires a more detailed dose–effect or substance sensitivity distribution for each constituent.

INTERACTIONS BETWEEN METALS IN A MIXTURE

Both reference models (CA and IA) depart from the idea of noninteractivity, that is, substances do not interact, and combined effects are described by IA or simple additive action. However, this assumption is not always fulfilled and substances often do interact when combined in a mixture. An interaction is believed to occur when a component influences the amount of another component accumulating at target site or its activity at that target site (Jonker et al., 2011). This interaction at the target site ultimately results in a different response than what can be expected based on the individual components. As a consequence, the interpretation of a mixture effect is dependent on the considered mixture reference model.

If, based on the reference model, the observed mixture effect is larger than expected, the mixture acts synergistically (also described as 'more than additive'). In contrast, antagonistic interactions occur if the observed mixture effects are smaller than those predicted with these models (also described as 'less than additive') (Jonker et al., 2005, 2011). It is important to note that the terms antagonistic and synergistic can only be determined after an analysis is made with either the CA or IA models, that is, the identification of the interaction depends on the chosen model. Unfortunately, many studies highlighting interactions do not specify which model was used to infer the interaction.

Although the toxicity of metal mixtures has been investigated for decades, no clear patterns have emerged from these studies. Two meta-analyses evaluated the mixture effects observed in metal mixture studies (Norwood et al., 2003; Vijver et al., 2011). Both authors found that interactive effects (i.e. antagonisms and synergism) were more commonly observed than noninteractive effects. Furthermore, they observed that interactive effects were highly variable and may depend on the test organism, metal combination, metal concentrations, and metal concentration ratio, water chemistry, exposed life stage, exposure duration, and endpoint considered.

The picture is further complicated by the paucity of chronic metal mixture toxicity data. The majority of studies reported metal mixture effects during acute exposure (Meyer et al., 2015,b), while regulatory frameworks such as the Water Framework Directive in Europe mostly rely on chronic toxicity data. Notably in these acute toxicity studies, combinations of metal concentrations were applied that are not environmentally realistic (too high).

Furthermore, it has been noted that a substantial number of metal mixture studies may have reported data obtained using nonsimultaneous toxicity testing of the individual metals and the metal mixture (Meyer et al., 2015). When mixture effects are experimentally evaluated using the above-mentioned reference models, it is however essential that the toxicity of the mixtures is investigated simultaneously with the toxicity of each of the individual components in the mixture. De Laender et al. (2009) have demonstrated that when the toxicity is not simultaneously assessed, erroneous conclusions on the interactive effects can be made.

Interactions between metals may occur at the different levels involved in the toxicokinetic process. Additionally, since metals compete with each other for binding sites on dissolved organic carbon, interactive effects may occur at the level of speciation. As a consequence, synergistic interactions at the dissolved level, may be actually noninteractive when expressed as the bioavailable free metal ion (Meyer et al., 2015). Alternatively, metals may compete at the transport sites of the cell membrane, thereby influencing each other's uptake. For instance, the uptake of nickel by *Daphnia magna* has been shown to be suppressed in the presence of zinc, which suggests that competitive effects at the uptake sites may take place (Komjarova and Blust, 2008). Once inside the organism, metals may affect each other's toxification and detoxification pathways by binding at target proteins. For instance, metallothionein concentrations in the midgut of shore crabs exposed to a mixture of cadmium–zinc was substantially larger than the concentrations in crabs exposed to either cadmium or zinc individually (Martín-Díaz et al., 2005). Since metal ions often show chemical and physical similarities, nonessential metals may bind to the binding site of essential metals in proteins and as such change the biochemical function of these proteins. It is the combination of all these processes that eventually leads to the mixture toxicity effects. Therefore, interactive mixture effects observed at the physiological level do not necessarily reflect toxicity effects (e.g. Sharma et al., 1999).

The recent, more systematic studies performed at environmentally more relevant concentrations (or at least sublethal, chronically toxic concentrations) have led to the general conclusion that in the large majority of cases, the combined effects of mixtures to single species are additive or antagonistic relative to the CA model for predicting mixture toxicity (e.g. Nagai and De Schamphelaere, 2016; Nys et al., 2017). The alternative IA model is on average more accurate, but more cases of synergisms are identified relative to this model (Nys et al., 2015), as opposed to the CA model. Synergisms are rare with the CA model and small, thus indicating that the CA model (at species level) is likely a conservative first tier in chronic toxicity prediction of metal mixtures. However, the validity of both CA and IA at the community level has rarely been evaluated, although this is the level that risk assessment frameworks ultimately aim to protect.

As a general, conservative approach, in absence of more detailed and consistent indications of specific interactions (antagonisms, synergism), concentration additivity can be applied for assessing combined effects of metals. It is noted here that this conclusion is related to inorganic substances.

7.3.2.2 Assessing the Combined Hazard of the Complex Inorganic Material

As a first step, the concentration range of the different constituents of the complex inorganic material (-group) is determined. The metal content in the material (ores and concentrates, scrap, UVCBs, and alloys) is usually determined because it represents commercial value or because the content needs to meet some standards (alloys).

In the second step, detailed ecotoxicity information on the metals/metal compounds contained in the complex inorganic material is collected. For hazard assessment, the ecotoxicity reference value (ERV) is needed for each of the constituting metals. ERVs are usually defined as the lowest value of observed effect concentrations from acute or chronic ecotoxicity datasets on a given substance. For the metals, ERVs are usually expressed as the 'dissolved metal ion' concentration.

ERVs for acute and chronic effect of different metals are, for example, presented in Tables 7.2. and 7.3. in sections 7.5.1.1. and 7.5.1.2.

The assessment of the hazard can be done using a 'default approach' or be further refined by considering solubility, as outlined in Fig. 7.4.

The default approach will be based on the worst-case composition of the hazardous constituents observed in the complex inorganic material, and consider all metals as fully soluble. The hazard will be determined by applying the mixture toxicity rules, that is, by comparing the concentration of each metal with its respective ERV and calculating the 'TU' for each of the metals contained.

According to the mixture toxicity rules, these TUs are added up in the combined toxicity assessment (see Eq. 1):

$$\text{Combined toxicity} = \sum_{i=1}^{n} \frac{[x_i]}{\text{ERV}_i} \tag{7.4}$$

where $[x_i]$ equals concentration of element i, $\text{ERV}i$ = reference toxicity concentration for element i.

Combined toxicity for aquatic life can be calculated for the acute and chronic time frame.

An example of this first-tier assessment is given in Tables 7.2 and 7.3 (see Section 7.5.1).

It is obvious that the default consideration of full solubility may not be realistic, for example, metallic alloys may only release a fraction of the metals contained. Therefore, in many cases it will be desirable to refine the default assessment. This can be done by using the 'Refined Approach' outlined in Fig. 7.4.

The refined approach may consist of:

- Considering the speciation of the metal compounds contained in the complex inorganic material. The hazard of different compounds from a same metal can be quite diverse, due to their different solubility, also when contained in a complex material. When the speciation of the metals in the material is known, their specific hazard (if documented) can be considered to refine the default approach, by considering real solubility instead of assuming full solubility. The consideration of specific hazard related to speciation is an option in the MeClas tool (http://www.meclas.eu) as described in Chapter 9.
- Testing the solubility of the metal compounds contained in the complex inorganic material. Since metal ecotoxicity is driven by metal ions, the ecotoxicity of metals and inorganic compounds is related to the concentration of metal ions that can go into solution, after being dissolved. This can be measured by testing the 'Transformation/Dissolution' (TD) of the metals contained in the selected representative material(s) (see grouping and selection of representative materials in Sections 7.2.1 and 7.2.2). A conservative estimate of metal solubility in the representative material(s), for example, the 90th percentile (P90) value, can subsequently be used for the hazard assessment of the complex inorganic materials group.

In the refined approach the hazard will as well be determined by applying the mixture toxicity rules, comparing in this case the experimentally defined **dissolved** concentration of each metal with its respective ERV and calculating the TU for each of the metals contained. The TUs will be added up in the combined toxicity assessment (see Eq. 1):

$$\text{Combined toxicity} = \sum_{i=1}^{n} \frac{[x_i]}{\text{ERV}_i} \tag{7.5}$$

where $[x_i]$ equals now the **dissolved** concentration of element i, $\text{ERV}i$ = reference toxicity concentration for element i.

THE TRANSFORMATION/DISSOLUTION APPROACH

The TD approach has been used in the EU since the 1990s for several regulatory processes (e.g. safety data sheets, EU CLP (2008), and EU REACH). It's use for assessing the ecotoxicity of metals, sparingly soluble metal compounds and metal alloys has been adopted by the GHS (United Nations, 2011). It is proposed to be used for the assessment of the solubility of metals contained in complex inorganic material in general.

By following the TD approach in a conservative way as described above, after grouping and selection of representative materials, there is no need to test the solubility of the constituents in every single specific material of the group. The worst-case TD values for each metal can be applied for defining the hazard of every sample of the complex inorganic materials group. Still, the specific solubility of the constituents of a given sample can always be determined by specific TD testing.

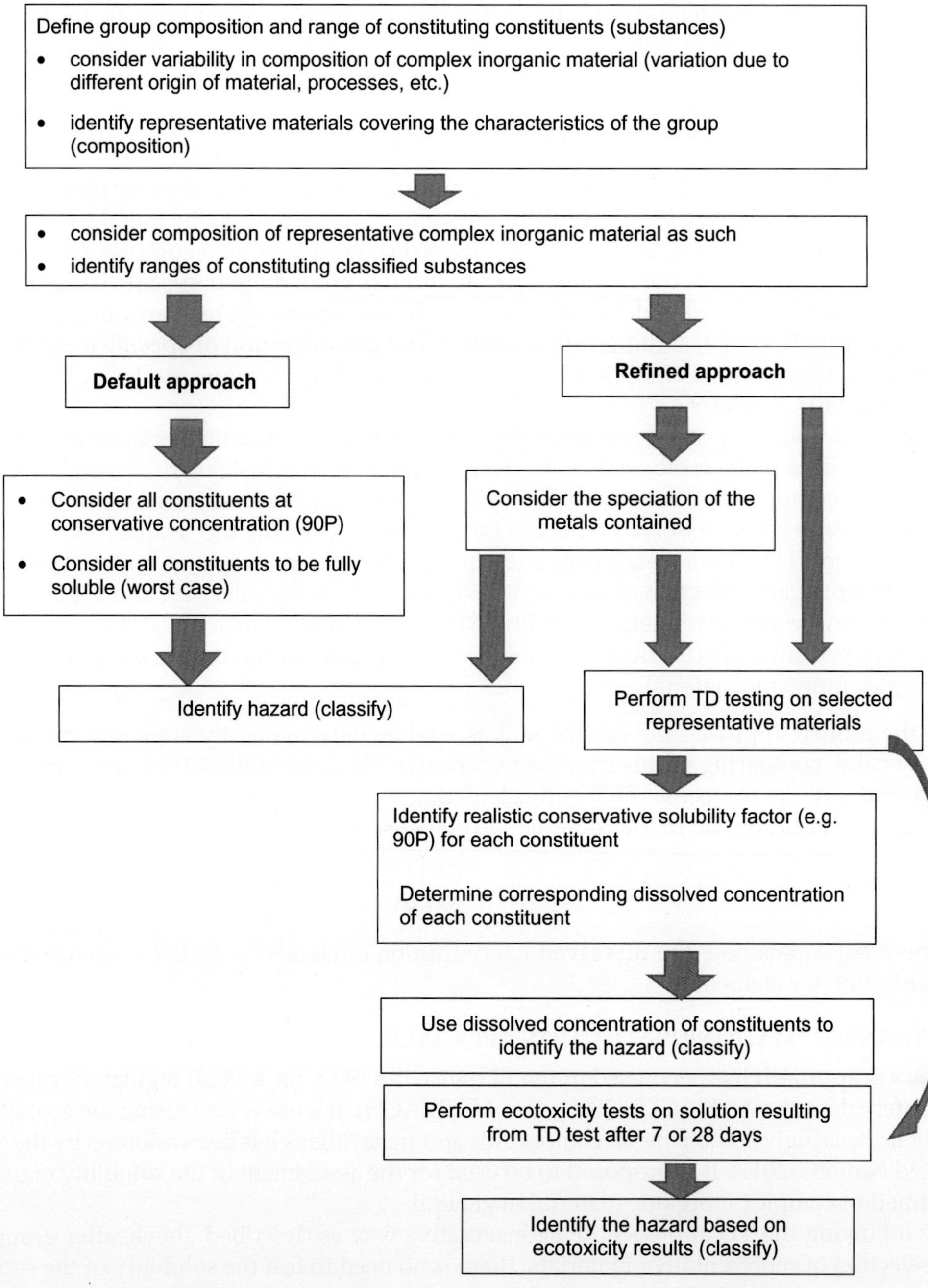

Define group composition and range of constituting constituents (substances)

- consider variability in composition of complex inorganic material (variation due to different origin of material, processes, etc.)

- identify representative materials covering the characteristics of the group (composition)

- consider composition of representative complex inorganic material as such
- identify ranges of constituting classified substances

Default approach

Refined approach

- Consider all constituents at conservative concentration (90P)

- Consider all constituents to be fully soluble (worst case)

Consider the speciation of the metals contained

Identify hazard (classify)

Perform TD testing on selected representative materials

Identify realistic conservative solubility factor (e.g. 90P) for each constituent

Determine corresponding dissolved concentration of each constituent

Use dissolved concentration of constituents to identify the hazard (classify)

Perform ecotoxicity tests on solution resulting from TD test after 7 or 28 days

Identify the hazard based on ecotoxicity results (classify)

FIG. 7.4   Scheme of default and refined approach for assessing the combined hazard of the complex inorganic material, based on the toxicity of the constituents.

The TD test approach can thus be used to generate relevant information on the ecotoxicity of the material, rather than performing ecotoxicity tests with the complex inorganic material on different test organisms directly. This approach has several advantages:

- by comparing TD data with the corresponding ERVs, information is generated that is relevant for all ecotoxicological endpoints, for example, algae, invertebrates, and fish, without need for testing on the organisms as such;
- the comparison of the measured solubility levels of the different metals contained in the complex inorganic material with their respective ERVs is providing a conservative assessment, because the ERVs are by definition conservative;
- due to the maximisation of the potential solubility of the metals by considering a wider pH range, the solubility assessment is maximised also, providing another element of conservatism;
- the TD approach is standardised and has proven intra- and interlaboratory repeatability (OECD, 2001). Its application avoids interlaboratory variability in ecotoxicity testing, following from inherent sensitivity of test organisms, differences in test protocol and test conditions between labs;
- for the complex inorganic material, the difficulties and possible error/variability by ecotoxicity testing of the materials are avoided by generating the required information in a conservative way; and
- last but not least, the TD approach allows to test in parallel a number of similar materials, whose varying composition is considered to cover the characteristics of the complex inorganic materials-entry in a relevant way. While interpreting the results of such parallel data, a conservative approach is adopted by considering a conservative dissolution value (e.g. the 90P) for the further assessment.

A technical description of the TD test is given in Sections 7.4.2 and 7.4.3.

In the last tier, specific testing on the solutions resulting from the TD test may still be performed (Fig. 7.4). Although the TD solutions do not fully correspond to the standard ecotoxicity test media, this approach may offer the advantage of providing a realistic soluble fraction of the metals for testing, and avoid possible artefacts related for example, to the presence of particles. In reality, given the variability amongst materials even originating from the same source, this option is not often used.

The general approach outlined above allows to determine in a conservative way the environmental hazard of the variable materials in a complex inorganic materials group (see Section 7.5.1, for an example). Hazards are usually driven by specific constituents with the highest toxicity (i.e. the 'drivers' of the classification).

Setting boundaries for the concentrations of the different combining (hazardous) elements in a group may be challenging; for the definition of the ranges, the focus is to be put on the most toxic constituent(s), of which different combinations are to be considered. In general, the similarity or sameness of complex inorganic material needs to be documented (source, process, and composition).

As mentioned above, grouping is done at the start of the assessment to facilitate and structure the work to be done. However, if the content or composition of hazardous constituents in a material varies too widely, or the presence of a different metal compound changes the hazard, grouping in different hazard categories may be appropriate. For alloys, see approaches further explained in Chapter 13.

A critical aspect of the TD approach outlined in this section is the identification of representative test samples covering the variability of the group. This aspect will be discussed in more detail in Chapter 11 for UVCBs.

7.3.2.3 *Assessing Combined Risks Based on the Constituents of the Material*

For the combined risk assessment of the complex inorganic material, the exposures (predicted environmental concentrations—PECs) resulting from the emissions of each metal contained in the material must be considered in combination (see also Fig. 7.3).

Combined exposure can be assessed in different ways. Generally, it is legally required for industrial sites to measure and document their local emissions; such data can most often be modelled into concentrations in the receiving environment. In some cases, metal concentrations in the receiving environment are monitored directly. If emission data are missing, specific environmental release factors ('SPERCs', see Verdonck et al., 2014, for relevant SPERCs for metals and also Chapter 9) can be used.

Secondly, detailed ecotoxicity information on the predicted no effect concentrations (PNECs) of the metals/metal compounds contained in the complex inorganic material is needed.

For most metals, these PNECs are also expressed on a soluble metal ion basis, and can be derived in a generic way (using ecotoxicity data as such) or by considering the bioavailability of the metals related to the physicochemical conditions of the receiving environment. Generic and specific local PNECs for a number of metals are listed in Table 7.5 (Section 7.5.2.2.).

DEFAULT APPROACH AND OPTIONS FOR REFINEMENT: PRINCIPLES

The combined risk assessment poses a number of scientific challenges:

- complex inorganic materials may contain many metals/metal compounds, with varying levels of ecotoxicity information
- all metals are naturally present at background concentration in the environment, which provides a 'combined background exposure' that can be significant
- literature on the combined effect of metals is limited
- metals may be antagonistic or synergistic, making the additivity principle either overly conservative, or not sufficiently conservative.

In general, there is lack of guidance on combined toxicity and risk. Owing to (i) the natural presence (background) of metals in the environment and (ii) the conservative approaches applied in PNEC setting (resulting for most metals in very low PNECs that are often close to the natural background), the contribution of background exposures may already be significant for a number of metals (see Section 7.5.2.1).

Considering this, the combined risk assessment may be refined by taking into account the metal-specific features that are applied for the risk assessment of single metals. This has resulted in a pragmatic, tiered approach that integrates recent scientific refinements in metal risk assessment. This generic approach should be applicable to any combination of metals in any type of complex inorganic material (e.g. alloys, UVCBs, and concentrates) and to all environmental compartments.

This tiered approach has been outlined further in Fig. 7.5.

In a first step, the relevant constituents of the material are identified (Fig. 7.5, tier 0—only constituents identified as hazardous to the environment should be taken into account). Based on measured or modelled concentrations of the metals, and comparison with their respective PNECs, a risk characterisation ratio (RCR) is calculated for each of them, and the risk ratios are added up in a 'concentration addition' approach (Fig. 7.5, tier 1) to assess the combined risk:

$$\text{Combined RCR} = \text{Sum}_{i=1}^{n}\text{RCR}_i \tag{7.6}$$

where RCRi is the risk characterisation ratio for metal i.

The existing evidence on combined metal toxicity at the species level shows that CA is generally a conservative approach, whereas for IA some cases of synergisms are identified, but these mostly occur at relatively high concentrations (Sharma et al., 1999; Nys et al., 2015). These observations in general do not justify the consideration of specific synergistic effects among the metals.

If relevant, in a next tier some specifics of metals, for example, their bioavailability and/or their background concentration can be considered to refine the assessment (Fig. 7.5, tier 2). Some further refinement by considering species sensitivity distributions (SSDs) rather than single data can be made, for example, in a 'multisubstance potentially affected fraction' ('msPAF') approach (De Zwart and Posthuma, 2005; Van Regenmortel et al., 2017). In cases where the emissions of different processes, some of which are not related to the complex inorganic material, are combined, some allocation of emissions and exposure can be made.

As the last step, toxicity tests can be performed (Fig. 7.5, tier 3). Toxicity testing is preferentially used as a 'risk management check' and effluents are considered the most relevant starting point for testing. Because of the mentioned issues on representativity of complex inorganic material samples when there is huge variability (e.g. UVCBs), testing the material itself to derive a safe threshold concentration (e.g. PNEC) for the material in the environmental compartments is not recommended.

BACKGROUND CONSIDERATIONS IN RISK ASSESSMENT

Metals occur naturally and are as such present in natural background concentrations in all environmental compartments. The natural backgrounds contribute to general exposure.

In a straightforward application of the default CA approach, using exposure based on monitored metal concentrations in the environment, the combined contributions of the natural background to the overall exposure of the respective metals may be significant and already add up to reach a risk level. This effect logically becomes stronger with an increasing number of metals considered in a CA approach. In Section 7.5.2.1, this problem will be illustrated by a worked-out example on EU soils.

This observation indicates the overconservative nature of the default CA approach when dealing with naturally occurring substances such as metals. It stresses the need for further refinement of the methodological approach for assessing combined metal toxicity.

7.4 IN PRACTICE

7.4.1 Covering the Complex Inorganic Material Variability

The content of metals and/or metal compounds in complex inorganic material represents value or is linked with some standards (alloys) and is therefore usually well defined. The composition may be specifically defined such as in alloys, but may also be more variable, when the materials are, for example, derived from different processes.

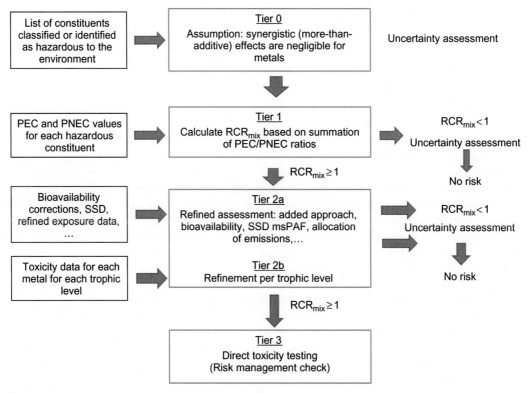

FIG. 7.5   Tiered approach for the refined risk assessment of the complex inorganic material.

When such materials of variable composition are to be assessed for their environmental hazard the major challenge is to identify/select the (range of) composition(s) that can be considered as representative for this material, and can be used as reference for the hazard assessments, or possibly also for testing.

For risk assessment, the challenge is to cover the proper range of processes and uses related to the complex inorganic material and to address the combined risk of exposure resulting from the related metal emissions to the environment. The risk assessment can be done either in a conservative way that is relevant for the exposure resulting from the industrial processes and practices of the complex inorganic material as a group, or can be performed on a local, site-specific scale.

It has been suggested to start with grouping of the materials. For the next steps (i.e. hazard and risk assessments), in particular when the composition of the materials is variable, it is crucial to select materials that are representative for the group. In summary, representative materials of the complex inorganic materials group are selected based on their composition of main constituents, granulometry (if relevant) and (mineral) speciation (if possible). Approaches on how to make such a selection for UVCBs are outlined in detail in Chapter 11. An example of hazard assessment of the group of the sulphidic zinc concentrates is provided in Section 7.5.1 and for other concentrates in Chapter 10.

7.4.2 Transformation/Dissolution (TD) Testing as Surrogate for Ecotoxicity Testing in Water

7.4.2.1 The TD Principles

The TD technique allows to generate information on the solubility and corresponding potential ecotoxicity of metals, sparingly soluble metal compounds and metal alloys (UN GHS, 2011; OECD, 2001). The solubility of inorganic substances is highly dependent on the medium in which they are dissolved. Factors such as pH, ionic strength, temperature, O_2, and CO_2 concentration are all important. The dissolved fraction is operationally defined as the fraction that is measured after filtering the solution through a 0.45 μm mesh and subsequent acidification with HNO_3.

A testing protocol was developed for quantifying the amount of metal (ions) that could be released from test materials under the conditions of existing standardised ecotoxicity testing media. Given that the pH is a major factor in metal solubility, a test range for pH was set (pH 6–8) to assess the maximal metal solubilisation within ecologically relevant conditions.

After solving the technical problems related to the setting and keeping of such lower pH-test conditions, a standard protocol for the TD technique was validated and defined by OECD (OECD, 2001). Under this protocol, the rate and extent of the dissolution of metal (ions) from given metal-containing materials can be measured under standardised conditions. The protocol was later translated in the Annex 10 of the GHS (UN GHS, 2011). Further technical guidance on the how to perform TD testing is given below.

In the TD approach, the solubility data that are generated (concentrations of dissolved metal, considered 'ion-form') are subsequently compared with the reference values for ecotoxicity of the substances under study. Such ERVs (based on EC50s for acute aquatic effect or NOECs/EC10 for long-term effect) are derived from datasets on readily soluble compounds of the metal, by preference generated under standardised ecotoxicity testing conditions, and under pH conditions between 6 and 8 (±0.5).

7.4.2.2 The TD Technique

In the TD test, a given mass loading of the test material (defined by the endpoint under study e.g. for chronic effect, 1 mg/loading is typically used) is brought in test solution at given pH (ranging between 6 and 8). The solutions are agitated under controlled conditions and samples are taken at given moments in time. By this procedure, the rate and extent of solubilisation of inorganic ions can be determined in a standardised way. Experimental variability and consistency is checked by using replicas. By convention, dissolution for the acute aquatic effects is assessed after 7 days, for the long-terms effects after 28 days (referring to the test time of standard ecotoxicity tests). Since the media are the same, the acute and chronic tests can be integrated. A scheme of the experimental set-up is presented in Fig. 7.6.

To ensure quality and repeatability, it is emphasised that the test protocol needs to be strictly followed. Under such conditions, the test has proven to be of high repeatability. Small deviations from the test protocol may influence the results significantly. The use of ultra-clean lab material and application of entire-procedure blanks to check for possible contamination, careful maintenance, and control of the experimental conditions, especially temperature and agitation rate, are crucial in this respect. Good-quality experiments are characterised by a consistent and repeatable dissolution pattern in time. For a standard set-up of particles with

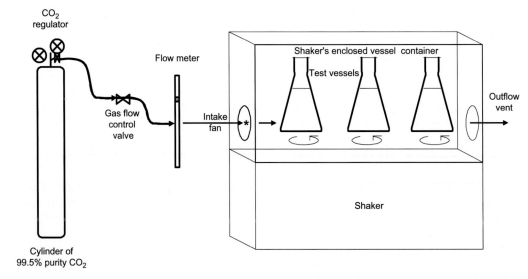

FIG. 7.6   Scheme of experimental setup for the TD test (CIMM 2008). *Reproduced with permission from CIMM.*

narrow range particles size, the within-vessel variation of TD data should be <10% and the between-vessel variation should be <20% (OECD, 2001). For some metal ions (e.g. Al, Fe, In, etc.), rapid precipitation may occur in the test. This reflects processes occurring in the real environment, and is taken into account in the interpretation of the results.

As a function of particle size, different set-ups are used:

- For fine inorganic compounds in powder form, the finest size and form available in the market is used for testing. The material is usually agitated by stirring, but rotary shaking is also an option. Care should be taken to avoid material sticking to the wall of the vessel.
- Some massive materials (e.g. metal alloys in massive form, materials in diameter ≥1 mm as put on the market), when stirred in solution, may be subject to significant abrasion effects. For those, massive forms can be mounted in an epoxy holder (Fig. 7.7). A well-defined cross-sectional surface of the material can as such be exposed to the solution and abrasion effects can be avoided. The solubilisation of ions from particulate material is driven by the surface exposed to the solution. These tests will thus be set up on the basis of exposed surface, and not mass loading. Results will be recalculated later to mass loading, to be compared with classification criteria. This latter set-up has been used with success for the hazard identification of a number of metals in massive form and as such directly applicable to the assessment of the metal alloys. When testing alloys, the concentration of the metals at the test surface should be checked for consistency with the composition of the alloy as such, since changes may occur when preparing the usually small cylindrical or wire forms of the alloy that are used for testing.

For the mass loading, reference is made to the classification criteria, for example, 0.1, 1, 10, and 100 mg/L, etc. However, given the potential degree of heterogeneity of the material, higher loading may be used to reduce variability. The assumption of linear ion release from the material within the loading range including the classification criterion should be documented.

Solvent exposed face

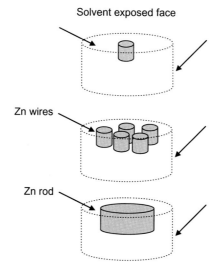

Zn wires

Zn rod

FIG. 7.7 Scheme of the mounting of solid surfaces in epoxy holder (CIMM, 2008). *Used with permission of CIMM.*

- *Comparison of TD data and ERVs*

As indicated above, TD should be determined over a pH range between pH 6 and 8. In principle, the highest solubility result, following from TD testing, should be used in comparison with the ERVs and for identifying the hazard. Two situations can occur in such comparison:

  ○ an ERV is available for a similar pH (or pH-range e.g. 7.5–8.2) as the pH at which the highest TD has been measured; in this preferred case, both values obtained at the same/similar pH are compared

  ○ no ERV is available at the pH where the highest TD value is observed (this may often be the case, since ecotoxicity tests are rarely performed at pH 6, where for several metals the highest TD is observed). In this case, the highest TD value is still compared with the ERV obtained at the different pH.

Under Section 7.5, an example case on sulphidic zinc concentrates will be worked out, showing the use of TD data to refine the combined hazard assessment.

7.4.3 Metal Solubility Assessment in Soils and Sediments

In the above sections, it has been outlined how the hazard or risk assessment of metals contained in complex inorganic materials can be refined and made more environmentally realistic by assessing the solubility and related bioavailability of the metals in the water environment. As a principle, a similar approach can be applied to assessing the risks of the material present in soils and the sediment.

No standard TD test has yet been accepted for soils. The caveat with soil is that adsorption of metal ions or metalloids already immobilises even the soluble components, often associated with slow reactions (ageing). Insoluble materials may gradually release soluble inorganic components in soil that, in turn, become gradually immobilised. For that reason, any TD in soil has to compare the solubility of the components in soil between soils amended with a complex inorganic material with a comparative set-up of a soil amended with its soluble metal or metal-loid compounds (often salts). An example of such comparative solubility tests has been worked out earlier (Smolders et al., 2007). A proposal of a TD test of sparingly soluble metal compounds in soil has been described before (McLaughlin et al., 2002) based on a toxicity assessment in which toxicity is compared between 7 and 60 days after mixing the compounds in soil.

For the soil compartment, exposure and effects data are generally expressed as total (or aqua regia-soluble) metal concentrations, regardless of the form and speciation of the metals present. Nevertheless, the source of metal contamination may affect the bioavailability and toxicity of the metal in the soil and this can be assessed by, for example, the isotopically exchangeable fraction of the total soil metal (Hamels et al., 2014). In many cases, predicted metal emissions on soil from current materials production and use are mainly due to application of sewage sludge on land and aerial deposition. Both sources can be considered relatively soluble compared to the original substances and, as such, a correction for solubility is generally not included in a prospective risk assessment of metals in soils next to bioavailability corrections for ageing reactions and varying soil properties (in contrast to retrospective risk assessment of, e.g. mining sites). There are obviously some metal sources in which the solubility of the metals or metalloids is limited. An example of that is the application of slags in construction works for which the release of toxic substances into the soil may be of concern.

The TD concept had already been used in tests for soils amended with blast furnace slags to address the issue of the vanadium (V) content. Vanadium is readily toxic on the short term to soil organisms when added to soil as a soluble form, but not as slags (Baken et al., 2012; Larsson et al., 2015). Comparative tests on V-bioavailability between soil treatments with aged or nonaged soluble V component and with the slag led to the conclusion that the measured V in soil solution explained the different biological effects, that is, a long-term solubility test of the complex inorganic material might reveal the risk relative to that of a soluble compound. The practical implementations of such studies in risk assessments have, however, not yet been adopted in a generic scheme.

For the sediments, the 'AVS-SEM' approach has recently been developed for considering metal bioavailability in risk assessment. This approach is in fact based on combined toxicity, since is compares the sulphide fraction in the sediment (AVS, acid volatile sulphide) with the concentration of the metals copper, zinc, lead, nickel, and cadmium present in the sediment, and extracted simultaneously with the same extractant (SEM, simultaneously extracted metals). Experimental evidence has shown that, due to the strong binding of metals to sulphides, the S-bound metals are not bioavailable for uptake by organisms and, consequently, are not toxic (ICMM, 2007). The AVS–SEM approach has been applied in EU-risk assessments on, for example, zinc and copper, to define the bioavailable fraction of these metals in sediment (JRC, 2010; ECHA, 2008b). It is noted that the AVS–SEM concept applies by definition a combined toxicity approach, since the sum of the bioavailable metals present is considered. For more information on the AVS–SEM concept and its application in sediment risk assessment, see the Metals Environmental Risk Assessment Guidance (ICMM, 2007). Further research is needed to also integrate other metals in this concept.

7.5 EXAMPLES OF IMPLEMENTATION

7.5.1 Example of Hazard Assessment of Zinc Concentrates

A procedure for a general approach for hazard assessment, relevant for a group of complex inorganic materials has been outlined in Fig. 7.4. A same approach can be followed for a specific complex inorganic material, where the assessment is done directly and solely on the specific material.

The approach for hazard assessment of complex inorganic material is illustrated using sulphidic zinc concentrates as an example. This example is considered representative for all types of complex inorganic materials (e.g. concentrates, alloys, and UVCBs).

The sulphidic concentrates are the main group of the zinc concentrates; they are derived from a similar source material that is, sulphidic zinc ore, or 'sphalerite'.

Metal sulphides are separated from the gangue material to upgrade the excavated metal sulphide ore to a metal sulphide concentrate with a concentration of 45%–55% zinc. Other metals (e.g. Pb, Cd, Cu, etc.) are typically present as metal sulphides in variable concentrations. Several of these metal sulphides have an identified acute and/or chronic hazard to aquatic life, for example, CdS, PbS. However, the metal sulphides are generally characterised by low solubility in water, so the aquatic effect hazard of this type of material will depend on the solubility of the metal compounds (or minerals) contained.

In Table 7.1, this group of materials is characterised for composition; the data illustrate the variability that is observed.

TABLE 7.1 Chemical Composition (wt%) of Sulphidic Zinc Concentrates

Concentrate	As	Cd	Cu	Pb	Zn	Ag	Hg	Ni
1	0.065	0.45	1.18	3.58	50.44	0.023	0.0003	0.0085
2	0.120	0.16	0.06	2.68	54.03	0.002	0.0041	0.0180
3	0.140	0.29	0.23	1.91	49.40	0.029	0.0130	0.0036
4	0.330	0.21	0.75	3.77	34.94	0.077	0.0099	0.0077
5	0.470	0.11	1.33	1.60	47.16	0.011	0.0520	0.0020
6	0.100	0.15	0.05	1.98	54.26	0.013	0.0019	0.0020
7	0.090	0.33	0.40	1.43	57.78	0.034	0.0019	0.0021
8	0.020	0.41	0.16	3.38	53.46	0.014	0.0076	0.0047
9	0.118	0.27	1.58	1.02	49.14	0.012	0.0071	0.0009
10	1.092	0.40	0.65	0.86	51.26	0.018	0.0015	0.0010
11	0.003	0.10	0.29	1.03	56.09	0.009	0.0008	0.0041
12	0.037	0.17	0.30	2.44	52.16	0.008	0.0032	0.0020
13	0.007	0.23	0.49	2.83	49.04	0.011	0.0026	0.0024
14	0.111	0.09	0.81	17.36	39.86	0.033	0.0010	0.0020
15	0.118	0.27	1.58	1.02	49.14	0.012	0.0071	0.0009
16	0.012	0.26	0.15	3.23	47.37	0.015	0.0116	0.0018
17	1.092	0.40	0.65	0.86	51.26	0.018	0.0015	0.0010
18	0.013	0.18	1.13	0.71	55.37	0.014	0.0159	0.0040
19	0.090	0.18	0.20	2.24	55.23	0.045	0.0033	
20	0.111	0.11	0.96	8.92	45.93			
90P	0.532	0.40	1.35	4.29	55.44	0.036	0.0136	0.0079

7.5.1.1 *Hazard Identification for Aquatic Life—Acute*

Following the 'default' approach (see Fig. 7.4), all contained metals in the complex inorganic material are assumed to be fully soluble. Considering the content of different metals in the material (Table 7.1), and the mixture toxicity rules (Eq. 4) the ERV values are reached/exceeded at a loading of 1 mg/L for several metal compounds; consequently, the default combined hazard is 'very toxic for aquatic life' (Table 7.2).

The hazard assessment can however be refined by considering the (mineral) speciation of the metal compounds contained in the material, and/or by testing the solubility of the compounds in a TD test (Section 7.4.2).

In a 'speciation' approach, the low hazard of, for example, the zinc sulphide and cadmium sulphide is considered, and the assessment would be more realistic. However, the 'lead compounds' and 'silver compounds' are among the major drivers for the hazard. The speciation of their different compounds may result in differences in solubility, but this is not considered in the 'group' hazard entry of these substances. It was therefore considered useful to perform TD testing on representative samples of the complex inorganic material. In the TD approach, the dissolved (bioavailable) fractions of the classified metals are experimentally determined.

The zinc concentrates in Table 7.1 were all tested for TD in the same testing facility following the same protocol (OECD, 2001), and resulting TD solutions were all analysed in the same accredited laboratory. Such an approach increases the consistency between results and their comparability. In a conservative approach, tests were performed at pH 6, since most of the metals concerned show highest solubility at this pH. To cover the variability, about 20 different sulphidic zinc concentrates from different geographical origins, representing the main world mining areas, were used for the generic assessment.

The results of the TD approach are summarised in Table 7.2. It follows from the data that the solubility (and related toxicity potential) of most metals in the complex inorganic material is rather limited.

For the generic assessment, 90P values of the respective parameter distributions are considered for the hazard assessment (Table 7.2). This is justified by the observation that metal-specific maximum values are not related to the same samples. By assessing the solubility of the metals, also the fraction resulting from the limited dissolution of the nonhazardous

TABLE 7.2   Assessment of Combined Toxicity—Acute

Metal	Content in Complex Inorganic Material (90P)	ERV Acute (μg/L)	Concentration Assumed Soluble at 1 mg/L Loading	TU at 1 mg/L Loading	T/D (% at 7 Days)	Soluble Concentration at 1 mg/L Loading	TU at 1 mg/L Loading
Zn	55.44	413	554.42	1.34	6.32	35.05	0.085
Pb	4.29	73.6	42.85	0.58	48.55	20.80	0.283
Cd	0.40	18	4.04	0.22	8.54	0.34	0.019
Cu	1.35	12.1	13.5	1.12	1.89	0.25	0.026
As	0.532	430	5.322	0.01	5.37	0.29	0.001
Ag	0.036	0.22	0.3636	1.65	<0.11	0.0004	0.002
			ΣTU	4.92		ΣTU	0.416

metal sulphides (e.g. zinc sulphide) will now be considered and added up in the combined toxicity hazard assessment. In this respect, this higher tier assessment is also more realistic.

Based on the solubility data in Table 7.2, a generic hazard—acute for aquatic life is identified at the level of >1–≤10 mg/L, since at that loading, the combined toxicity of all constituting elements of the complex inorganic material results in a combined TU of >1. Calculations of the hazard (and corresponding classifications) can be done with the MeClas tool (Chapter 9).

7.5.1.2 *Hazard Identification to Aquatic Life—Chronic*

As for the assessment of acute hazard outlined above, the chronic hazard can be assessed at different levels of refinement. Using the sulphidic zinc concentrates again as an example, it follows from their composition that, when it is assumed that all metals are fully soluble in the complex inorganic material, a chronic hazard is identified at the level of >0.01–<0.1 mg/L (Table 7.3). This default assessment was refined by doing TD testing.

For the chronic assessment, the solubility of the constituting elements of the complex inorganic material is determined after 28 days. Based on the results, the hazard to aquatic life—chronic of the complex inorganic material is now identified at the level >0.1–<1 mg/L (Table 7.3).

The results of this analysis are relevant for the sulphidic zinc concentrates as a complex inorganic material group. Considering the variability of this group, and the fact that these concentrates are constantly produced and blended in variable composition, the purpose of the work presented in this example was to derive a generic hazard assessment that is relevant for the group as a whole. For this reason, a conservative approach was followed. It is noted

TABLE 7.3 Assessment of Combined Toxicity—Chronic

Metal	Content in Complex Inorganic Materials (90P)	ERV Chronic (µg/L)	Concentration Assumed Soluble at 1 mg/L Loading	TU at 1 mg/L Loading	T/D (% at 28 Days)	Soluble Concentration at 1 mg/L Loading	TU at 1 mg/L Loading
Zn	55.44	82	554.42	6.76	13.85	76.78	0.936
Pb	4.29	17.8	42.85	2.41	65.45	28.05	1.576
Cd	0.40	0.21	4.04	19.24	13.44	0.54	2.571
Cu	1.35	11.4	13.53	1.19	2.53	0.34	0.030
As	0.532	40	5.32	0.13	6.55	0.35	0.009
Ag	0.036	0.12	0.36	3.0	<0.11	0.0004	0.004
			ΣTU at 1 mg/L	32.73		ΣTU at 1 mg/L RD[a]	5.103
						ΣTU at 1 mg/L NRD[b]	0.011
			ΣTU at 0.1 mg/L	3.27		ΣTU at 0.1 mg/L	0.62
			ΣTU at 0.01 mg/L	0.33		ΣTU at 0.01 mg/L	0.062

[a]Metals equivalent to being rapidly degradable.
[b]Metals equivalent to being not rapidly degradable.

that the approach as outlined above can also be applied for the identification of the hazard to specific concentrates, by using specific data on composition and TD results. It is emphasised further that this generic assessment is only relevant for the sulphidic concentrates; for the zinc concentrates derived from other, for example, carbonate, ores, then specific TD testing is required to check the hazard of the specific material.

In this example, the scheme presented in Fig. 7.4 for 'refined approach' is applied up to the use of the TD results. No ecotoxicity testing was performed on the TD solutions (last step in Fig. 7.4), and the hazard assessment was concluded based on the conservative analysis of the solubility of the constituting elements, determined by TD testing.

7.5.2 Examples of Risk Assessment

In the following, some examples of mixed exposure to metals and combined risk assessment, occurring in practical cases of production and use of complex inorganic material will be presented and discussed. Since the risk assessments are based on the combined exposure of metals released during the production and/or use, the approach is relevant for all complex inorganic materials (e.g. UVCBs, alloys).

7.5.2.1 Combined Metal Backgrounds Showing Risk by Default

A case of combined risk assessment of metals present in soil is based on data for measured aqua regia-soluble metal concentrations and major soil properties in 2108 arable land soils across Europe as documented in the GEMAS project (Reimann et al., 2014). All sites were under normal agricultural production at the time of sampling and were not subjected to significant local metal emissions. Widespread metal toxicity on soil organisms was not expected. The metal concentrations measured in these soils must be considered as ambient concentrations, that is, the sum of the natural background of the elements combined with the diffuse anthropogenic input in the past or present, for example, through the use of fertilisers including manure (ECHA, 2008c).

However, no indications were detected at the continental scale in support of a significant impact of diffuse contamination on the regional distribution of element concentrations in the European agricultural soil samples. At this European scale, the variation in the natural background concentration of these metals in the agricultural soil samples is much larger than any anthropogenic impact (Reimann et al., 2017).

In this example, all soil PNEC values were normalised with regard to the soil conditions on each individual site. The RCR is calculated as the ratio of the measured metal concentration in soil over the corresponding normalised PNEC value. The combined risk via CA is calculated as the sum of the RCRs for the individual metals while the combined effect via the IA approach is calculated according to Eq. (3) and the potentially affected fractions predicted for the prevailing metal concentrations by the normalised SSDs for each metal.

It is noted that the PNEC values used for these metals are already using advanced refining techniques, that is, all metals have an extensive dataset for their effects on soil organisms, allowing a species sensitivity approach to derive their respective PNEC. Moreover, all these metals have bioavailability models in place to correct for varying soil properties on their bioavailability and toxicity in soil. These refinements have been included in the assessment.

Based on the individual metals, 12.7% of the sites sampled are predicted to be at risk, that is, they have a RCR ≥ 1 for 1 or more metals for their effects on soil organisms. When the effects of these five essential metals are combined, the CA approach however predicts that 55.2% of the sites are at risk (Table 7.4). This means that in 42.5% of the soils (i.e. 55.2%–12.7%), risk is predicted because of the combined exposure to these five metals while the individual metals do not indicate toxic effects on soil organisms. In contrast, the IA approach predicts a risk for 16.9% of the sites and consideration of combined toxicity according to this approach results in a prediction of risks due to the combination of metals when no single metal was above the threshold for only 4.2% (16.9%–12.7%) of the sites.

As indicated above, those observations are mainly attributed to natural background levels of these metals in soil because of the generally relatively small contribution of diffuse anthropogenic contamination to the measured metal concentrations (Reimann et al., 2017). This example indicates the overconservative nature of the default CA approach when dealing with naturally occurring substances such as metals. Compliance with the CA approach for metal exposure in soil would most likely only be reached at deficiency levels for these essential metals. This illustrates that the refinements developed for hazard and risk assessment for the individual metals to account for metal specific issues, for example, essentially and bioavailability do not ensure a sound combined risk assessment of these metals.

7.5.2.2 Risk Assessment on Measured Emissions at Local Site

In the following example, a risk assessment for complex inorganic material produced at a zinc production site is presented. The exposure estimates are based on typical measured emissions to water for such sites. The receiving water is a large river (e.g. river Rhine) with high flow rate ($10^6 \, m^3$/day). Generic PNEC values for the individual metals in surface water are taken from the corresponding European REACH dossiers. The generic RCR values for all individual metals are below 1, but the combined risk for exposure to these six metals predicted according to CA shows a potential risk for aquatic organisms (Table 7.5).

The PNEC values for zinc, lead, and copper, however, can be normalised towards the typical abiotic conditions for large rivers in Europe (pH 7.8, hardness 217 mg $CaCO_3$/L and 2.8 mg/L DOC) according to available bioavailability correction models (http://bio-met.net).

TABLE 7.4   Risks Calculated for Individual Metals and Combined Exposure in 2108 European Agricultural Soils

	Co	Cu	Mo	Ni	Zn	CA	IA
10th–90th percentile of measured metal concentrations (mg/kg soil)	1.7–17	4.6–38	0.2–1.2	3.1–43	17–87		
10th–90th percentile normalised PNEC values (mg/kg soil)	13–72	43–124	11–168	18–80	71–276		
% sites at risk[a]	2.5	1.5	0.1	10.3	2.1	55.2	16.9

[a]$RCR \geq 1$ or $E(mix) \geq 5\%$.

TABLE 7.5 Example of Typical Emission Data of Large Zinc Producers and Corresponding Exposure and First Tier Combined Risk Assessment With and Without Bioavailability Corrections

Constituent	Emissions to Water (kg/year)	Concentration in Receiving Water (μg/L)	$PNEC_{generic}$ (μg/L)	$RCR_{generic}$	$PNEC_{bioav}$ (μg/L)	RCR_{bioav}
Zn	3652	10.0	20.6	0.486	29.1	0.344
Cd	22	0.06	0.19	0.317	0.25	0.24
Pb	165	0.45	3.1	0.146	6.5	0.07
Cu	53	0.15	7.8	0.019	9.3	0.016
As	23	0.06	17.1	0.004	17.1	0.004
Hg	4.4	0.01	0.06	0.210	0.06	0.210
Sum				1.181		0.884

This results in larger PNEC and smaller RCR values compared to the generic PNEC scenario and the absence of risk prediction when bioavailability corrected RCR values for these six metals are combined. These data show that normalisation of the PNEC to local environmental conditions can significantly affect the outcome of the (combined) risk assessment and should be done wherever possible to avoid both over- and underconservative assessments based on generic PNEC values.

KEY MESSAGES

The management of the risks posed by complex inorganic materials on the environment requires the identification of their possible hazards and exposure. Ecotoxicity information on complex inorganic materials is usually scarce and testing encounters several limitations such as the low solubility of the materials and the variability of their composition.

In view of the number of materials and the need to fulfil data requirements, it is proposed, as a first step, to group materials by (their) expected physicochemical and ecotoxicological properties and to carry out a generic assessment on representative material(s) of the group.

The latter will be based preferably on the assessment of the constituents, making use of available data on metals and metal compounds. Data gathering should include data on solubility, speciation and particle size, all key factors in toxicity.

The assessment can be further refined by applying practical approaches that quantify the solubility and bioavailability of the metals contained in the material:

- For hazard assessment: solubility of the metal compounds or minerals contained can be quantified with the Transformation/Dissolution (TD) test. This test provides a realistic, conservative estimate of the solubility of the contained metals, which can be directly related to the ERVs of the related metals.
- For risk assessment: existing models for metal bioavailability can be applied to refine the conservative default approach that considers all metal to be soluble and bioavailable. In addition, natural background can be considered.

The constituents-based approach is however confronted with the challenge of having to predict the combined exposure and effects of the constituents. A way forward by using tiered approaches with different levels of refinement, is proposed.

The approaches for refinement are most advanced for the water environment. Still, a number of issues remain, for example, the proper accounting for natural background in exposure assessments based on monitored data, the better understanding of metal interactions in the different environmental media namely, that is, water, soil, and sediment. These aspects require further research.

References

Allen HE, Hall RH, Bristin TD: Metal speciation: effects on aquatic toxicity, *Environ Sci Technol* 14:441–443, 1980.

Altenburger R, Backhaus T, Boedeker W, Faust M, Scholze M: Simplifying complexity: mixture toxicity assessment in the last 20 years, *Environ Toxicol Chem* 32:1685–1687, 2013.

Baken SMA, Larsson MA, Gustafsson JP, Cubadda F: Smolders: ageing of vanadium in soils and consequences for bioavailability, *Eur J Soil Sci* 63(6):839–847, 2012.

Bliss CI: The toxicity of poisons applied jointly, *Ann Appl Biol* 26:585–615, 1939.

CIMM: *Proposal for zinc massive transformation/dissolution tests: pH and exposed surface loading effects*, Report to International Zinc Association, 2008.

De Laender F, Janssen CR, De Schamphelaere KAC: Nonsimultaneous ecotoxicity testing of single chemicals and their mixture results in erroneous conclusions about the joint action of the mixture, *Chemosphere* 76(3):428–432, 2009.

De Zwart D, Posthuma L: Complex mixture toxicity for single and multiple species: proposed methodologies, *Environ Toxicol Chem* 24:2665–2676, 2005.

ECHA: *Guidance on information requirements and chemical safety assessment*, QSARS and grouping of chemicals (chapter R.6). Available from: https://echa.europa.eu/guidance-documents/guidance-on-information-requirements-and-chemical-safety-assessment, 2008a.

ECHA: *Copper voluntary risk assessment report*, Available from: https://echa.europa.eu/copper-voluntary-risk-assessment-report, 2008b.

ECHA: *Guidance on information requirements and chemical safety assessment*, Appendix R.7.13-2: Environmental risk assessment for metals and metal compounds. Available from: http://echa.europa.eu/documents/10162/13632/information_requirements_r7_13_2_en.pdf, 2008c.

ECHA: *Guidance on information requirements and chemical safety assessment*, Evaluation of available information (Chapter R.4). Available from: https://echa.europa.eu/guidance-documents/guidance-on-information-requirements-and-chemical-safety-assessment, 2011. ECHA-2011-G-13-EN.

ECHA: *Transitional guidance on mixture toxicity assessment for biocidal products for the environment*, Available from: http://echa.europa.eu/documents/10162/15623299/biocides_transitional_guidance_mixture_toxicity_en.pdf, 2014.

EU CLP Classification: *Labelling and packaging of substances and mixtures regulation*, EC N° 1272/2008, Annex VI. Available from: http://eur-lex.europa.eu/legal-content/EN/TXT/?uri=CELEX:02008R1272-201701012008.

Hamels F, Malevé J, Sonnet P, Kleja DB, Smolders E: Phytotoxicity of trace metals in spiked and field-contaminated soils: linking soil-extractable metals with toxicity, *Environ Toxicol Chem* 33:2479–2487, 2014.

ICMM: *Metals environmental risk assessment guidance (MERAG)*, Available from: https://www.icmm.com/merag, 2007.

Jonker MJ, Gerhardt A, Backhaus T, Van Gestel CAM: Test design, mixture characterization and data evaluation. In van Gestel CAM, Jonker MJ, Kammenga JE, Laskowski R, Svendsen C, editors: *Mixture toxicity: linking approaches from ecological and human toxicology*, Pensacola, FL, 2011, CRC Press, pp 121–155.

Jonker M, Svendsen C, Bedaux JJM, Bongers M, Kamenga JE: Significance testing of synergistic/antagonistic, dose-level-dependent, or dose-ratio-dependent effects in mixture dose-response analysis, *Environ Toxicol Chem* 24(10):2701–2713, 2005.

JRC—European Commission: *European Union risk assessment report on Zinc metal, part I—environment*, Luxembourg, 2010, Publications Office of the European Union.

Komjarova I, Blust R: Multi-metal interactions between Cd, Cu, Ni, Pb and Zn in water flea *Daphnia magna*, a stable isotope experiment, *Aquat Toxicol* 90:138–144, 2008.

Larsson MA, Baken S, Smolders E, Cubadda F, Gustafsson JP: Vanadium bioavailability in soils amended with blast furnace slag, *J Hazard Mater* 296:158–165, 2015.

Loewe S, Muischnek H: Effect of combinations: mathematical basis of problem, *Arch Exp Pathol Pharmakol* 114:313–326, 1926.

Martín-Díaz ML, Vellane-Lincoln A, Bamber S, Blasco J, DelValls T: An integrated approach using bioaccumulation and biomarker measurements in female shore crab, *Carcinus maenas, Chemosphere* 58:615–626, 2005.

McLaughlin MJ, Hamon RE, Parker DR, et al: Soil chemistry. In Fairbrother A, et al, editors: *Test methods to determine hazards of sparingly soluble metal compounds in soils*, Pensacola, FL, 2002, SETAC Press, pp 5–16.

Meyer JS, Farley KJ, Garman ER: Metal mixtures modelling evaluation project: 1. Background, *Environ Toxicol Chem* 34:726–740, 2015.

Meyer JS, Ranville JF, Pontasch M, Gorsuch JW, Adams WJ: Acute toxicity of binary and ternary mixtures of Cd, Cu, and Zn to *Daphnia magna, Environ Toxicol Chem* 34:799–808, 2015.

Nagai T, De Schamphelaere KAC: The effect of binary mixtures of zinc, copper, cadmium and nickel on the growth of the freshwater diatom *Navicula pelliculosa* and comparison with mixture toxicity model predictions, *Environ Toxicol Chem* 35:2765–2773, 2016.

Norwood WP, Borgmann U, Dixon DG, Wallace A: Effects of metal mixtures on aquatic biota: a review of observations and methods, *Hum Ecol Risk Assess* 9(4):795–811, 2003.

Nys C, Asselman J, Hochmuth JD, et al: Mixture toxicity of nickel and zinc to *Daphnia magna* is noninteractive at low effect sizes, but becomes synergistic at high effect sizes, *Environ Toxicol Chem* 34:1091–1102, 2015.

Nys C, Janssen CR, De Schamphelaere KAC: Development and validation of a metal mixture bioavailability model (MMBM) to predict chronic toxicity of Ni-Zn-Pb mixtures to *Ceriodaphnia dubia, Environ Pollut* 220:1271–1281, 2017.

OECD: Environmental health and safety publications, Series on testing and assessment No. 29. In *Guidance document on transformation/dissolution of metals and metal compounds in aqueous media*, 2001.

OECD: Environmental health and safety publications, Series on testing and assessment No. 165. In *Report of an OECD workshop on metal specificities in environmental risk assessment*, 2012.

OECD: Environmental health and safety publications, Series on testing and assessment No. 194. In *Guidance on grouping of chemicals*, ed 2 2014.

Rasmussen K, Pettauer D, Vollmer G, Davis J: Compilation of EINECS: descriptions and definitions used for UVCB substances: complex reaction products, plant products, (post-reacted) naturally occurring substances, micro-organisms, petroleum products, soaps and detergents, and metallic compounds, *Toxicol Environ Chem* 69(3/4):403–416, 1999.

Reimann R, Birke M, Demetriades A, Filzmoser P, O'Connor P, editors: Chemistry of Europe's agricultural soils—part A: methodology and interpretation of the GEMAS data set. In *Geologisches Jahrbuch*, Hannover, 2014, Schweizerbartvol. B 102.

Reimann C, Fabian K, Birke M, et al.: GEMAS: establishing geochemical background and threshold for 53 chemical elements in European agricultural soil. Accepted by Appl Geochem 2017. (in press). Available from: http://www.sciencedirect.com/science/article/pii/S0883292716304577.

SCHER, SCCS, SCENIHR: *Opinion on the toxicity and assessment of chemical mixtures*, Available from: http://ec.europa.eu/health/scientific_committees/environmental_risks/docs/scher_o_155.pdf, 2012.

Sharma SS, Schat H, Vooijs R, Van Heerwaarden LM: Combination toxicology of copper, zinc, and cadmium in binary mixtures: concentration-dependent antagonistic, nonadditive, and synergistic effects on root growth in *Silene vulgaris, Environ Toxicol Chem* 18:348–355, 1999.

Smolders E, McGrath S, Fairbrother A, et al: Hazard assessment of inorganic metals and metal substances in terrestrial systems. In Adams WJ, Chapman PM, editors: *Assessing the hazard of metals and inorganic metal substances in aquatic and terrestrial systems*, Boca Raton, FL, 2007, CRC Press, pp 113–130.

UN GHS: *Globally harmonized system of classification and labelling of chemicals*, Annexes 9 and 10. Available from: https://www.unece.org/fileadmin/DAM/trans/danger/publi/ghs/ghs_rev04/English/ST-SG-AC10-30-Rev4.pdf, 2011.

US EPA: *Toxic Substances Control Act Inventory Registration for Products containing two or more substances: formulated and statutory mixtures*, Available from: http://www.epa.gov/opptintr/newchems/mixtures.txt, 2005a.

US EPA: *Toxic Substances Control Act Inventory Registration for Combinations of two or more substances: complex reaction products*, Available from: http://www.epa.gov/opptintr/newchems/rxnprods.txt, 2005b.

Van Regenmortel T, Nys C, Janssen R, Lofts S, De Schamphelaere KAC: Comparison of four methods for bioavailability-based risk assessment of mixtures of Cu, Zn, and Ni in freshwater. Accepted in Environ Toxicol Chem DOI: https://doi.org/10.1002/etc.3746. 2017 (in press).

Verdonck F, Van Assche F, Hicks K, Mertens J, Voight A, Verougstraete V: Development of realistic environmental release factors based on measured data: approach and lessons from the EU metal industry, *Integr Environ Assess Manag* 10(4):529–538, 2014.

Vijver MG, Elliott EG, Peijenburg WJGM, de Snoo GR: Response predictions for organisms water-exposed to metal mixtures: a meta-analysis, *Environ Toxicol Chem* 30:1482–1487, 2011.

Further Reading

ARCHE: *MeClas tool for classification of metals*, Available from: http://www.meclas.eu/, 2017.

ASTM E1833-07a: *Standard practice for sampling of blister copper in cast form for determination of chemical composition*, 2009.

CEN TR 1510: *Characterization of waste. Sampling of waste materials*, 2006.

ECHA: *Guidance for identification and naming of substances under REACH and CLP*, Available from: https://echa.europa.eu/guidance-documents/guidance-on-reach, 2016a.

ECHA: *Guidance on the application of the CLP criteria*, Annex IV metals and inorganic metal compounds. Available from: https://echa.europa.eu/documents/10162/13562/clp_en.pdf2016b.

ISO 12743: *Copper, lead, zinc and nickel concentrates—sampling procedures for determination of metal and moisture content*, 2006.

ISO 14488: *Particulate materials—sampling and sample splitting for the determination of particulate properties*, 2007.

ISO 3082: *Iron ores—sampling and sample preparation procedures*, 2000.

Human Health (Toxicity) Assessment of Complex Inorganic Materials

Violaine Verougstraete[*], *Rayetta Henderson*[†], *Carol Mackie*[‡], *Tony Newson*[§], *Adriana Oller*[¶]

[*]Eurometaux, Brussels, Belgium [†]ToxStrategies, Wilmington, NC, United States [‡]Copper Compound Consortium, Loanhead, United Kingdom [§]Consultant, Rotherham, United Kingdom [¶]NiPERA Inc., Durham, NC, United States

8.1 INTRODUCTION

While some metals are used as engineering materials or in consumer products in their elemental form or as simple metal compounds, the majority of metals in commerce are used in the form of more complex materials like alloys. Alloys, which are not simple mixtures, have been defined by the Globally Harmonised System of Classification and Labelling of Chemicals (UN GHS) as "a metallic material, homogeneous on a macroscopic scale, consisting of two or more elements so combined that they cannot be readily separated by mechanical means" (UN GHS, 2015). Consumers are extensively exposed to alloys present in their daily environment by way of articles with specific shapes and functions. For example, a wide variety of stainless steels are used in cookware (e.g. stainless steel grades 304 and 316) (Nickel Institute, 2005). Workers are exposed at the workplace during the production of and industrial uses of alloys.

Also, at and around industrial sites, metals can be contained in numerous other types of complex materials (e.g. ores and concentrates, slimes and sludges, flue dust, and other residues), which are used and generated during primary and secondary production of metals, ceramics, and inorganic pigments. These materials usually qualify as Unknown or Variable composition, Complex reaction products or Biological materials (UVCBs) (Rasmussen et al., 1999; US EPA, 2005a,b), due to their variability in composition and physical form. Such substances cannot be sufficiently identified by parameters like the International Union of Pure and Applied Chemistry (IUPAC) nomenclature and/or other name or identifiers or by its molecular, structural information, or chemical composition (ECHA, 2016). The use and production of such materials at industrial sites may be associated with the exposure of workers and the general population living around the sites.

As for all substances/mixtures, an appropriate assessment of the human health hazards of complex inorganic materials is required to ensure their adequate risk management. Such an assessment involves the collection of sufficient and relevant data to evaluate all potential toxicological endpoints and routes of exposure, taking into account the unique properties of the material. The outcomes of the hazard assessment will be confronted with the evaluation of exposures so as to evaluate health risks.

Chemicals management systems (e.g. the EU REACH, UN GHS, and US TSCA) usually set very clear information requirements to guide the collection of information and identify possible data gaps in the knowledge of the hazards of a substance or mixture.

The toxicity data on complex inorganic materials are scarce. To gather the information needed to perform the hazard assessment, one may therefore consider direct testing of the material. However, testing is confronted with at least two important limitations: first, the number of complex materials to be tested would be high and certainly go against animal welfare considerations and second, the variability inherent to some of these materials will make the selection of a representative sample for testing somewhat challenging. Also, in some regions (like the EU), legislation prevents the testing of mixtures for some endpoints, for example, carcinogenicity, mutagenicity, and reproductive toxicity (EU CLP, 2008). These limitations are discussed in more detail in Section 8.2.

An alternative to direct testing of the material is to use available data on the constituents of the material. Yet, for materials like alloys, the toxicity may not be directly related to and deduced from the metal content due to the inclusion of the metals in a 'matrix' that may modify their properties.

The challenge is therefore to find alternative ways to generate relevant information on the intrinsic properties of the complex material to allow for an appropriate hazard assessment for human health endpoints, while minimising animal testing.

This chapter describes the principles behind hazard identification of complex inorganic materials and discusses recent refinements that can help fulfil information requirements taking into account the unique properties of the material, while reducing the need for toxicity testing.

8.2 MEETING HEALTH HAZARD INFORMATION REQUIREMENTS

A number of data should be available to perform the appropriate hazard and risk assessment of the complex inorganic materials. The list of human health hazard endpoints (see example in Table 8.1) is no different for complex inorganic materials than for 'simple substances' like metals and metal compounds; however, the difficulties in addressing those increase when dealing with mixtures, alloys, UVCBs, or multiconstituent substances where the metal or metal compound is only one of the ingredients or constituents. This is especially true when dose–responses and effect levels need to be established for the material.

TABLE 8.1   List of Human Health Hazard Endpoints to be Addressed for Registration Under REACH (Related Information Requirements Are Detailed in REACH Annexes VII-X) and UN GHS

EU REACH	UN GHS
Acute toxicity	Acute toxicity
Skin irritation and corrosion	Skin corrosion/irritation
Sensitisation	Respiratory or skin sensitisation
Eye irritation and corrosion	Serious eye damage/eye irritation
Repeated dose toxicity	
Mutagenicity/genotoxicity	Germ cell mutagenicity
Carcinogenicity	Carcinogenicity
Reproductive toxicity	Reproductive toxicity
	Specific target organ toxicity-single exposure
	Specific target organ toxicity-repeated exposure
	Aspiration hazard

8.2.1 Grouping of Complex Inorganic Materials

The number and the variability of existing complex inorganic materials make it challenging or impossible to assess each and every material specifically. A possibility for structuring the assessments is to group materials of similar composition and expected properties or other characteristics. Alloys, for example, are often already pregrouped in numerous national and international standards, according to their composition and technical function (e.g. AFNOR, AISI, DIN, ASTM, JISI, and UNS). Within a group, a *generic* assessment can be made on representative material(s) which is considered to be relevant for all the materials of the group. The aim is to define substances within the group that are typical for its characteristics as a whole ('representative') and to keep at hand the reason behind the selection of the representative material. This is further described for alloys in Section 8.4. This grouping does not need to be performed when the assessment is limited to a single complex inorganic material.

8.2.2 Using Toxicity Data on the Complex Inorganic Material

In some cases, toxicity data may be available on the complex inorganic material. For example, in vivo toxicity data are available on some slags (e.g. REACH Registration dossier for slags steelmaking converter; Stettler et al., 1988) or specific grades of steel (e.g. Stockmann-Juvala et al., 2013; Summer et al., 2007; Liden et al., 1996), which have been tested for some specific hazard endpoints. Overall, however, these types of data are limited and do not fulfil all of the hazard data requirements.

Additional data could be generated by launching a test on the complex material but testing is confronted with several limitations such as the following:

- The significant number of materials to be tested goes against practicality and animal welfare considerations. If, for example, the different alloys existing on the market or the inorganic UVCBs used/generated during metal manufacturing processes were each to be tested for the various hazard endpoints to be investigated (in particular, the long-term effects endpoints, e.g. carcinogenicity), the number of toxicological tests needed for assessing the safety of these materials would require a huge number of laboratory animals and represent significant resource (monetary and time) investments in in vivo testing. This volume of testing goes against society's interest to minimise the use of animal tests in regulatory compliance and to support the development of alternative in vitro methods for testing the safety of materials (3Rs, Russell and Burch, 1959). This minimisation of animal testing is embedded in several regulations (e.g. REACH), connected with research to develop suitable in vitro and in silico methods.

- Also, when proceeding with testing, one should test a representative sample of what human populations are exposed to over the lifecycle of the material. Identifying a representative sample for a complex inorganic material like a UVCB can be complicated as it will typically have a varying composition (contain varying amounts of metal, metal compounds, and minerals) resulting in different particle size distributions. This will translate into variability in exposure and the level of toxic response. In some instances, the change in hazard profile results in a change of the classification under UN GHS/ EU CLP. The observed variation in the composition of such materials is related to their source, for example, the inherent temporal and geographical heterogeneity of ores in the primary production (ICMM, 2010), the variability of the feed material in recycling, and/ or the differences in the industrial processes generating the materials. This variability in composition will affect the representativeness of the UVCB sample. More precisely, it will be difficult to identify/select the (range of) composition(s) that can be considered relevant for this material and conditions of use, and those composition(s) that can be used as starting point for the assessment. It will often not be possible to find 'one' sample that would be representative of all the hazard profiles of UVCBs. To cover the whole spectrum of the variability, a sample could theoretically be artificially created, using, for example, extremes in composition of some of the constituents. While the use of such a worst-case composition that maximises the hazardous constituents would ensure that the approach is precautionary, its relevance can be questioned, as that worst-case material may actually not exist in the context of the applied industrial processes. These processes normally maximise the content of one constituent while simultaneously minimising the content of another one. Creating an artificial sample also alters the relative composition of constituents and thus alters any potential interactive effects of the constituents. Furthermore, different samples may be needed for different toxicological endpoints because one constituent of the UVCB may be more hazardous for a particular endpoint whereas another constituent may elicit different toxic effects. It therefore quickly becomes clear that the number of representative samples needed to fulfil the regulatory toxicological endpoints for a single UVCB material would lead to an unreasonable and unjustified number of studies. In addition, this would not resolve the uncertainty regarding the selection of the most representative combination of constituents.

8.2.3 Using Read-Across and Bridging

In the absence of data on the complex inorganic material, the assessor may consider the possibility to 'extrapolate' the available hazard knowledge on 'comparable' materials. This extrapolation of information shall follow a number of principles and be transparent so as to allow its understanding and reproducibility by another assessor.

Read-across is a technique used to predict endpoint information for one substance by using data from the same endpoint from another chemical which is considered to be 'similar' in some way(s) that can serve as the basis for performing such an assessment (OECD, 2014). Grouping or 'chemical grouping' describes the general approach to assess more than one chemical at the same time. It can include the formation of a chemical category (group of chemicals whose physicochemical and toxicological properties are likely to be similar or follow a regular pattern as a result of structural similarity) or rely on the identification of a chemical analogue (one chemical considered to be 'similar' to the chemical of interest), so as to apply read-across. The principles for grouping of metals and metal compounds have been outlined in several documents, for example, the OECD Guidance on Grouping Chemicals (2014) and a proposed read-across strategy has to be fully justified. An example of read-across is provided in Section 8.4 for a series of copper materials.

For mixtures, the UN GHS and EU CLP classification systems enable suppliers to derive health hazard classifications of materials based on available data on similar, previously tested materials and on the ingredients or constituent substances, when information on the material is not available; this approach is called bridging (UN GHS, 2015; EU CLP, 2008). Bridging ensures that the classification process, when performed based on the outcome of a hazard assessment, uses the available data to the greatest extent possible in characterising the hazard properties of the mixture without the necessity for additional testing in animals (UN GHS, 2015). Bridging can be used when a hazard classification exists for a 'source' complex inorganic material and sufficient data are available to demonstrate that the 'target' material is predicted to have a toxicity similar to the 'source' material (e.g. comparable bioavailability). Bridging can also be performed with data on the ingredients/constituents of the source material when hazard classifications/data on the source material itself are not available. The results of bridging need to be considered in a verified weight-of-evidence approach. An example of bridging of alloys is provided in Section 8.4.

8.2.4 Use of Data on the Constituents/Ingredients of the Complex Inorganic Material

A last alternative for assessing the human health hazard of complex inorganic materials is to use the data on the constituents or ingredients of the materials. This approach is further detailed in Chapters 11–13, for UVCBs, inorganic pigments, and alloys, respectively. Generically, the assumption is that the toxicity of the material is driven by the toxicity of its constituents/ingredients. Using this assumption as a starting point for the hazard assessment forces the assessor to have access to reliable and complete datasets on the ingredients/constituents (e.g. from high-quality databases) so as to build a matrix of information covering all constituents/ingredients and endpoints in order to perform the hazard assessment. Examples of information sources where such data can be found are further detailed in Chapter 6.

Still, one may be confronted with several issues when extrapolating toxicity data on the ingredients/constituents to the complex inorganic material like differences in speciation or physical form between the metal or metal compound used as a starting point and the speciation/physical form of the metal or compound included in the complex inorganic material or the inclusion in a matrix or crystalline structure. Finally, if the toxicity of the ingredients/constituents is considered to drive the toxicity profile of the complex inorganic material, the question of possible interactions (additivity, potentiation, synergy, and antagonism), and hence combined toxicity between the different ingredients/constituents shall be considered. These aspects are described below with some possible ways forward.

8.3 POSSIBLE WAYS FORWARD

The number of complex inorganic materials to assess and the limitations in generating test data de novo for these materials will tend to lead the assessor to either use a bridging/read-across approach or gather data on constituents/ingredients as much as possible. In every case, the aspects of speciation, physical form, bioavailability, and combined toxicity need to be taken into account.

8.3.1 Speciation

IUPAC defines 'chemical species' as a specific form of an element defined as isotopic composition, electronic or oxidation state, and complex or molecular structure (Duffus et al., 2009). For the risk assessment of a given metal, the term 'speciation' refers to its different chemical species, including its physicochemical characteristics that are relevant to bioavailability (see also Chapter 3). It is well known that for some metals, the species present will have a direct impact on the mechanism of action (e.g. chromium VI vs chromium III, Katz and Salem, 1993) or on toxicokinetics (dissolution kinetics in the lung, absorption, etc.) including bioavailability (extent to which a substance is taken up by an organism and is available for metabolism and interaction) (e.g., Goodman et al., 2011). For example, the oral uptake of zinc is recognised to vary as a function of chemical speciation: soluble zinc compounds (zinc chloride, zinc sulphate) have been reported to have a gastrointestinal uptake rate of 40% while less soluble forms of zinc (zinc metal and zinc oxide) have been assigned lower default uptake rates of 20%, based on human observational studies (HERAG, 2007). On the other hand, it should be acknowledged that the information on the speciation of the ingredients/constituents in some complex inorganic materials is not always available. In alloys, it is expected that the metal is present in its metallic form or as a surface oxide. However, in UVCB materials such as ores, concentrates, slags, and scrap the speciation may be different. For UVCBs, as indicated previously, while the elemental composition is generally known (as it has market implications and presents a high variability in concentrations), the chemical speciation of the constituents is often not known. Expert judgement can define it based on the knowledge of the industry processes; it can also be assessed from sequential extraction/metal analysis (XRD, XANES, EXAFS, FTIR, and chemical fractionation and analysis). The latter provides information on the form in which each elemental constituent is present in the UVCB (e.g. oxide, sulphide, etc.). Depending on the level of knowledge, two different situations can be distinguished: the chemical speciation of the constituent in the UVCB is known and can be used for hazard identification and classification, or the information on chemical speciation is not complete. In the latter case, the worst-case speciation for the purpose of classification and hazard assessment is assumed for each toxicological endpoint, that is, the speciation that would lead to the most severe classification or to the lowest effect level. This is further detailed in Chapter 11.

8.3.2 Physical Form and Particle Size

Physical form and particle size play a role in metals toxicity, and may need to be considered when extrapolating the data from one metal/metal compound having a defined particle size to a complex inorganic material having a different particle size. For example, when preparations of metallic lead of different particle sizes ranging between 0 and 250 μm were incorporated into rat diets, Barltrop and Meek, 1979 found an inverse relationship between particle size and lead absorption determined by measurement of tissue lead concentrations. This relationship was mostly marked in the 0–100 μm range. Therefore, factors such as speciation and particle size need to be considered with care when using data on constituents to predict data on the complex material of which they are a part.

In some cases, there is a difference in toxicological properties between the material and the ingredients/constituents, due to the inclusion of the latter in a crystalline structure or a matrix, which affects their bioavailability and hence their toxicity (see also Section 8.3.3). The hazard assessment of such materials requires specific approaches to consider the differences in these properties. Alloys are a typical example of 'matrix materials'. They are designed to have properties different from those of their ingredient metals, for example, to make them less reactive, more durable, have greater strength, or to improve machinability. Sometimes substantial changes in the intrinsic properties introduced during alloying often make translation of properties such as corrosion resistance and release rates of the metal ion from the individual ingredients (e.g. copper) to the alloy (e.g. brass) inaccurate (Leygraf et al., 2016).

8.3.3 Bioavailability and Its Use to Support Hazard Assessment

Both the speciation and the physical form/particle size and/or inclusion in a matrix affect the bioavailability of the metal ion. Bioavailability is a widely accepted concept based on the implicit knowledge that before an organism may accumulate or show a biological response to a chemical, that element or compound must be available to the organism. Since the toxicity of metals and their inorganic metal compounds is primarily associated with the release of soluble metal ions (or ion complex including a hydrated metal ion), their bioavailability is defined as the extent to which the soluble metal ion will be available at the target organ/site (Klaassen, 2001).

The information on bioavailability is usually obtained from toxicokinetic studies for all relevant routes of exposure and all relevant chemical forms or physical states of a metal-containing substance.

In situations where the bioavailability of a substance/material is not known or where it is not feasible to determine this in vivo, bioaccessibility may be used to conservatively estimate bioavailability. Bioaccessibility is defined as the "fraction of a substance that dissolves under surrogate physiological conditions" and therefore is "potentially available" for absorption into systemic circulation (Heaney, 2001).

Bioelution refers to the in vitro extraction methods used to measure the degree to which a substance is released in artificial biological fluids (e.g. synthetic sweat, simulated gastric juice, etc.), that is, the bioaccessibility of a substance (Henderson et al., 2014). For metals, metal compounds, and complex inorganic materials, bioelution tests can be used to estimate a substance or a metal-containing material's bioaccessibility in the form of released metal (e.g. metal ion release). Such in vitro tests are physicochemical tests, which are comparable to a water solubility test; their main strengths are that they are affordable, reproducible, and rapid.

The bioavailability of a metal can indeed be affected by a number of factors, such as its chemical speciation, its physical form, and its inclusion in a matrix or crystalline structure. Therefore, the bioavailability of a metal may vary considerably depending on the complex inorganic material in which it is present. Bioelution tests allow this variability to be assessed by measuring relative metal ion releases, that is, relative bioaccessibility (e.g. Hillwalker and Anderson, 2014).

Other important benefits of bioelution tests to predict bioavailability are as follows:

- Bioelution tests provide an in vitro alternative, allowing a reduction in animal testing (toxicokinetic and toxicity studies).
- The results of such tests are usually conservative. By measuring the ion released in an artificial body fluid (i.e. bioaccessibility), the test estimates the potential bioavailability of the metal. The absolute releases determined in a bioelution test setting may overestimate the bioavailability potential in vivo; substances migrating to the fluids might not be (fully) absorbed into the body (e.g. gastric fluid and digestive tract). Bioelution can thus usually be considered as a worst-case estimation for biological availability (Brock and Stopford, 2003; Skowronski et al., 2001; Collier et al., 1992).
- The assessment of the toxicity of a metal requires careful consideration of the route of exposure, as absorption and bioavailability are specific to each route. Bioelution tests can be tailored to provide data for specific exposure pathways. For example, the dissolution in artificial sweat can be used to estimate the metal ion bioaccessibility in the surface layer in contact with the skin to support predictions for both the sensitisation potential and systemic effects following dermal uptake. The dissolution in various artificial gastric/intestinal/saliva fluids can be used to estimate the relative metal ion bioaccessibility to support predictions of systemic effects following oral exposure (Henderson et al., 2012). Data from dissolution in simulated lung/lysosomal fluids can provide support for predictions of systemic inhalation effects (and supporting data for assessing local effects) (Stockmann-Juvala et al., 2013).

The main potential applications of bioaccessibility testing for metals, metal compounds, and metal-containing complex inorganic materials in the context of hazard assessment are grouping/read-across and bridging for hazard identification and classification.

In view of these possible applications and the requirement to minimise animal testing, bioelution testing—to estimate bioaccessibility and predict bioavailability—has become an active research area with researchers working to develop and validate bioelution protocols. Bioaccessibility, as a concept, has already been integrated in some regulatory arenas to assess the potential bioavailability of metals in environmental matrices and articles. For example, bioelution methods have been formalised as standards for product testing: EN 71 BS EN 71-1:2011+A3, 2014 and ASTM F-963, 2003 (methods to measure metal releases and standards for determining toy safety which specifies migration limits for metals from toy materials), EN 1811: 2011 + A:2015 (assesses nickel release from consumer articles intended for prolonged and direct skin contact), and ASTM D5517-17, 2007 (measures extractability of metals from art materials). In the above examples, absolute metal releases are compared with reference values stipulated in the legislation or standardised method. However, in most instances the relative bioaccessibility results are of interest (e.g. for read-across/grouping). For example, such results have been reported in the Bioaccessibility Research Group of Europe (BARGE, n.d.) studies and US EPA soil studies, which investigated the human bioavailability of priority contaminants in soils such as arsenic, lead, and cadmium via the gastrointestinal tract.

Guidance on how to conduct bioelution tests can be obtained from several regulatory authorities (e.g. US EPA, 2004; RIVM, 2005, 2006). However, to date there is no internationally standardised protocol, where all parameters that could possibly impact the reliability and reproducibility of the bioelution test results have been addressed and defined. There is ongoing work in the EU to validate a protocol, to be submitted to OECD for further standardisation. In addition, Henderson et al. (2014) have reported a proposed Standard Operating Procedure for testing metals, metal compounds, alloys, and a concentrate in several different synthetic fluids that can be used in the development of an internationally agreed protocol.

Until an OECD harmonised guideline has been published, and all possible factors affecting the ion release in bioelution tests have been identified and standardised, it is still possible to use bioaccessibility results in both the comparative and the weight-of-evidence approach when performing assessments.

'Comparative' means that in all cases, the tested substance or material is compared with at least one reference substance or material. In generic terms, the reference material should be so far as chemically (and physically, e.g. particle size distribution) as possible similar to the form of the metal present in the material and have a well-documented hazard profile. Since the selection of the reference material is usually done on a case-by-case basis, the rationale for its selection needs to be clearly documented. The results should be expressed as mass of metal released per liter or per mg of sample, and/or bioaccessible concentration (BC).

For example: if an alloy containing metal (X) undergoes bioelution testing, the pure metal form of X should be used as reference material and undergo the comparable bioaccessibility testing. The release data generated are then compared, so that the ion release of a given metal X from the alloy is assessed relative to the ion release of the same metal from its pure form, under the same bioelution conditions. The BC is the concentration of the bioaccessible metal, that is, the actual concentration that could be released and potentially absorbed, in the alloy based on results from bioelution tests.

It can be calculated as follows:

$$BC = \frac{\text{Release from alloy}}{\text{Release from pure metal}} \times 100\% \tag{8.1}$$

Example: an alloy contains 10% of a classified metal. Overall, 2 g of the alloy sample are added to 1 L of gastric fluid. The pure metal is tested in parallel under the same conditions. After 2 h, 0.0001 g soluble metal/L are detected for the alloy sample and 0.1 g soluble metal/L are detected for the pure metal (100% metal) sample. The BC of the metal in the alloy is

$$BC = \frac{0.0001 \text{g metal} / \text{L from alloy}}{0.1 \text{g metal} / \text{L from pure metal}} \times 100\% = 0.1\% \qquad (8.2)$$

In this example, even if the alloy contains 10% of the metal, the alloy does not behave as a simple mixture (a simple mixture would have yielded 0.01 g metal/L); rather the alloy behaves as if it has 0.1% of the classified metal.

The bioelution test can provide information on the comparative, or relative, potential bioavailability and hence toxicity of each metal constituent and provide the basis for, for example, grouping, performing read-across.

This information should be used as part of a weight-of-evidence approach to assess the relative (or sometimes absolute) bioavailable fraction, or 'potentially absorbable dose' following human exposure. The other components of the weight-of-evidence approach will depend on the purpose (read-across/grouping, bridging, classification) but will usually include information regarding the physical size and form of the materials, toxicokinetics, pattern of exposures, and other relevant details regarding the chemical nature of the specific material of interest. An example of application is provided in Section 8.4.

It is acknowledged that the use of bioelution results still relies on expert judgement and interpretation, but the available evidence and experience are continuously increasing, hence augmenting the robustness of the conclusions drawn. It is important at this stage to document explicitly the type of data that have been used in the weight-of-evidence judgement and to stress clearly the limitations/boundaries of these tests. An important word of caution is provided here. For the inhalation route, given its complexity, a more mechanistic approach should be applied, using information on inhalability as well as particle deposition, for example. Also, for local effects, factors other than the concentration of the ion at the target site may be important in determining toxicity in the lung (e.g. particle effect, lung overload, redox reactions, oxidative stress, and changes in pH). Although bioelution tests may give a first estimate of the 'persistence'/dissolution of the substance in the lung, these outcomes should not be used in isolation to predict toxicity.

Bioelution tests focus on the release of the soluble metal ion. In some cases, it is also important to consider the counter-ion present especially if this could be considered biologically active (OECD, 2014). This is particularly important in comparison to acute local toxicity effects where the counter-ion may result in the substance being classified as corrosive. Finally, where there is a valence issue for a particular metal in the complex inorganic material (e.g. chromium and vanadium) it is important that the most toxicologically relevant ion is considered in the weight-of-evidence assessment.

8.3.4 Combined Toxicity

Complex inorganic materials do not just include one single metal. When the hazard assessment relies on the combined knowledge of the different constituents/ingredients assessed one by one, the assessor may question whether the simultaneous presence of various metals/metal compounds (combined exposure) could result in antagonistic, additive, or even synergistic effects. The possibility of combined toxicity should be addressed in some way.

Information on combined toxicity is, however, not readily available for complex inorganic materials, except for a few metals where interactions have been documented. For example, in situations where the toxicity of workers exposed to particular mixtures has been studied or the studies assessing the effect of one metal on the absorption of another (e.g. zinc intake is associated with a lower cadmium burden, Vance and Chun, 2015). Regulatory guidance to assess and include combined toxicity in the hazard assessment of a material is under development in several jurisdictions. Several guidance documents already provide a framework that can be used in developing robust argumentation (IGHRC, 2009; SCHER, SCCS, SCENIHR, 2012; ECHA, 2014). The methods currently proposed to assess mixtures can take account of additive effects, such as dose/concentration addition or response–effect addition.

More specifically the European Scientific Committees (SCHER, SCCS, SCENIHR, 2012) propose to make a distinction between chemicals based on knowledge about their mode of action:

- Chemicals for which the mode of action is known are as follows:
  ○ Chemicals with common modes of action/target organs will act jointly to produce combination effects that are greater than the effects of each mixture component applied singly. These effects can be described by dose–concentration addition.
  ○ Chemicals with different modes of action (acting independently): no robust evidence is available to confirm that exposure to a mixture of such substances is of health or environmental concern if the individual chemicals are present at or below their respective threshold effect levels.
- Chemicals for which the mode of action is unknown: in this case, the dose–concentration addition method should be preferred over the independent action approach. Prediction of possible interaction requires expert judgement and hence needs to be considered on a case-by-case basis.

The European Scientific Committees also stated that interactions (including antagonism, potentiation, and synergies) usually occur at medium or high dose levels (relative to the lowest effect levels). At low-exposure levels, they are either unlikely to occur or are toxicologically insignificant. With regard to the assessment of chemical mixtures, such as complex inorganic materials, a major knowledge gap at the present time is the lack of exposure information and a rather limited number of chemicals for which there is sufficient information on their mode of action. There is a need for an agreed inventory of mode of actions/target organs for metals and a defined set of criteria on how to characterise or predict a mode of action for data-poor chemicals (SCHER, SCCS, SCENIHR, 2012).

In Chapter 11, a specific approach is proposed for the risk assessment of UVCB substances, by refining the assessment of effects (using knowledge on target organs as a surrogate for information on mechanism of action), exposure (refinement of exposure levels based on measurements in representative workplaces), and risk management.

8.4 IN PRACTICE

How to use data on the constituents of a UVCB or an alloy will be further detailed in Chapters 11 and 13, respectively. This section will focus on read-across/grouping and bridging to highlight the main principles.

8.4.1 Read-Across and Bridging

As mentioned earlier, read-across is a technique used to predict endpoint information for one substance by using data from the same endpoint from another chemical which is considered to be 'similar' in some way (OECD, 2014). This can be done by using data from one chemical to predict endpoint information for another chemical considered as 'analogue'. This can also be done by assessing more than one chemical at the same time, by forming a category. The hypothesis when starting grouping is that all chemicals in a category and all endpoints will be covered, meaning that the conclusions will be valid for all members of the category in the absence of substance-specific data. It is key that grouping of chemicals in a category is based on credible scientific and verifiable hypotheses. The assumptions made and the conclusions drawn should be appropriately documented and key data easily identifiable, so as to allow other assessors and authorities to evaluate the robustness and reliability of the read-across.

A number of documents exist that support the use of grouping and read-across in general (e.g. ECHA Guidance, 2008; OECD, 2014). The Read-Across Assessment Framework (RAAF) published by ECHA (2015) not only provides a systematic method to assess whether the proposed justifications for the extrapolation of data are compliant under EU REACH but also where those can be improved. For metals and inorganic metal compounds, the hypothesis is that toxicological properties are likely to be similar or follow a similar pattern as a result of the presence, solubility in relevant fluids, and bioavailability of the common metal ion (or ion complex including a hydrated metal ion). Organometallic compounds will generally have a different mode of action since the metal ion is not likely to be present in the same form as in inorganic compounds. In such cases, read-across between inorganic and organometallic compounds is not recommended, although read-across may well be appropriate between different organometallic compounds (OECD, 2014). It is the bioavailability of the metal ion at target sites that in most cases determines the occurrence and severity of the health effects to be assessed during read-across of metal substances (e.g., OECD, 2014; Henderson et al., 2012; Goodman et al., 2011).

The information on physicochemical properties of the metal/metal compound (e.g. water solubility), particle size and structure, in vivo data on systemic effects, and toxicokinetics will be used to complement bioaccessibility in a weight-of-evidence approach to support the grouping and read-across (OECD, 2014). The use of bioaccessibility data as a supportive tool in the context of grouping and read-across of metal toxicity has been illustrated for nickel compounds by Henderson et al. (2012).

How can this be applied to complex inorganic materials?

The example below illustrates a read-across of acute oral toxicity data from a water-soluble copper compound ('source') to various copper-bearing materials such as copper massive, copper powder, and minerals ('target').

The starting point in this assessment is the acute oral toxicity data (LD50) for a water-soluble copper compound, copper sulphate pentahydrate ($CuSO_4 \cdot 5H_2O$) (source compound). This compound has a large database available. The LD50 for this compound in rats was 481 mg/kg body weight (bw) or 122 mg Cu/kg bw (Lheritier, 1994).

Under the assumption that the acute oral toxicity of $CuSO_4 \cdot 5H_2O$ is driven by the bioavailable Cu ion, bioelution tests in gastric fluid were conducted to assess the relative bioaccessibility of Cu ion as predictor of the relative in vivo Cu bioavailability from various Cu-containing materials. Release/dissolution of copper ions from copper materials and copper compounds in biological fluids simulating oral exposure followed the ASTM D 5517-07 protocol, using HCl 0.07 N (pH 1.5) as a gastric mimetic fluid (Rodriguez et al., 2010a,b). The results from these bioelution tests are illustrated in Table 8.2.

Using the bioaccessibility data from coated copper flakes, chalcocite, copper massive, and copper powder in gastric fluid relative to that of copper sulphate pentahydrate (Table 8.3, column 2), predicted LD50s were calculated as mg Cu/kg bw or by mg copper-containing material/kg bw (knowing the compositions) (Table 8.3, column 3). The predicted values were compared to measured data when available (Table 8.3, column 4). The higher the predicted LD50 value is, the lower the oral toxicity is predicted to be, the less severe the classification will be. These predicted values are expected to be conservative (lower) than the available in vivo LD50s (copper flakes) as not all bioaccessible copper will be absorbed from the digestive tract.

8.4.1.1 Bridging of Mixtures Like Alloys

Alloys are often already pregrouped in numerous national and international standards, according to their composition and technical function (e.g. AFNOR, AISI, DIN, ASTM,

TABLE 8.2 Composition, Water Solubility and Bioelution Test Results for Various Copper-Containing Materials After 2 h in Gastric Fluid

Material Tested	Composition	Water Solubility	Bioaccessibility of Cu Ion
$CuSO_4 \cdot 5H_2O$ ('source')	25.45% Cu	Highly soluble	100
Coated copper flakes	93.7% Cu, 2.6% Cu_2O	Soluble	56.5
Chalcocite (Cu_2S)	79.9% Cu	Insoluble	3.3
Copper powder	99.7% Cu, 0.3% Cu_2O	Insoluble	1.1
Copper massive	>99.9% Cu	Insoluble	0.096–0.105

TABLE 8.3 Bioaccessibility Results, Predicted LD50 and Proposed Classifications

Cu Material	% Bioaccessibility Cu	LD50 (mg Cu/kg bw)	LD50 (mg Material/kg bw)	Proposed Classification
$CuSO_4 \cdot 5H_2O$ 25.45% Cu	100	Measured: 122	Measured: 481 (Lheritier, 1994)	Harmful
Coated copper flakes 93.7% Cu, 2.6% Cu_2O	56.5	Predicted: 215	Predicted: 230 Measured: 400 (Sanders, 2001)	Harmful
Cu2S 79.9% Cu	3.3	Predicted: 3700	Predicted: $4600 = (3700 \times 0.799)$ Measured: >2000 (Bradshaw, 2012)	Not classified
Copper powder 99.7% Cu	1.1	Predicted: 11,090	Predicted: $11,123 = (11,090 \times 0.997)$	Not classified
Copper massive >99.9% Cu	0.10	≫	≫	Not classified

JISI, and UNS). In addition, they can also be pregrouped with respect to their technical performance (e.g. corrosion resistance, wear resistance, etc.) and metallurgical structure. However, these types of groupings are in most cases not sufficient to allow bridging in order to fulfil hazard information requirements and cover existing data gaps.

It is well known that for most alloys, their toxicity will be influenced by the rate of metal ion release and the concentration of this ion in the human body fluids under biologically relevant conditions (i.e. bioavailability of the metal ion). As discussed earlier in this chapter, these release rates and concentrations can, however, not necessarily be adequately predicted based solely on the bulk composition of the alloy or its surface properties (Odnevall Wallinder et al., 2006; Herting et al., 2008a,b).

Thus, grouping of alloys based on comparable release rates in bioelution tests provides a more toxicologically relevant approach for bridging. In the first step, a few representative alloys are selected to define the 'groups'. These representative materials can be selected based on the 'pregrouping' referred above (standards, composition, engineering applications, and corrosion). The groups can then be further populated on the basis of bioelution tests, which will measure metal release from equivalent amounts of a representative alloy and from those alloys one would consider adding to the group. This will be done using appropriate bioelution protocols and artificial biological fluids relevant to the routes of exposure based on the toxicity profile of the source alloy (or of the source alloy constituents). Alloys which release hazardous metals similar to that of a representative alloy in similar bioelution conditions will be allocated to the same group.

Once the grouping of the materials to be assessed has been performed, the assessor may select one 'source alloy' and use the hazard data available for this representative material to predict the hazard of the other members of the group (e.g. 'target alloy'). This requires an understanding (and documentation) of all the relevant factors that contribute to the metal toxicity behaviour of the 'source alloy' and their comparison with those of the other members of the group ('target alloy'). Compiling a table reporting not just bioaccessibility data, but also providing information on the physicochemical properties (e.g. surface properties), knowledge on effects and hazard classifications, and other relevant elements that may contribute to the metal release behaviour, will increase the transparency or the decision-making process.

Several options and approaches are available for the verification of the appropriateness of the applied bridging and these include: (1) assembling and evaluating available toxicity data on complex materials (e.g. alloys) in comparison with analogous toxicity studies of the metal ingredients/constituents for specific endpoints, (2) conducting in vivo bioavailability or toxicokinetic studies in representative animal species, and (3) conducting in vitro toxicity assays or in vivo toxicity studies of complex materials in comparison with their metal ingredients/constituents.

It should be stressed that, at this stage, some routes of exposure may have been more closely examined than others when it comes to the predictability of in vitro bioelution tests for in vivo bioavailability. In particular, for the oral route, there are multiple sources of available data that could be used to verify the assumptions made (Drexler and Brattin, 2007; Ruby et al., 1999; Brattin et al., 2013; Denys et al., 2012; Rodriguez and Basta, 1999; Henderson et al., 2012). However, care must be taken when evaluating other routes of exposure (e.g. inhalation, dermal) where fewer test data are available to verify the in vivo bioavailability predictive role of bioelution tests. One example where such data exists is nickel, for which an artificial sweat test exists (EN 1811: 2011 + A:2015, 2011) which can be used to determine the skin sensitising potential of an alloy for classification purposes (EU CLP, 2008).

In addition, a repeated inhalation study in rats has confirmed that the low toxicity of a stainless steel alloy is much better predicted by the BC of nickel in lung fluids than by its content in the alloy (Stockmann-Juvala et al., 2013). In general, when evaluating the inhalation and dermal routes of exposure one must consider that solubilisation of a substance or material may only be one factor in predicting toxicity, whereas other factors may play an equal or even more significant role.

KEY MESSAGES

The assessment of the human health hazards of complex inorganic materials requires the collection of sufficient and relevant data to evaluate all potential toxicological endpoints and routes of exposure, taking into account the unique properties of the material. The outcomes of the hazard assessment will be confronted with the evaluation of exposures to evaluate health risks. Toxicity data on complex inorganic materials allowing to fulfil the information requirements and derive dose–responses/effect levels are usually scarce and testing is confronted with practical difficulties like the number of materials to test, their variability, and in some regions, the legislation preventing testing of mixtures for some endpoints.

It is therefore proposed to structure the assessment by grouping materials of comparable composition and according to expected properties, and to select representative materials on which a generic assessment is carried out.

This assessment will most often use the data on the constituents/ingredients of the material ('constituents-based approach'), based on the assumption that the toxicity of the material is driven by the toxicity of its constituents. This assumption needs, however, to be refined by the consideration of speciation, the physical form, and the particle size, affecting bioavailability and hence toxicity of the metal ion released by the material. Information on bioavailability is usually obtained from toxicokinetic studies for all relevant routes of exposure and all relevant chemical forms or physical states of a metal-containing substance. In situations where the bioavailability of a substance/material is not known or where it is not feasible to determine this in vivo, bioaccessibility (determined in vitro) may be used to conservatively estimate bioavailability.

The constituent-based approach is confronted with the question of possible interactions and hence combined toxicity between the different constituents. Information, except for a few metals where interactions have been documented, is limited. Several guidance documents or regulatory bodies provide a framework that can be used as way forward, to be completed with information on exposure and an inventory of mode of actions/target metals.

References

ASTM D5517-17: *Standard test method for determining extractability of metals from art materials*, D5517-94, Philadelphia, PA, 2007, American Society for Testing and Materials. Available from: www.astm.org.

ASTM F-963: *Standard consumer safety specification for toy safety*, West Conshohocken, PA, 2003, ASTM International. https://doi.org/10.1520/C0033-03. Available from: www.astm.org.

BARGE, n.d. Available from: http://www.bgs.ac.uk/barge/home.html.

Barltrop D, Meek F: Effect of particle size on lead absorption from the gut, *Arch Environ Health* 34(4):280–285, 1979.

Bradshaw J: *Dicopper sulphide: Acute oral toxicity in the rat – fixed dose method*. In *Testing laboratory: Harlan Laboratories Ltd. Report no.: 41201347. Owner company: Dicopper sulphide group of the Copper Compound Consortium*, 2012.

Brattin W, Drexler J, Lowney Y, Griffin S, Diamond G, Woodbury L: An in vitro method for estimation of arsenic relative bioavailability in soil, *J Toxic Environ Health A* 76:458–478, 2013.

Brock T, Stopford W: Bioaccessibility of metals in human health risk assessment: evaluating risk from exposure to cobalt compounds, *J Environ Monit* 5(4):71N–76N, 2003.

Collier CG, Pearce MJ, Hodgson A, Ball A: Factors affecting the in vitro dissolution of cobalt oxide, *Environ Health Perspect* 97:109–113, 1992.

Denys S, Caboche J, Tack K, et al: In vivo validation of the unified BARGE method to assess the bioaccessibility of arsenic, antimony, cadmium, and lead in soils, *Environ Sci Technol* 46:6252–6260, 2012.

Drexler JW, Brattin WJ: An in vitro procedure for estimation of lead relative bioavailability: with validation, *Hum Ecol Risk Assess* 13:383–401, 2007.

Duffus J, Nordberg M, Templeton D: Glossary of terms used in toxicology, second edition IUPAC recommendations 2007, *Pure Appl Chem* 79(7):1153–1344, 2009.

ECHA (European Chemicals Agency): *Guidance on information requirements and chemical safety assessment*, QSARs and grouping of chemicals (chapter R.6). Available from: https://echa.europa.eu/documents/10162/13632/information_requirements_r6_en.pdf/77f49f81-b76d-40ab-8513-4f3a533b6ac9, 2008.

ECHA (European Chemicals Agency): *Transitional guidance on the biocidal products regulation transitional guidance on mixture toxicity assessment for biocidal products for the environment*, Available from: https://echa.europa.eu/documents/10162/15623299/biocides_transitional_guidance_mixture_toxicity_en.pdf, 2014.

ECHA (European Chemicals Agency): *Read-across assessment framework (RAAF)*, Available from: https://echa.europa.eu/documents/10162/13628/raaf_en.pdf/614e5d61-891d-4154-8a47-87efebd1851a, 2015.

ECHA (European Chemicals Agency): *Guidance for identification and naming of substances under REACH and CLP Version 2.0*, 2016, ISBN: 978-92-9495-711-5.

EN 1811: 2011+A: 2015: *Reference test method for release of nickel from products intended to come into direct and prolonged contact with the skin*, CEN, Ref No EN 1811:1998 E, 2011.

EN 71 BS EN 71-1:2011+A3: *Safety of toys mechanical and physical properties*, 2014.

EU CLP: Classification, Labelling and Packaging of Substances and Mixtures, EC Regulation (EC) No. 1272/2008, 2008.

Goodman JE, Prueitt RL, Thakali S, Oller AR: The nickel ion bioavailability model of the carcinogenic potential of nickel-containing substances in the lung, *Crit Rev Toxicol* 41(2):142–174, 2011.

Heaney RP: Factors influencing the measurement of bioavailability, taking calcium as a model, *J Nutr* 131(4):1344–1348, 2001.

Henderson RG, Cappellini D, Seilkop SK, Bates HK, Oller AR: Oral bioaccessibility testing and read-across hazard assessment of nickel compounds, *Regul Toxicol Pharmacol* 63:20–28, 2012.

Henderson RG, Verougstraete V, Anderson K, et al: Inter-laboratory validation of bioaccessibility testing for metals, *Regul Toxicol Pharmacol* 70:170–181, 2014.

HERAG: *Gastrointestinal uptake and absorption & catalogue of toxicokinetic models*, Available from: http://hub.icmm.com/document/264, 2007.

Herting G, Odnevall Wallinder I, Leygraf C: Metal release rate from AISI 316L stainless steel and pure Fe, Cr and Ni into a synthetic biological medium—a comparison, *J Environ Monit* 10:1092–1098, 2008a.

Herting G, Odnevall Wallinder I, Leygraf C: Corrosion-induced release of chromium and iron from ferritic stainless steel grade AISI 430 in simulated food contact, *J Food Eng* 87:291–300, 2008b.

Hillwalker W, Anderson K: Bioaccessibility of metals in alloys: evaluation of three surrogate biofluids, *Environ Pollut* 185:52–58, 2014.

ICMM: *An industry approach to EU Hazard Classification, M2020*, Available from: http://hub.icmm.com/document/749, 2010.

IGHRC: *Chemical mixtures: a framework for assessing risk to human Health (CR14)*, Cranfield, 2009, Institute of Environment and Health, Cranfield University.

Katz SA, Salem H: The toxicology of chromium with respect to its chemical speciation: a review, *J Appl Toxicol* 13:217–224, 1993.

Klaassen CD: *Casarett and Doull's Toxicology: the basic science of poisons*, ed 6, New York, 2001, McGraw-Hill.

Leygraf C, Odnevall Walinder I, Tidblad J, Graedel T: *Atmospheric corrosion*, ed 2, New Jersey, 2016, Wiley, ISBN: 9781118762271.

Lheritier M: *Test to evaluate the acute toxicity following a single oral administration (LD50) in the rat*, 1994, Pharmakon Europe. Report no. 44193.

Liden C, Menné T, Burrows T: Nickel-containing alloys and platings and their ability to cause dermatitis, *Br J Dermatol* 134:193–198, 1996.

Nickel Institute: *The effective use of nickel in stainless steels*, Available from: http://www.nickelinstitute.org/en/KnowledgeBase/TrainingModules/EffectiveUseofNickelinStSt.aspx, 2005.

Odnevall Wallinder I, Bertling S, Berggren Kleja D, Leygraf C: Corrosion induced release and environmental interaction of chromium, nickel and iron from stainless steel, *Water Air Soil Pollut* 170:17–35, 2006.

OECD: *Guidance on grouping of chemicals*, ed 2, 2014. Series on testing & assessment No. 194. Available from: http://www.oecd.org/officialdocuments/publicdisplaydocumentpdf/?cote=ENV/JM/MONO(2014)4&doclanguage=en.

Rasmussen K, Pettauer D, Vollmer G, Davis J: Compilation of EINECS: descriptions and definitions used for UVCB substances: complex reaction products, plant products, (post-reacted) naturally occurring substances, micro-organisms, petroleum products, soaps and detergents, and metallic compounds, *Toxicol Environ Chem* 69(3,4):403–416, 1999.

RIVM: *Consumer product in vitro digestion model: bioaccessibility of contaminants from toys and application in risk assessment*, Bilthoven, The Netherlands, 2005, National Institute for Public Health and the Environment (RIVM). Report no. 320102004/2005.

RIVM: *How can information on oral bioavailability improve human health risk assessment for lead-contaminated soils? Implementation and scientific basis*, Bilthoven, The Netherlands, 2006, National Institute for Public Health and the Environment (RIVM). Report no. 711701042/2006.

Rodriguez RR, Basta NT: An in vitro gastrointestinal method to estimate bioavailable arsenic in contaminated soils and solid media, *Environ Sci Technol* 33:642–649, 1999.

Rodriguez P, Arbildua J, Opazo M, Urrestarazu P: *Copper and copper compounds: Bio-elution in gastric*. In *European Copper Institute Report Projects ENV 99332b*, 2010a.

Rodriguez P, Arbildua J, Opazo M, Urrestarazu P: *Copper concentrates and minerals: Bio-elution in gastric mimetic fluids*. In *European Copper Institute Report, Projects ENV 99332b*, 2010b.

Ruby MV, Schoof R, Brattin W, et al: Advances in evaluating the oral bioavailability of inorganics in soil for use in human health risk assessment, *Environ Sci Technol* 33(21):3697–3705, 1999.

Russell WMS, Burch RL: *The principles of humane experimental technique*, London, 1959, Methuen, ISBN: 0900767782.

Sanders A: *Copper powder: acute oral toxicity in the rat—acute toxic class method*, 2001, SafePharm Laboratories. Report no. 1451/001.

SCHER, SCCS, SCENIHR: *Opinion on the toxicity and assessment of chemical mixtures*, Available from: http://ec.europa.eu/health/scientific_committees/environmental_risks/docs/scher_o_155.pdf, 2012.

Skowronski GA, Seide M, Abdel-Rahman MS: Oral Bioaccessibility of trivalent and hexavalent chromium in soil by simulated gastric fluid, *J Toxicol Environ Health Part A* 63(5):351–362, 2001.

Stettler LE, Proctor JE, Platek SF, Carolan RJ, Smith RJ, Donaldson HM: Fibrogenicity and carcinogenic potential of smelter slags used as abrasive blasting substitutes, *J Toxicol Environ Health* 25(1):35–56, 1988.

Stockmann-Juvala H, Hedberg Y, Dhinsa NK, et al: Inhalation toxicity of 316L stainless steel powder in relation to bioaccessibility, *Hum Exp Toxicol* 32(11):1137–1154, 2013.

Summer B, Fink U, Zeller R, et al: Patch test reactivity to a cobalt-chromium-molybdenum alloy and stainless steel in metal-allergic patients in correlation to the metal ion release, *Contact Dermatitis* 57(1):35–39, 2007.

UN GHS: *UN Globally harmonised system of classification and Labelling of Chemicals (GHS)*, Sixth revised edition. United Nations, New York and Geneva. ST/SG/AC.10/30/Rev.6. https://www.unece.org/fileadmin/DAM/trans/danger/publi/ghs/ghs_rev06/English/ST-SG-AC10-30-Rev6e.pdf, 2015.

US EPA: *Estimation of relative bioavailability of lead in soil and soil-like materials using in vivo and in vitro methods*, Draft Washington, DC, 2004, U.S. Environmental Protection Agency, Office of Solid Waste and Emergency Response.

US EPA: *Toxic Substances Control Act Inventory Registration for Combinations of two or more substances: complex reaction products*, Washington, DC, 2005a, U.S. Environmental Protection Agency, Office of Solid Waste and Emergency Response.

US EPA: *Toxic substances control act inventory registration for chemical substances of unknown or variable composition, complex reaction products and biological materials: UVCB substances*, Washington, DC, 2005b, U.S. Environmental Protection Agency, Office of Solid Waste and Emergency Response.

Vance TM, Chun OK: Zinc intake is associated with lower cadmium burden in US adults, *J Nutr* 145(12):2741–2748, 2015.

Further Reading

HERAG: *Alloys fact sheet: hazard identification and classification of alloys for human health endpoints*, 2014 (draft).

UN GHS: *UN globally harmonized system of classification and labelling of chemicals (GHS)*, First revised edition, New York/Geneva, 2005, United Nations. ST/SG/AC.10/30/Rev.1 Health Hazards. https://www.unece.org/fileadmin/DAM/trans/danger/publi/ghs/ghs_rev01/English/03e_part3.pdf.

UN GHS: *UN Globally harmonized system of classification and labelling of chemicals (GHS)*, Second revised edition, New York/Geneva, 2007, United Nations. ST/SG/AC.10/30/rev.2 Health Hazards. https://www.unece.org/fileadmin/DAM/trans/danger/publi/ghs/ghs_rev02/English/03e_part3.pdf.

UN GHS: *UN Globally harmonized system of classification and labelling of chemicals (GHS)*, Third revised edition, New York/Geneva, 2009, United Nations. ST/SG/AC.10/30/rev.3 Health Hazards. https://www.unece.org/fileadmin/DAM/trans/danger/publi/ghs/ghs_rev03/English/03e_part3.pdf.

UN GHS: *UN Globally harmonized system of classification and labelling of chemicals (GHS)*, Fourth revised edition, New York/Geneva, 2011, United Nations. ST/SG/AC.10/30/Rev.4. https://www.unece.org/fileadmin/DAM/trans/danger/publi/ghs/ghs_rev04/English/ST-SG-AC10-30-Rev4e.pdf.

UN GHS: *UN Globally harmonized system of classification and labelling of chemicals (GHS)*, Fifth revised edition, New York/Geneva, 2013, United Nations. ST/SG/AC.10/30/Rev.5. https://www.unece.org/fileadmin/DAM/trans/danger/publi/ghs/ghs_rev05/English/ST-SG-AC10-30-Rev5e.pdf.

US EPA: *Estimation of relative bioavailability of arsenic in soil and soil-like materials by in vivo and in vitro methods*, Review draft, Washington, DC, 2005c, US Environmental Protection Agency. Region 8.

Specific Methodologies/Tools to Support Assessment

Patrick Van Sprang*, Frederik Verdonck*, Hugo Waeterschoot†, Isabelle Vercaigne*, Daniel Vetter‡, Jutta Schade‡, Kevin Rader§, Kevin Farley¶, Richard Carbonaro§, Koen Oorts*, Violaine Verougstraete†, Graham Merrington**, Adam Peters**, Rüdiger Vincent Battersby‡

*ARCHE Consulting, Ghent, Belgium †Eurometaux, Brussels, Belgium ‡EBRC Consulting GmbH, Hannover, Germany §Mutch Associates, Ramsey, NJ, United States ¶Manhattan College, Riverdale, NY, United States **WCA Environment Ltd, Faringdon, United Kingdom

9.1 MECLAS, AN ONLINE TOOL FOR HAZARD IDENTIFICATION AND CLASSIFICATION OF COMPLEX INORGANIC MATERIALS

9.1.1 Introduction to the Tool

MeClas (currently version 4.2) is a web-based tool (www.meclas.eu) used to generate (eco) toxicity hazard categories and corresponding classification and labelling information of inorganic metal-containing complex materials such as ores, concentrates, intermediates, or alloys for which the manual application of the UN GHS/EU CLP rules is very complex and requires a high level of consistency (Verdonck et al., 2017).

The tool comprises several tiers, aimed at the progressive refinement of classification through the recognition of specific mineral content, speciation up to bioavailability corrections. The better the quality of the information that is available, the higher the tier, reducing thereby the conservatism of the assessment and increasing its precision (see Fig. 9.1).

Tier 0 is solely based on the quantification of the elemental concentrations of the constituents/ingredients in the complex inorganic materials. Tier 1 requires more knowledge on speciation and is based on the determination of the metal speciation/mineralogy. In Tier 2, hazard classification of inorganic materials can be further refined by comparing the released

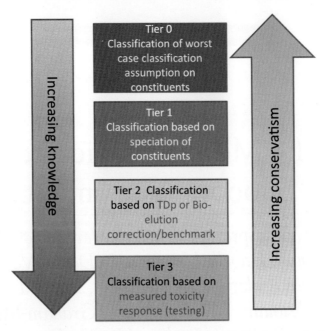

FIG. 9.1   Schematic overview of the tiered classification approach in MeClas for complex inorganic materials.

bioavailable metal concentrations after specific time periods with ecotoxicological endpoints, using metal release data from Transformation/Dissolution, or from bioelution tests for some human health endpoints. More information can be found in Verdonck et al. (2017).

MeClas addresses the UN GHS (UN GHS, 2015) and EU CLP (EU CLP, 2008) human health and environmental hazard endpoints and is based on an unambiguous algorithm defined under GHS/CLP and has a well-defined domain of applicability and robust predictability.

The tool allows for a consistent approach across companies that have to implement classification in line with UN GHS ruling and guidance (and regional implementations), while considering the metal specificities and the most up-to-date (eco)toxicological hazard information on self-classifications and ERV/TRV. Where relevant in a regional jurisdiction (EU and US), mandatory classification references are used complementary to high-quality (eco)toxicity reference values (ERV/TRV) and self-classifications.

The main benefit of using MeClas is its unique access to the most up-to-date reference values on the constituents of the complex inorganic materials. To date this is a feature unique to this tool. The continuous updating process on both ruling and ecotox references is managed by providing a platform for relevant data sharing between the metal consortia and companies collecting these data for the EU REACH.

Moreover, since the tool is web based and tracks versions/tool changes, it ensures consistency across the metal industry.

The main limitations of MeClas are that the system is applicable for inorganic materials only (not for organics and metalloorganic compounds) and for environmental and human health hazards only (and not for physical hazards). The up-to-date character of the database also depends on the willingness and resources of the metal sector to share promptly any changes in classifications. To mitigate this latter point, international databases are also screened for changes in the classifications. Lastly, some modules such as the use of bioelution testing for human health hazard characterisation are still under discussion with some regulatory authorities. This is clearly indicated in MeClas as a warning, with the possibility to remain at a lower tier.

9.1.2 User Interface

The user has to input the information (s)he has on the composition, and if available, its solubility characteristics, and after saving, (s)he can immediately calculate the classification. If useful, more information on the underlying calculations can be retrieved. The elemental and speciation composition of an inorganic complex material or mixture can be defined in the 'add/edit composition' user interface. This is shown as an example in Fig. 9.2:

In a material, 0.2% Ag is 100% present as a massive Ag. The relative releases according to the Transformation Dissolution protocol test (at 7 and 28 days) can be specified in Tier 2, when available. In this example, 0.01% of Ag is released after 7 days and 0.02% after 28 days.

The calculated output is the classification and labelling of the material according to UN GHS and different regional formats such as the EU CLP and US OSHA (US, 2012) for each toxicological and ecotoxicological endpoint.

Fig. 9.3 presents the result of the classification of a hypothetical inorganic complex material obtained from the MeClas tool.

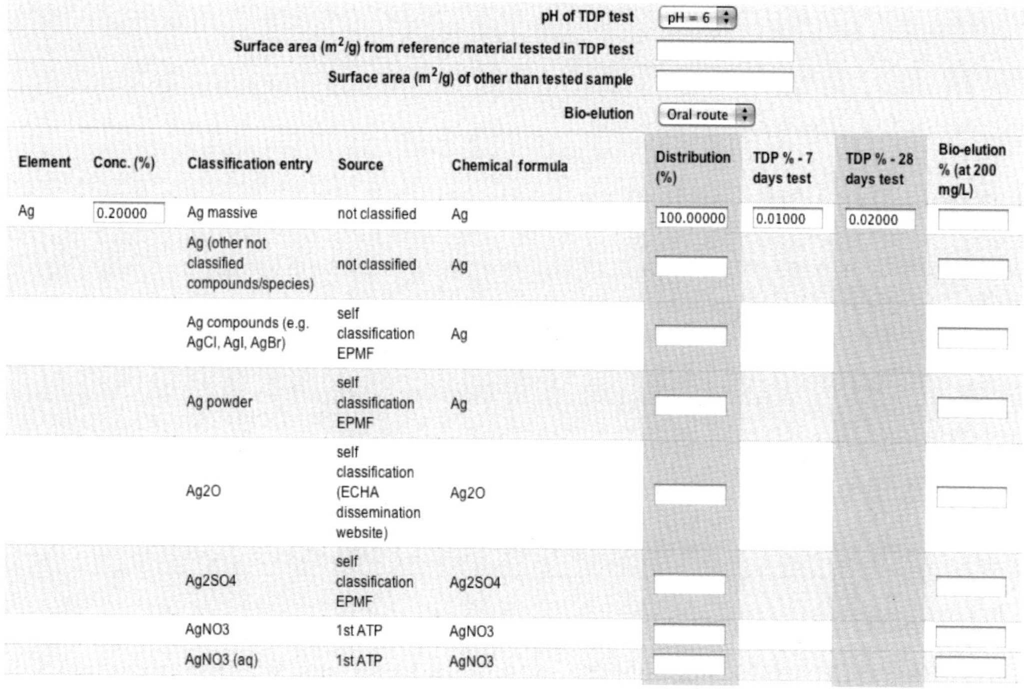

						Distribution (%)	TDP % - 7 days test	TDP % - 28 days test	Bio-elution % (at 200 mg/L)

pH of TDP test — pH = 6
Surface area (m²/g) from reference material tested in TDP test
Surface area (m²/g) of other than tested sample
Bio-elution — Oral route

Element	Conc. (%)	Classification entry	Source	Chemical formula		Distribution (%)	TDP % - 7 days test	TDP % - 28 days test	Bio-elution % (at 200 mg/L)
Ag	0.20000	Ag massive	not classified	Ag		100.00000	0.01000	0.02000	
		Ag (other not classified compounds/species)	not classified	Ag					
		Ag compounds (e.g. AgCl, AgI, AgBr)	self classification EPMF	Ag					
		Ag powder	self classification EPMF	Ag					
		Ag2O	self classification (ECHA dissemination website)	Ag2O					
		Ag2SO4	self classification EPMF	Ag2SO4					
		AgNO3	1st ATP	AgNO3					
		AgNO3 (aq)	1st ATP	AgNO3					

FIG. 9.2 MeClas screenshot of input user interface.

Endpoint	Self-cassification (GHS)	EU (CLP)	US (OSHA)
Acute toxicity-oral	Cat. 5; H303	Not classified	Cat. 5; H303
Acute toxicity-dermal	Not classified	Not classified	Not classified
Acute toxicity-inhalation	Not classified	Not classified	Not classified
Skin corrosion/irritation	Not classified	Not classified	Not classified
Serious eye damage/eye irritation	Not classified	Not classified	Not classified
Respiratory or skin sensitisation	Skin Sens. Cat. 1; H317	Skin Sens. Cat. 1; H317	Resp./Skin Sens. Cat. 1; H334/H317
Germ cell mutagenicity	Not classified	Not classified	Not classified
Carcinogenicity	Cat. 1B; H350	Cat. 1B; H350	Cat. 1A; H350
Reproductive toxicity	Cat. 1A; H360	Cat. 1A; H360	Cat. 1A; H360
Specific target organ toxicity - single exposure	Not classified	Not classified	Not classified
Specific target organ toxicity - repeated exposure	Cat. 1; H372	Cat. 1; H372	Cat. 1; H372
Aspiration hazard	Not classified	Not classified	Not classified
Hazardous to aquatic environment - ACUTE	Acute Cat. 1; H400	Acute Cat. 1; H400	
Hazardous to aquatic environment - CHRONIC	Chronic Cat. 2; H411	Chronic Cat. 2; H411	

FIG. 9.3   MeClas screenshot of output interface.

The three outputs (UN GHS, EU CLP, and US OSHA) give similar results except for acute toxicity, sensitisation, and carcinogenicity. The difference for acute toxicity is due to the non-existence of Category 5 under the EU CLP. The difference for sensitisation is due to the more stringent cut-off limit in the mixture ruling under US OSHA. In addition, the difference for carcinogenicity can be explained because US OSHA is referring to the—in this case—more precautionary IARC database that is used to base the classifications for carcinogenicity.

More information on the MeClas tool can be found in Verdonck et al. (2017).

9.2 SPERCs: REALISTIC ENVIRONMENTAL RELEASE FACTORS FOR METALS TO BE USED FOR THE ASSESSMENT OF ENVIRONMENTAL EXPOSURE

9.2.1 Introduction to the Tool

The assessment of environmental exposure and risks associated with the production or use of a substance on an industrial site includes the estimation of the releases to the environment. In the absence of measured release data on the specific substance, a risk assessor would rely on default release factors to the environmental compartments as developed in international, national, or regional context. As a wide variety of substances, processes, and uses has to be covered, default release factors are as a rule conservative, usually leading to significant overprediction of releases and hence to overpredicted environmental exposure concentrations and risks.

However, in practice, unrealistic and worst-case predictions do not support a more efficient management of releases and risk.

In response to the conservatism included in the release factors of the environmental release categories (ERCs) in the EU, a significant number of sector groups of the chemical industry and their downstream user industries have developed the so-called SPERCs (Sättler et al., 2012). SPERCs are intended to be used as a refinement of the ERC, as an advanced tier instrument in environmental safety assessment, increasing the realism and accuracy of the resulting environmental emissions and exposure estimates. SPERCs are based on a standardised calculation approach.

Verdonck et al. (2014) have proposed a more realistic approach to characterise the environmental releases from the manufacture, processing, and downstream uses of the metals and their compounds, including in or from complex inorganic materials. Although developed in the European Union (EU), this approach can also be used in other regions and in other chemical management systems addressing metals or complex inorganic materials containing metals.

A database consisting of more than 1300 recent (1993–2010) site-specific measured release factors to air and water for 18 different metals from 21 EU Member States was compiled and used to calculate average and reasonable worst-case release factors for multiple metal manufacturing and industrial use processes. The parameters influencing releases to water were found to depend predominantly on life cycle step (manufacture and/or use), the sector and/or the solid–water partition coefficient (K_d). The derived SPERCs can be downloaded from http://www.arche-consulting.be/Metal-CSA-toolbox/spercs-tool-for-metals.

The release factors can be used as an advanced tier instrument in environmental safety assessments for regulations such as REACH, increasing the realism of the estimates while still keeping a sufficient level of conservatism. As the metal and metal compound release factors are based on real-world measured release data, the estimations are more realistic and relevant for present-day industrial operations in the EU and are therefore considered as an improvement compared with the existing defaults (e.g. ERCs). The proposed metal release factors still provide a conservative estimate of environmental emissions of the metals industry because metal and metal compounds release factors are based on the 90th percentiles, are integrative of all on-site processes and all metal species rather than focusing on the single substance being assessed, and are based on recent (2007–12) and older (1993–2007) data (considering the overall tendency of decreased emissions over time). The use of these conservatively derived default environmental release factors allows for a realistic environmental exposure and risk assessment for a substance life cycle, in a simplified approach. These refined defaults allow not only to meet the increasing data requirement needs of regulations (e.g. REACH) but also a more efficient risk management, if needed.

9.2.2 Overview of the Metal Release Factors

An overview of the metal-specific typical (50th percentile) and reasonable worst-case (90th percentile) release factors to water and air for several metal life cycle steps (Verdonck et al., 2014) is provided in Table 9.1.

More information on the SPERCs can be found in Verdonck et al. (2014).

TABLE 9.1   Overview of the Metal-Specific Typical and Reasonable Worst-Case Release Factors to Air and Water for Several Metal Life Cycle Steps

Life cycle step (category)	Subcategory (K_d)	Typical (50th Percentile) Release Factor (After On-site Treatment) to		Reasonable Worst-Case (90th Percentile) Release Factor (After On-site Treatment) to	
		Water (%)	Air (%)	Water (%)	Air (%)
Manufacture and recycling of	Massive metal and metal powder	0.01	0.002	0.03	0.03
	25,000–60.000 L/kg	0.002		0.01	
	60,000–190,000 L/kg	0.0005		0.005	
	190,000–400,000 L/kg	0.0001		0.002	
Manufacture of	Metal compounds	0.02	0.003	0.2	0.03
	10,000–25,000 L/kg	0.02		0.2	
	25,000–60,000 L/kg	0.005		0.04	
	60,000–100,000 L/kg	0.002		0.01	
	100,000–190,000 L/kg	0.0007		0.005	
	190,000–250,000 L/kg	0.0003		0.002	
	250,000–400,000 L/kg	0.0002		0.001	
Formulation	In alloys	0.0002	0.0001	0.005	0.005
Industrial use: shaping	Massive metal	0.00001	0.0001	0.003	0.02
Industrial use in batteries	Metals (compounds)	0.0001	0.0006	0.003	0.003
Industrial use in metallic coating	Metal (compounds)	0.02	0	0.2	0.5
Formulation of	Metal compounds	0	0	0.0002	0.005
	Plastics and rubber				
	Paints and coatings	0.001	0.0005	0.01	0.005
	Other sectors	0.1[a]	0.00002	2[a]	0.01
Industrial use of	Metal compounds	0	0	0.001	0.001
	Plastics and rubber				
	Textile	0.002	0	0.007	0.001
	Glass	0.007[a]	0.004[a]	0.5[a]	2[a]

In the first data cell of the Manufacture and recycling of row, the subcategory is "10,000–25,000 L/kg". In the Manufacture of Metal compounds first row, the subcategory is "1000–10,000 L/kg".

[a]Before on-site risk management measure sewage treatment plant.

9.3 MEASE, A FIRST TIER MODELLING TOOL FOR THE ASSESSMENT OF WORKPLACE EXPOSURE

9.3.1 Introduction to the Tool

The exposure assessment tool MEASE (currently version 1.02.01) has been developed for the estimation and assessment of the exposure to metals and inorganic substances. It combines approaches from the EASE expert system (HSE, 1997), the ECETOC targeted risk assessment (TRA) tool (ECETOC, 2009), the ECHA (2015) and from the health risk assessment guidance for metals project (HERAG, 2007). It represents a first tier screening tool for the estimation of occupational inhalation and dermal exposure to metals and inorganic substances.

A screenshot of the MEASE interface is shown in Fig. 9.4.

This tool and the related documentation can be freely downloaded from http://www.ebrc. de/industrial-chemicals-reach/projects-and-references/mease.php.

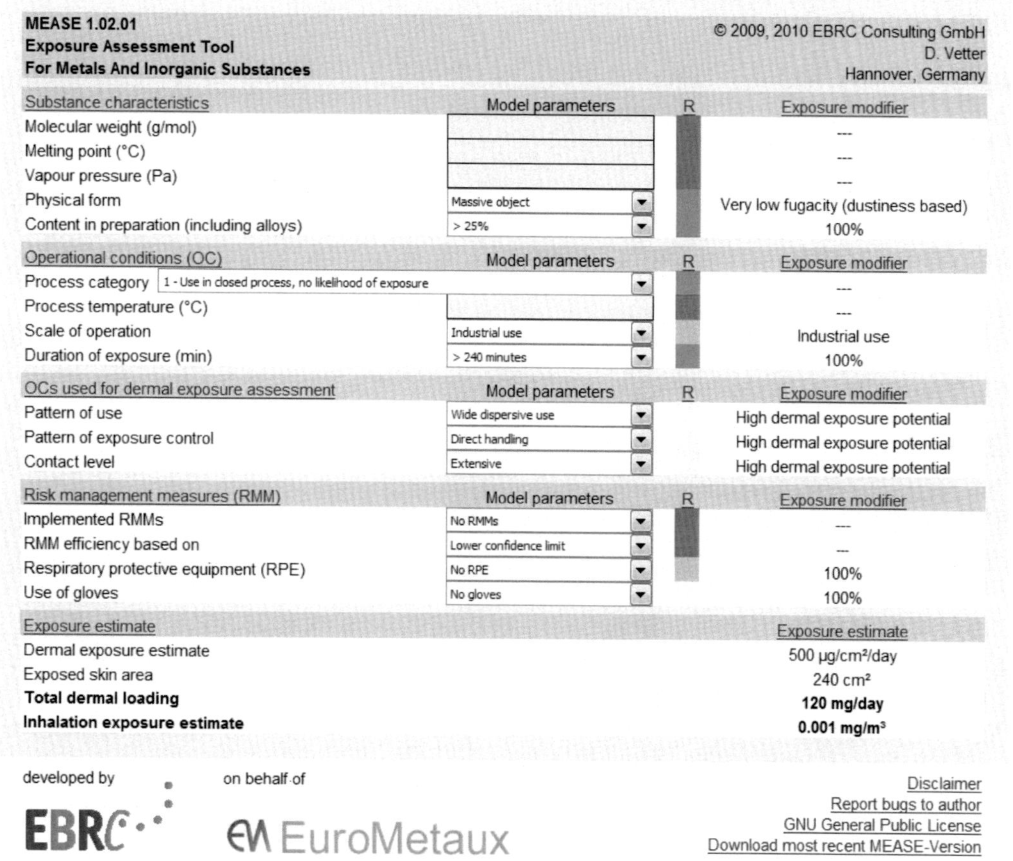

FIG. 9.4   Screenshot of MEASE.

For inhalation exposure, the tool follows the process-specific categories (PROC)—specific approach of the ECETOC TRA tool reported in the EU and selects initial exposure estimates from three so-called 'fugacity classes'. For some metal-specific PROCs 21: low-energy manipulation and handling of substances bound in/on materials or articles, 22: manufacturing and processing of minerals and/or metals at substantially elevated temperature, 23: open processing and transfer operations at substantially elevated temperature, 24: high (mechanical) energy work-up of substances bound in/on materials and/or articles, 25: other hot work operations with metals, 26: handling of solid inorganic substances at ambient temperature, 27a: production of metal powders (hot processes), the initial exposure estimates in MEASE are based on air monitoring data from the metals' industry. The initial exposure estimates can be furthered with the tool by reflecting operational conditions and risk management measures. Efficiency values for risk management measures are based on the publication of Fransman et al. (2008).

For dermal exposure, MEASE is based on the classification system of the former broadly used EASE system. The exposure estimates are however based on real measured data for several metals. These measurements have been collated and plotted against the EASE exposure classes in the 'dermal factsheet' of the HERAG project (HERAG, 2007). The factsheet concluded on a system of exposure bands, which are proposed to replace the EASE estimates within its own classification system. These conclusions have been implemented in MEASE. MEASE has been taken up in the ECHA guidance for the assessment of occupational exposure under EU REACH (ECHA, 2016).

9.3.2 Notes on the Application of MEASE for the Assessment of Exposure to Complex Inorganic Materials

MEASE has been developed for the estimation of exposure to given substances (metals and inorganic substances). If such a substance only represents a part of a mixture, it is possible to refine the assessment by reflecting the reduced concentration in the tool. For mixtures, MEASE has foreseen a categorical approach, which provides an exposure reduction according to four concentration ranges:

- 0% reduction for 25%–100%,
- 40% reduction for 5%–25%,
- 80% reduction for 1%–5%,
- 90% reduction for >1%.

However, if exposure is only to be assessed for a specific constituent of a given complex inorganic material, this approach may be considered as highly conservative. In particular, for processes conducted at ambient temperature, a linear relationship between the constituent concentration and the exposure reduction can safely be assumed. It is therefore suggested to use the reasonable worst-case concentration of a constituent in a given complex inorganic material and to manually reduce the exposure estimate on a linear basis. For hot metallurgical processes, where different emission potentials for constituents from a molten complex inorganic material apply, the original MEASE approach may be followed as a precautionary measure.

9.3.3 Validation Status of MEASE and Development of MEASE 2

MEASE has been validated in the recently finalised ETEAM project (Lamb et al., 2015) as one of the main exposure assessment tools used under the EU REACH Regulation. MEASE has been found to generally produce conservative exposure estimates. Some of ETEAM findings have suggested that the between-user variability may be increased in MEASE compared with the other tools validated in the ETEAM project. To address the ETEAM recommendations, a version 2 of MEASE will be released in 2017.

9.4 GUIDANCE ON THE ASSESSMENT OF OCCUPATIONAL EXPOSURE TO METALS BASED ON MONITORING DATA

On the basis of experiences acquired along with the risk assessments of metals and metal compounds under the EU Existing Substances Regulation (Council Regulation EEC N° 793/93), exposure assessments of the metals required under REACH and various site-specific assessments carried out for the European metals industry, Vetter et al. (2016) have developed a guidance document aiming at harmonising occupational exposure assessments based on monitoring data (Fig. 9.5). This document can be downloaded from http://www.ebrc.de/tools/occupational-exposure.php.

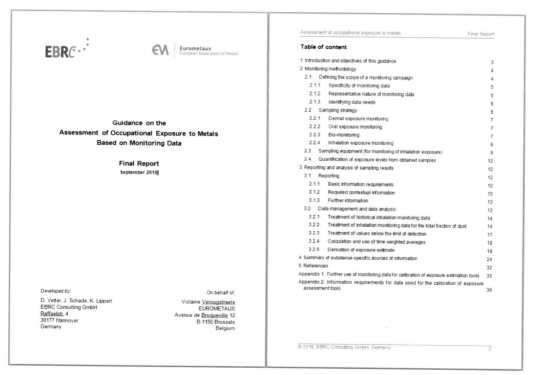

FIG. 9.5   Title page and table of content of the exposure assessment guidance.

This guidance addresses the aspects of monitoring methodology and data reporting relevant for occupational exposure assessments within the entire supply chain of metals (and their compounds), including handling and manufacturing of complex inorganic materials. Although it is acknowledged that exposure in the workplace may occur via inhalation, dermal contact, and via (inadvertent) ingestion, the guidance focusses on the monitoring of inhalation exposure as the main route of exposure.

The information included in this document addresses generic issues related to the monitoring of occupational exposure to metals on purpose. However, references to information sources for aspects outside the scope of this guidance and for metal-specific guidance have been included, when available, in the document.

9.5 TOOLS TO ASSESS BIOAVAILABILITY TOOLS FOR METALS

9.5.1 Bioavailability in Freshwater

Metals released from complex materials might enter the environment and may harm freshwater organisms when specific metal threshold concentrations are exceeded. It is generally recognised by regulators, industry, and academia that such concentrations of metals, expressed as total or dissolved concentrations, are not good predictors of toxic effects and that metal bioavailability needs to be taken into account to accurately assess the potential impact of metals on aquatic ecosystems (Bergman and Dorward-King, 1997; Janssen et al., 2000; Paquin et al., 2002; Fairbrother et al., 2007). Indeed, numerous detailed studies have demonstrated the effects of complexation by dissolved organic carbon (DOC), and competing cations such as Ca^{2+}, Mg^{2+}, Na^+, and H^+ on the toxicity of metals to fish, crustaceans, and algae. This research has culminated into the development of the biotic ligand model (BLM), which can accurately predict the toxicity of some metals as a function of the physicochemistry of the water. On the basis of the increasing knowledge of factors which influence metal bioavailability, different approaches were developed according to Rüdel et al. (2015), as illustrated in Table 9.2.

Full or complex BLMs provide sound assessment of metal bioavailability and toxicity but require a high number of measured water quality parameters as input such as temperature, pH, DOC, humic acid (HA), Ca, Mg, Na, K, SO_4, Cl, S, and alkalinity.

BLMs are either developed for acute or chronic endpoints. In the United States, the full acute BLM is currently accepted by the US EPA as an acceptable model for use by risk assessors to develop aquatic life criteria for copper in freshwater (US EPA, 2007). This model developed by Windward can be downloaded from http://www.windwardenv.com/biotic-ligand-model/. Other acute BLMs have been developed for aluminium, silver, cadmium, cobalt, nickel, and lead, and can be downloaded from the same website. The tool further calculates for lethal concentration 50 (LC50) values for specific water chemistries. Furthermore, an HC5 (hazardous concentration for 5% of the species) option is available to calculate both acute (criterion maximum concentration, CMC) and chronic (criterion continuous concentration, CCC) water quality criteria from the fifth percentile of the SSD.

Although chronic BLMs have been developed for several metals such as nickel, copper, cobalt, or zinc and successfully used in a regulatory context, such as REACH and the EU

TABLE 9.2 Description of Different Approaches for Consideration of Metal Bioavailability in Aquatic Systems (Rüdel et al. 2015)

Approach	Basis	Monitoring Requirements	Example
Full or complex BLM (partly including chemical speciation models)	Consideration of metal speciation and interactions between metal ions, organic matter content, particles, and abiotic and biological ligands	Dissolved target metal concentration, pH, concentrations of major cations and anions, DOC, suspended particles, temperature, others	Acute and Chronic Cu—BLM Acute and Chronic Pb—BLM
User-friendly BLM-based bioavailability tools	Simplification of the complex BLM, based on calculations with complex BLM or based on transfer functions; validated against the complex BLM	Dissolved target metal concentration, pH, hardness (e.g. as Ca concentration) and DOC concentration	Biomet M-BAT PNEC.pro
Simplified bioavailability approaches	For example, DOC-related assessments or hardness banding	Dissolved target metal concentration, hardness, or DOC concentration depending on approach	Hardness banding of Cd (directive 2008/105/EC, EC, 2008)

Water Framework Directive, no downloadable tools were made available for these metals. However, for lead (Pb), a chronic BLM has recently been developed by Ghent University (Belgium), ARCHE Consulting (Belgium), KTH Royal Institute of Technology (Sweden), and ILA (International Lead Association, formerly known as International Lead Zinc Research Organization, USA). The lead BLM normalisation tool is a free resource (http://www.ila-lead.org/responsibility/lead-blm-tool) for anybody interested in using bioavailability-based approaches for assessing the long-term hazard of lead in the freshwater aquatic environment. For each site, the best-fitting distribution and lognormal distribution is displayed, and standard deviation (or relevant fitting parameters) and HC_{5-50} with 90% confidence interval ($HC_{5,5}$ and $HC_{5,95}$) are calculated.

Fig. 9.6 shows the example of HC_{5-50} calculation for a site (site 1) with the following characteristics: pH: 7.7; Ca: 0.34 mmol/L and DOC: 2.5 mg/L. From the best-fitting distribution (i.e. log-logistic), a Pb-specific HC_{5-50} of 8.56 µg/L was calculated, while from the lognormal distribution a Pb-specific HC_{5-50} of 7.03 µg/L was generated.

Bioavailability models have been developed and validated for water chemistry ranges that are critically important for bioavailability normalisation (notably Ca and pH). It is important to note that different validation ranges apply to the different models (Table 9.3). Normalisations of toxicity data to target waters outside these pH and Ca water chemistry ranges should be interpreted with care.

However, the currently available chronic BLM software tools for bioavailability calculations are data-demanding (about 10 physicochemical input parameters are required) and often insufficiently user-friendly and time-consuming. To address these barriers, user-friendly tools were developed.

Currently, according to Rüdel et al. (2015), there are three readily available user-friendly tools:

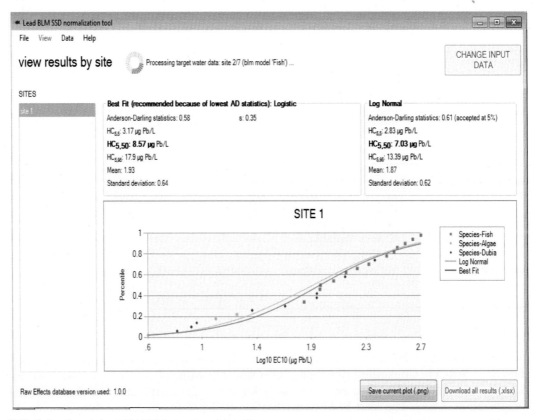

FIG. 9.6   Chronic lead (Pb) BLM output user interface.

TABLE 9.3   Validity Range for the Chronic BLMs

Metal	pH	Ca (mg/L)
Zn	6.0–8.0	5.0–160
Ni	6.5–8.7	2.0–88
Cu	6.0–8.5	3.1–93
Pb	6.3–8.4	3.6–204

- **Biomet** (current version 3.5)—is a lookup table-based tool in MS Excel spreadsheet originally developed by ARCHE Consulting/WCA-environment (www.bio-met.net). This tool calculates local site-specific EQS values for copper, nickel, zinc, and lead based on the chronic BLMs. Fig. 9.7 shows, for example, the copper HC_{5-50} calculation for two different waters (i.e. water 1 and 2) with the following characteristics: pH: 7.0; Ca: 10 mg/L and DOC: 1.0 mg/L for water 1 and pH: 7.0; Ca: 10 mg/L, and DOC: 1.0 mg/L for water 2. Biomet subsequently calculates local copper EQS values (i.e. HC_{5-50}) of 4.95 (water 1) and 10.0 µg/L (water 2), the bioavailability factor (i.e. BioF), the bioavailable

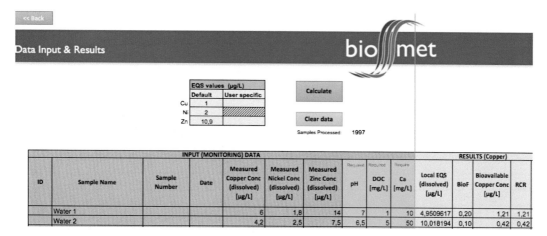

FIG. 9.7 Biomet input and results interface.

metal concentration and the risk characterisation ratios. The BioF is the reference EQS divided by the local (site-specific) EQS, the bioavailable metal concentration is provided by the product of dissolved metal concentration by the BioF and the risk characterisation ratios corresponding to the dissolved metal concentration divided by the local (site-specific) EQS or the bioavailable metal concentration divided by the reference EQS. To provide these outputs, the tool requests to input DOC, pH, and Ca concentrations.

- **M-BAT**—an algorithm-based tool in MS Excel developed by WCA-environment (UK) (http://www.wfduk.org/tagged/bioavailability-assessment-tool). This tool calculates local site-specific predicted no effects concentrations (PNEC) values for copper, nickel, zinc, and manganese based on the chronic BLMs. Fig. 9.8 shows the copper HC_{5-50} calculation for two different waters (i.e. water 1 and 2) with the following characteristics: pH: 6.8; Ca: 10 mg/L and DOC: 3.2 mg/L for water 1 and pH: 7.5; Ca: 52 mg/L, and DOC: 8.9 mg/L for water 2.

M-BAT subsequently calculates water site-specific copper PNEC values (i.e. HC_{5-50}) of 10.4 (water 1) and 40.7 μg/L (water 2), the bioavailability factor (i.e. BioF), the bioavailable

Metal Bioavailability Assessment Tool (M-BAT)

				INPUT DATA					RESULTS (Copper)			
ID	Location	Waterbody	Date	Measured Cu Concentration (dissolved) [μg l⁻¹]	pH	DOC	Ca	Site-specific PNEC Dissolved Copper [μg l⁻¹]	BioF	Bioavailable Copper Concentration [μg l⁻¹]	Risk Characterisation Ratio	
1		Water 1		5.2	6.8	3.2	10	10.39	0.10	0.50	0.50	
2		Water 2		3.6	7.5	8.9	52	40.23	0.02	0.09	0.09	

FIG. 9.8 M-BAT input and results interface.

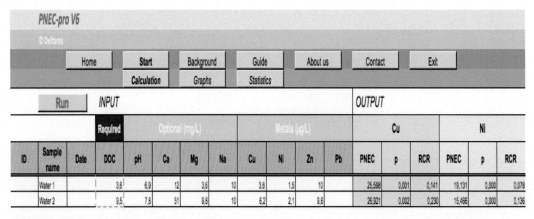

FIG. 9.9 PNEC.Pro input and results interface.

metal fraction and the risk characterisation ratios. This tool requests the input of DOC, pH, and Ca concentrations.

- **PNEC.Pro** (current version 6)—an algorithm-based tool in MS Excel developed by Deltares (The Netherlands) (www.pnec-pro.com). It calculates local, water type-specific no-effect concentrations (PNEC) of copper, nickel, zinc, and lead based on BLMs. As an example Fig. 9.9 shows the copper and nickel HC_{5-50} calculation for two different waters (i.e. water 1 and 2) with the following characteristics: pH: 6.9; Ca: 12 mg/L, and DOC: 3.6 mg/L for water 1 and pH: 7.8; Ca: 51 mg/L, and DOC: 9.5 mg/L for water 2.

PNEC.Pro subsequently calculates water site-specific PNEC values of 25.6 (copper)/19.1 (nickel) μg/L for water 1 and 26.9 (copper)/15.5 (nickel) μg/L for water 2, the bioavailability factor (i.e. BioF), and the risk characterisation ratios.

This tool requests the input of DOC only, while the input of pH/Ca/Mg/Na concentrations is optional.

For some metals, such as cadmium, no chronic BLMs are available. However, water hardness appears to be the major factor influencing Cd toxicity and therefore simple regression relating chronic Cd toxicity to hardness has been developed by US EPA (2016). Bioavailability correction for cadmium to derive the site-specific CCC in the United States relies on the following equation:

$$CCC = e^{(0.7977 \times \ln(\text{hardness}) - 3.897)} \tag{9.1}$$

More information on the cadmium aquatic life ambient water quality criteria is provided on https://www.epa.gov/sites/production/files/2016-03/documents/cadmium-final-report-2016.pdf. In the EU, hardness corrected PNEC values as follows:

$$\text{Site-specific PNEC} = 0.09 (\text{hardness} / 50)^{0.7409} \tag{9.2}$$

More on this approach can be retrieved from https://echa.europa.eu/documents/10162/4ea8883d-bd43-45fb-86a3-14fa6fa9e6f3.

9.5.2 Bioavailability in Soil

Metal toxicity not only depends on the total metal dose, but also on the time elapsed since contamination and physicochemical soil properties. Variation in soil properties can indeed result in a more than 10-fold variation in toxicity and therefore it is important that an ecological risk assessment is based on soil-specific quality standards (to avoid over- or under-protective scenarios).

Risk assessments for metals in the framework of the EU ESR and REACH Regulations have triggered research on bioavailability and effects of metals and metalloids on terrestrial organisms (plants, invertebrates, and microbial processes) for the last two decades. As a result, a wealth of ecotoxicity data was generated and correction factors were developed in order to account for (i) the differences in toxicity among soils due to varying soil properties (so-called normalisation models) and (ii) the difference in bioavailability and toxicity of metal(loid)s between laboratory and field conditions (so-called leaching-ageing factors). This has further lead to the development of bioavailability correction models for several metals, allowing the derivation of site-specific ecological soil quality standards based on scientific understanding of the behaviour and ecotoxicity of metals in soils (Smolders et al., 2009).

An excel spreadsheet, freely available from http://www.arche-consulting.be/en/our-tools/soil-pnec-calculator/calculates the predicted ecological risks of several metals (Cu, Ni, Zn, Pb, Cd, Mo, and Co) in soil, based on their PNECs to soil organisms as reported in the EU REACH datasets for these metals. An update of the spreadsheet and its extension to Ag is foreseen for 2017. All available data were compiled and implemented in this tool, so as to allow straightforward calculations of soil-specific ecological quality standards and resulting risk characterisation ratios for various metals and make this information available to potential users (industry or regulators).

The input parameters are dependent on the metal under consideration, and soil parameters are likely to be determined in routine soil analyses (metal background concentrations, pH, % organic matter, % clay, and eCEC).

Fig. 9.10 shows the example of copper PNEC calculation for a specific site with the following soil characteristics: pH: 7.0; eCEC: 12.04 cmol/kg; clay content: 10%, and organic carbon content: 2%. From these input data, a copper PNEC value of 52.8 mg/kg using the added risk approach and 71.1 mg/kg using the total risk approach is calculated.

9.6 TICKET-UWM: A TOOL TO ASSESS THE ENVIRONMENTAL FATE OF COMPLEX INORGANIC MATERIALS IN THE AQUATIC COMPARTMENT

9.6.1 Introduction to the Tool

Assessment of the fate of complex inorganic materials in surface waters requires consideration of many different processes, such as chemical and transport processes. Important chemical processes to consider include: kinetic reactions, such as degradation reactions (biological or abiotic) and transformation reactions (e.g. dissolution of metals from solids); phase changes, such as precipitation and condensation; and speciation reactions including complexation and adsorption (equilibrium or kinetic). Key transport processes include advection, dispersion/diffusion, settling, resuspension, burial of particulate (or particle-associated) forms of the

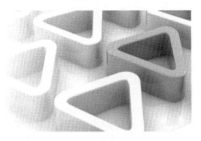

PNEC and Risk Characterisation Ratio (RCR) for:

Copper				Back To Input

Results for site specific information:
eCEC= 12.04 cmolc/kg, pH= 7, Org. C= 2 %, Clay= 10 %

	PNEC (mg/kg)	PEC (mg/kg)	RCR /	PAF %
ADDED approach				
PNECadded,site specific	52.8	75	1.42	11.5
TOTAL approach				
PNECtotal,site specific	71.1	100	1.41	12.0

FIG. 9.10 Soil PNEC calculator output user interface.

chemicals, and transient storage. For metal substances, the need to quantify bioavailability to quantify risk adds an additional layer of complexity to the assessment since it is closely linked to chemical processes taking place in surface waters.

While models do exist to quantify the above-described processes in great detail, there is a need for relatively simple, screening models. Ideally, these models would be capable of providing information on fate in prototypical scenarios, with varying levels of chemical and transport complexity, in a package useable by all stakeholders.

There is a family of screening-level models known as Unit World Models (UWMs). Initially developed for organic chemicals, the UWM approach evaluates the environmental exposure and fate of chemicals in the idealised environmental settings by partitioning them into various compartments (Mackay, 1979, 1991; Mackay and Paterson, 1991). The Tableau Input Coupled Kinetics Equilibrium Transport Unit World Model (TICKET-UWM) was developed originally as a screening-level model for organics and metals for a simplified one-dimensional, two-layer lake or impounded river (Farley et al., 2011).

Although created to consider idealised environmental settings (i.e. the unit-world), further developments of the TICKET-UWM also allow it to be used to assess risks on a site-specific basis or for a set of prototypical site types under various loading scenarios/time scales, while considering the key transport and chemical features described above.

It shall be noted that several 'unit world-type models' have been made available for metals (Bhavsar et al., 2004; Gandhi et al., 2011; Harvey et al., 2007). However, TICKET integrates state-of-the-art metal speciation and bioavailability models, cycling of carbon and sulphide, and is highly flexible owing to its tableau framework. One particular feature that sets it apart is its ability to perform simulations with metal forms that are sparingly soluble and kinetically release metal into the dissolved phase.

9.6.2 Model Description

The conceptual framework for TICKET-UWM is based on a well-mixed water column with an underlying oxic or anoxic sediment layer. In the model, metals can enter the water column via surface runoff, storm water flow, groundwater flow, atmospheric deposition, and/or direct discharges. The form of the metal released into the environment (e.g. soluble salt, metal powder, massive) is also considered, by specifying kinetic expressions to describe dissolution rates in the water column and underlying sediment layer.

Processes that are modelled include the following:

- Dissolved and particulate phase transport by flow in and out of the water column, diffusive exchange of dissolved metal between the water column and sediment, settling of particulate matter in the water column, and burial of solids in the deeper sediments.
- Metal complexation to natural organic matter (NOM)—both dissolved and particulate—and inorganic ligands using WHAM V, WHAM VI, or WHAM VII (Tipping, 1998; Tipping and Hurley, 1992; Tipping et al., 2011).
- Metal binding to iron and manganese hydroxides (HFO and HMO) using surface complexation sub-models (Dzombak and Morel, 1990; Tonkin et al., 2004).
- Precipitation of metal sulphides, hydroxides, and carbonates in the water column and sediment using the seed/solid solution approach.
- A description of biogeochemical cycling of carbon, sulphur, iron, and manganese in the water column and sediment.
- Competitive interactions of metals (i.e. multimetal) and major cations on the biological site of action for water and sediment dwelling organisms as described by BLM (Di Toro et al., 2001; Paquin et al., 2002; Santore et al., 2001).

These processes are illustrated in Fig. 9.11.

The TICKET-UWM algorithm was constructed as a general solver based upon the Tableau Input Coupled Kinetics Equilibrium Transport (TICKET) numerical engine (Farley et al., 2008). Chemical species, chemical equilibrium constants, and kinetic rate coefficients are stored in external databases and can be edited by the user.

The model is currently able to perform the following calculations:

- Steady-state response: Metal concentrations resulting from an annual average metal load.
- Time variable response: Concentrations of metals in the water column and sediment as a function of time.

Specific model applications include the following:

- Critical loads: Calculation of the allowable metal load which would not exceed a water quality standard or critical biotic ligand concentration.
- Removal time assessment: Calculation of metal removal from the water column as a function of time after an instantaneous release.
- BLM with metal precipitation: Calculation of metal concentrations on the biotic ligand using WHAM V, VI, or VII to describe metal-NOM partitioning and the MINTEQA2 database to describe precipitation of solids.

It can be found on http://www.unitworldmodel.net/.

9.6.3 Illustrative Simulation

9.6.3.1 *Input Details*

Farley et al. (2011) demonstrated use of TICKET-UWM to determine the long-term fate of metal powders released into a lake environment in a continuous load (e.g. in kg/year). The

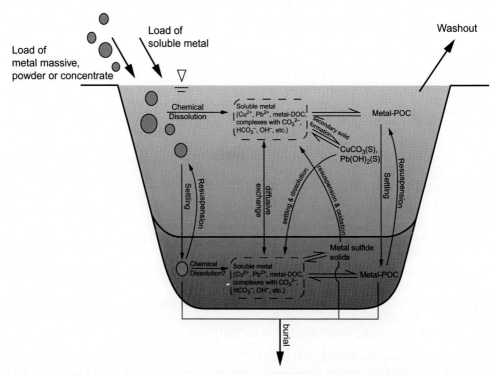

FIG. 9.11 Conceptual description of the processes affecting fate of metals during release of a metal concentrate and/or soluble metal to a lake. *Arrows* represent transport processes modelled by TICKET-UWM. *Dashed line* represents speciation calculations performed in the water column and sediment.

illustrative simulation discussed below examines short-term behaviour of a generic metal concentrate released into the water column of a lake as a single dose (i.e. a spike load). The generic concentrate is 15% copper (Cu) by mass and 10% lead (Pb) by mass. The dissolution of Cu and Pb from the concentrate, typically determined using transformation/dissolution protocols (Skeaff et al., 2006), are assumed to be first-order processes with rate constants of 0.005 per day and 0.01 per day, respectively. The settling velocity of the generic concentrate is estimated to be 50 m per day, calculated using Stokes' Law (average particle diameter = 25 μm, concentrate particle density = 3.38 g/cm^3, concentrate particles assumed to be spherical). The release of the concentrate to the lake is modelled by specifying an initial concentration of concentrate in the lake.

The illustrative simulation uses 'Shield Lake', a generalised lake in the Sudbury area of the Canadian Shield with an underlying anoxic sediment layer as described by Farley et al. (2011) and Mackay et al. (2001). The relevant transport and chemistry parameters for the simulation are illustrated in Table 9.4.

9.6.3.2 Results

WATER COLUMN

Following addition to the water column, the concentrate Cu and Pb concentrations decrease rapidly as the concentrate settles and enters the sediment (Fig. 9.12, green lines). Prior

TABLE 9.4 TICKET Unit World Model Input Parameters for Shield Lake (Farley et al. 2011; Mackay et al. 2001)

Parameter	Value	Parameter	Value
Simulation time, day	30	pH	7.10
Surface area of water column, km^2	20	Alkalinity, mg/L as CaCO$_3$	15.2
Depth of water column, m	10	Calcium, mg/L	6.9
Active sediment depth, cm	1	Magnesium, mg/L	2.1
Residence time of water column, year	2.28	Sodium, mg/L	16.3
Settling rate, m/day	1	Potassium, mg/L	0.7
Burial rate, cm/year	0.073	Sulphate, mg/L as SO$_4$	9.7
Resuspension rate, cm/year	0.073	Chloride, mg/L	21.9
Diffusive exchange, cm/day	0.96	Sediment solids conc., g/L	240
Suspended solids, mg/L	10	Porosity	0.9
POC, mg/L	0.6	Sediment f_{oc}	0.05
DOC, mg/L	2.7	AVS, $\mu mol/g_{dry}$	20
		Sediment pH, cations, and anions	Same as water column

to leaving the water column, the concentrate releases some soluble Cu and Pb which is referred to as 'solubilized Cu and Pb'. The solubilised metal is defined as the sum of total dissolved metal (dissolved free metal ions and aqueous inorganic and organic complexes) and particulate metal (precipitated and adsorbed). The release causes an initial rapid increase in solubilised Cu and Pb concentrations until a maximum concentration is achieved at approximately 1 day (black dashed lines in Fig. 9.12). Maximum solubilised Cu and Pb concentrations are less than 0.3 μg/L. Solubilised Cu and Pb concentrations then begin to decrease.

On the basis of TICKET-UWM speciation output, most of the solubilised Cu and Pb are bound to be dissolved and particulate organic carbon (POC); very little Cu or Pb is present in the free ionic or inorganically complexed forms (Fig. 9.12, pie charts). Therefore, the gradual removal of Cu and Pb from the water column is due to adsorption to POC in the water column and settling to the sediment.

SEDIMENT

In this illustrative simulation, it is assumed that the mechanism of Cu and Pb release from the concentrate is oxidative dissolution. Dissolution, therefore, would not be expected to occur in an anoxic sediment. Concentrate builds up in sediment as it settles through the water column (Fig. 9.12). Since dissolution does not occur in the sediment, the small increase in solubilised Cu and Pb in the sediment comes from transfer with the water column (settling of POC-sorbed metal and diffusive transport). Model speciation results indicate that essentially all solubilised Cu and Pb in the sediment precipitate as metal sulphide solids (Fig. 9.12, pie charts).

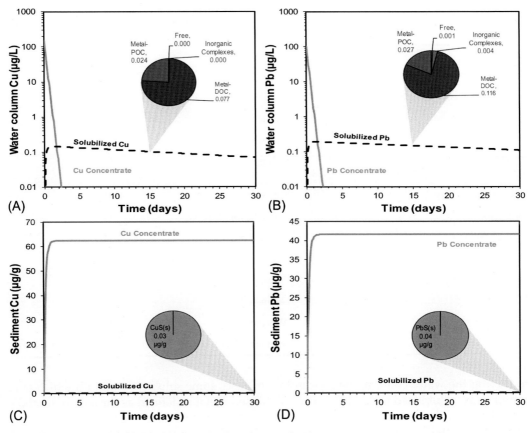

FIG. 9.12 (A, B) Water column concentration of concentrate Cu and Pb and solubilized Cu and Pb. (C, D) Sediment concentrations of concentrate Cu and Pb and solubilized Cu and Pb.

SENSITIVITY ANALYSES

The model can be used to conduct sensitivity analysis to assess the impact associated with uncertainty/variation in concentrate properties such as settling rate and dissolution rate. Fig. 9.13 shows the results of two analyses compared with the based case for Cu.

In the first sensitivity analysis scenario, the dissolution rate of the Cu in the concentrate is increased by a factor of 10. The concentrate settles rapidly from the water column as in the base case simulation, but more copper is able to dissolve during this time and consequently solubilised Cu reaches a maximum value of just above 1 µg/L.

In the second sensitivity analysis scenario, the size of the concentrate particles was lowered from 25 to 5 µm. As a result of the smaller particle size, the settling velocity decreases from 50 to 2 m/day. The lower settling velocity increases the concentrate residence time in the water column and more Cu is solubilised compared with the base case. A higher maximum Cu concentration is attained relative to the base case which occurs later in the 30-day simulation.

In the final sensitivity analysis, the Cu in the concentration is allowed to dissolve at the faster rate (0.05 per day) and dissolution is allowed to occur in the sediment. The water column Cu behaviour is essentially the same as in the first sensitivity analysis scenario (data not shown). The concentrate Cu rapidly increases in the sediment over the first day as it

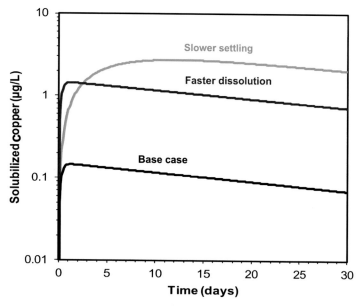

FIG. 9.13 Water column copper concentrations for sensitivity analysis simulations with the Cu dissolution rate increased from 0.005 to 0.05 per day *(blue lines)* and with the concentration settling rate decreased from 50 to 2 m/ day *(green line)*.

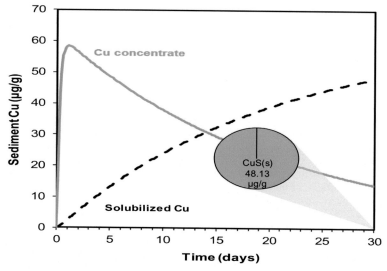

FIG. 9.14 Sediment copper concentrations for sensitivity analysis simulations with the Cu dissolution rate increased from 0.005 to 0.05 per day *(blue lines)* in the water column and sediment.

settles in the water column, reaches a maximum value, and then decreases steadily thereafter as it dissolves (Fig. 9.14, green line). Consequently, solubilised Cu in the sediment increases with time (Fig. 9.14, black dashed line), but all of the Cu is precipitated as a sulphide solid as a result of the initial sulphide specified in the sediment layer (Table 9.4).

9.6.4 Summary

This illustrative simulation and associated sensitivity analyses provide an indication of types of assessments that can be conducted with the TICKET-UWM. As indicated above, the water quality and transport parameters of the model can be set to represent either a prototypical surface water environment to match site-specific parameters of a particular system under consideration. The discussion above was focused on a complex inorganic material (i.e. metal concentrate). However, the mode framework is flexible and can also consider soluble metals and organic chemicals. Although not highlighted in the above discussion, this model framework can easily integrate metal bioavailability assessments into transport and fate simulations making TICKET-UWM a valuable risk assessment tool for complex inorganic materials such as metal concentrates and other sparingly soluble metal species.

KEY MESSAGES

Several IT tools have been developed to facilitate the evaluation of complex inorganic materials, with the aim to support the assessor during specific steps of the assessment.

The Metals Classification Tool (MeClas) will facilitate the hazard identification and classification of complex inorganic materials, working with several tiers and aimed at the progressive refinement of classification through the recognition of specific mineral content, speciation up to bioavailability corrections.

MEASE and SPERCs generate exposure estimates in the absence of reliable measured data. Tools such as Biomet, PNEC.Pro, M-BAT, and the TICKET-UWM have been developed and made available by the scientific, regulatory, and industry communities to consider metal specificities such as bioavailability in the environment and fate.

All these tools are easily accessible, based on current knowledge and are regularly updated to take into account new metal science and user-friendliness.

References

Bergman HL, Dorward-King EJ: *Reassessment of metals criteria for aquatic life protection*, Pensacola, FL, 1997, SETAC Press.

Bhavsar SP, Diamond ML, Evans LJ, Gandhi N, Nilsen J, Antunes P: Development of a coupled metal speciation-fate model for surface aquatic systems, *Environ Toxicol Chem* 23:1376–1385, 2004.

Di Toro DM, Allen HE, Bergman HL, Meyer JS, Paquin PR, Santore RC: Biotic ligand model of the acute toxicity of metals. 1. Technical basis, *Environ Toxicol Chem* 20:2383–2396, 2001.

Dzombak DA, Morel FMM: *Surface complexation modeling: hydrous ferric oxide*, New York, 1990, Wiley.

EC, 2008. Directive 2008/105/EC of the European Parliament and of the Council of 16 December 2008 on environmental quality standards in the field of water policy, amending and subsequently repealing Council Directives 82/176/EEC, 83/513/EEC, 84/156/EEC, 84/491/EEC, 86/280/EEC and amending Directive 2000/60/EC of the European Parliament and of the Council. Official Journal of the European Union, L 348/84, 24.12.2008.

EU CLP (Classification, labelling and packaging of substances and mixtures), EC Regulation No. 1272/2008, 2008.

ECETOC Targeted Risk Assessment (TRA)—worker exposure estimation—V 2.0, ECETOC 2009.

ECHA: *Guidance on information requirements and chemical safety assessment*, Chapter R.14: occupational exposure assessment, version 3.0, 2016, ISBN: 978-92-9495-081-9.

ECHA: *Guidance on information requirements and chemical safety assessment*, Chapter R.12: Use description, version 3.0 2015ISBN: 978-92-9247-685-4.

Fairbrother A, Wenstel R, Sappington K, Wood W: Framework for metals risk assessment, *Ecotoxicol Environ Safety* 68:145–227, 2007.

Farley KJ, Carbonaro RF, Fanelli CJ, Costanzo R, Rader KJ, Di Toro DM: TICKET-UWM: A coupled kinetic, equilibrium, and transport screening model for metals in lakes, *Environ Toxicol Chem* 30:1278–1287, 2011.

Farley KJ, Rader KJ, Miller BE: tableau input coupled kinetic equilibrium transport (TICKET) model, *Environ Sci Technol* 42:838–844, 2008.

Fransman W, Schinkel J, Meijster T, Van Hemmen J, Tielemans E, Goede H: Development and evaluation of an exposure control efficacy library (ECEL), *Ann Occup Hyg* 52(7):567–575, 2008.

Gandhi N, Bhavsar SP, Diamond ML: Critical load analysis in hazard assessment of metals using a unit world model, *Environ Toxicol Chem* 30:2157–2166, 2011.

Harvey C, Mackay D, Webster E: Can the unit world model concept be applied to hazard assessment of both organic chemicals and metal ions? *Environ Toxicol Chem* 26:2129–2142, 2007.

HERAG: *(Health Risk Assessment Guidance) Fact Sheet: Assessment of occupational dermal exposure and chemical absorption for metals and inorganic metal compounds*, http://www.ebrc.de/downloads/HERAG_FS_01_August_07.pdf, 2007.

HSE: *EASE for Windows 2.0: A system for the estimation and assessment of substance exposure (EASE)*, Version 2.0, developed by the health and safety executive (HSE), UK in conjunction with AIAI, Edinburgh, 1997.

Janssen C, De Schamphelaere K, Heijerick D, et al: Uncertainties in the environmental risk assessment of metals, *Hum Ecol Risk Assess* 6:1003–1018, 2000.

Lamb J, Hesse S, Miller BG, et al: *Evaluation of tier 1 exposure assessment models under REACH (ETEAM) project—final overall project summary report. Endbericht zur Evaluierung von Tier 1-Modellen, 1, Auflage, Dortmund: Bundesanstalt für Arbeitsschutz und Arbeitsmedizin 2015*, 2015, ISBN: 978-3-88261-160-1. Projektnummer: F 2303.

Mackay D: Finding fugacity feasible, *Environ Sci Technol* 13:1218–1223, 1979.

Mackay D: *Multimedia environmental models: the fugacity approach*, Chelsea, MI, 1991, Lewis Publishers.

Mackay D, Paterson S: Evaluating the multimedia fate of organic chemicals: a level III fugacity model, *Environ Sci Technol* 25:427–436, 1991.

Mackay D, Sharpe S, Cahill T, Gouin T, Cousins I, Toose L: *Assessing the environmental persistence of a variety of chemical substances including metals*, CEMC Report 200104, Peterborough, ON, 2001, Canadian Environmental Modelling Centre, Trent University.

Paquin PR, Gorsuch JW, Apte S, et al: The biotic ligand model: a historical overview, *Comparative Biochemistry and Physiology Part C: Toxicology & Pharmacology* 133:3–35, 2002.

Rüdel H, Díaz Muñiz C, Garelick H, et al: Consideration of the bioavailability of metal/metalloid species in freshwaters: experiences regarding the implementation of biotic ligand model-based approaches in risk assessment frameworks, *Environ Sci Pollut Res Int* 22:7405–7421, 2015.

Santore RC, Di Toro DM, Paquin PR, Allen HE, Meyer JS: Biotic ligand model of the acute toxicity of metals. Application to acute copper toxicity in freshwater fish and Daphnia, *Environ Toxicol Chem* 20:2397–2402, 2001.

Sättler D, Schnöder F, Aust N, et al: Specific environmental release categories—a tool for improving chemical safety assessment in the EC—report of a multi-stakeholder workshop, *Integr Environ Assess Manag* 8:580–585, 2012.

Skeaff JM, Ruymen V, Hardy DJ, et al: The standard operating procedure for the transformation/dissolution of metals and sparingly soluble metal compounds, Natural Resources Canada, CANMET-MMSL Division Reports MMSL 06-085 (TR). 555. CANMET-MMSL, Booth St., Ottawa, Canada. In *K1A OG1*, 2006.

Smolders E, Oorts K, Van Sprang P, et al: Toxicity of trace metals in soil as affected by soil type and ageing after contamination: using calibrated bioavailability models to set ecological soil standards, *Environment Toxicol Chemistry* 28(8):1633–1642, 2009.

Tipping E: Humic ion-binding model VI: an improved description of the interactions of protons and metal ions with humic substances, *Aquat Geochem* 4:3–47, 1998.

Tipping E, Hurley MA: A unifying model of cation binding by humic substances, *Geochim Cosmochim Acta* 56:3627–3641, 1992.

Tipping E, Lofts S, Sonke JE: humic ion-binding model VII: a revised parameterization of cation-binding by humic substances, *Environmental Chemistry* 8:225–235, 2011.

Tonkin JW, Balistrieri LS, Murray JW: Modeling sorption of divalent metal cations on hydrous manganese oxide using the diffuse double layer model, *Appl Geochem* 19:29–53, 2004.

Verdonck F, Van Assche F, Hicks K, Mertens J, Voigt A, Verougstraete V: Development of realistic environmental release factors based on measured data: approach and lessons from the EU metal industry, *Integrated Environmental Assessment and Management* 10(4):529–538, 2014.

Verdonck F, Waeterschoot H, Van Sprang P, et al: MeClas: an online tool for hazard identification and classification of complex inorganic metal-containing materials, *submitted*, 2017.

Vetter D, Schade J, Lippert K: Guidance on the assessment of occupational exposure to metals based on monitoring data, Final Report, 2016.

UN GHS (Globally harmonized system of classification and labelling of chemicals) Sixth revised edition, ST/SG/AC.10/30/Rev.6. United Nations, ST/SG/AC.10/30/Rev.6, 2015, New York and Geneva, https://www.unece.org/fileadmin/DAM/trans/danger/publi/ghs/ghs_rev06/English/ST-SG-AC10-30-Rev6e.pdf.

US EPA: *Aquatic life ambient freshwater quality criteria—copper*, EPA-822-R-07-001. Washington, DC, USA, 2007, United States Environmental Protection Agency (US EPA), Office of Science and Technology.

US Environmental Protection Agency (EPA): *Aquatic life ambient freshwater quality criteria—cadmium*, EPA-820-R-16-002 Washington, DC, 2012, Office of Water.

US EPA: *Aquatic life ambient freshwater quality criteria—cadmium*, EPA-820-R-16-002. Washington, DC, USA, 2016, United States Environmental Protection Agency (US EPA), Office of Science and Technology.

Further Reading

US OSHA: *HA hazard communication standard (HAZCOM) 1910.1200 FR 77: 17574*, http://www.ishn.com/articles/105552-osha-hazard-communication-standard-hazcom-19101200, 2012.

Hazard Assessment of Ores and Concentrates

Johannes A. Drielsma*, Katrien Delbeke[†], Patricio H. Rodriguez[‡], José Jaime Arbildua[‡], Frank Van Assche[§]

*European Association of Mining Industries, Metal Ores and Industrial Minerals, Brussels, Belgium [†]European Copper Institute, Brussels, Belgium [‡]Universidad Adolfo Ibanez, Santiago, Chile [§]International Zinc Association, Brussels, Belgium

10.1 INTRODUCTION

Ores and concentrates are naturally occurring substances of variable composition. Ores are composed of one or more metal-bearing minerals, of sufficient quantity and quality to be mined for profit. Once mined, ores are most often physically processed to remove minerals of no economic interest (gangue minerals), allowing value minerals to accumulate within a concentrate. This allows for more efficient transport of the valuable minerals to market. Metal concentrates are therefore routinely stored, loaded, transported long distances, unloaded and finally used for the production of metals, metal compounds, and alloys.

Human health and environmental risks during mining, processing, storage, and transport are mostly regulated through national or provincial regulations. The EU REACH Regulation as entered into force in 2007 and the EU CLP (2008) consider metal ores and concentrates as naturally occurring substances requiring hazard classification, but exempt from registration. This is justified by low levels of exposure together with the need to keep regulatory frameworks proportionate and workable (European Parliament, 2006). Similarly, most international guidelines and jurisdictions relevant to handling, use and transport of ores and concentrates are, as outlined below, focussing on hazard identification/classification and resulting risk reduction.

United Nations and OECD guidelines on harmonised hazard identification, classification criteria, and testing procedures are complemented by various hazard communication tools, such as the Globally Harmonised System of Classification and Labelling of Chemicals (UN GHS, 2015), the EU Classification and Labelling Regulations (EU CLP, 2008), and the UN Model Regulations on the Transport of Dangerous Goods (DGs).

Risk Management of Complex Inorganic Materials
https://doi.org/10.1016/B978-0-12-811063-8.00010-0

The UN, through its committees and regional economic commissions (United Nations Economic Commission for Africa, United Nations Economic Commission for Europe, United Nations Economic Commission for Latin America, United Nations Economic and Social Commission for Asia and the Pacific), administers a set of agreements that aim at effective implementation of these mechanisms as far as transport of DGs by road, rail, and inland waterways are concerned [e.g., the European Agreement concerning the International Carriage of Dangerous Goods by Road (ADR, 2016), the European Regulations concerning the International Transport of Dangerous Goods by Rail (RID, 1996), and the European Agreement concerning the International Carriage of Dangerous Goods by Inland Waterways (ADN, 2012), respectively]. These agreements further inspire regional/national regulations of transport, disaster prevention, and emergency response (e.g., EU transport regulations and the EU Seveso Directive).

International rules for marine transport of goods are agreed and administered by the International Maritime Organisation (IMO). Environmental risk mitigation measures are foreseen for substances that are 'Harmful to the Marine Environment (HME)' under the International Convention for the Prevention of Pollution from Ships (MARPOL) Annex V. Human health safety measures for DGs and Materials Hazardous only in Bulk (MHB) are foreseen by the International Convention on Safety of Life at Sea (SOLAS) in the International Maritime Solid Bulk Cargoes (IMSBC) Code and the International Maritime Dangerous Goods Code (IMDG) (International Maritime Dangerous Goods Code, 2016) Code. The HME criteria (IMO, 2012) and MHB criteria (IMSBC, 2013) are based on the UN GHS criteria.

Tables 10.1 and 10.2 summarise the UN GHS environmental and human health classification criteria that lead to the classifications of DGs and require various risk reduction measures for packaged and bulk cargoes during storage and transport. It can be seen that substances that meet the criteria for environmental classification under UN GHS as Aquatic Acute 1 and Aquatic Chronics 1 and 2 are considered as DGs (Marine Pollutants class 9—miscellaneous dangerous substances) under the UN Maritime Transport Regulations. Substances that meet the UN GHS criteria for human health hazard classification as acute toxicity (Categories 1, 2, and 3) are considered as dangerous goods (DG class 6—Toxic) under the UN Maritime Transport Regulations.

TABLE 10.1   Environmental Hazard Classification Categories in Various International Systems

Environment Hazard endpoint	UN GHS	UN TDG (EU ADR/ADN/RID)	IMO-IMDG	IMO-MARPOL Annex V	EU SEVESO[a]
Aquatic acute	Category 1	DG Class 9	DG Class 9	HME	E1
	Category 2	–	–	–	–
	Category 3	–	–	–	–
Aquatic chronic	Category 1	DG Class 9	DG Class 9	HME	E1
	Category 2	DG Class 9	DG Class 9	HME	E2
	Category 3	–	–	–	–

[a]*Risk management measures depend on the tonnages handled.*
ADR, International Carriage of Dangerous Goods by Road; ADN, European Agreement concerning the International Carriage of Dangerous Goods by Inland Waterways; IMO, International Maritime Organisation; IMDG, International Maritime Dangerous Goods Code; MARPOL, International Convention for the Prevention of Pollution from Ships; RID, European Regulations concerning the International Transport of Dangerous Goods by Rail; UN GHS, United Nations Globally Harmonised System of Classifications and Labelling of chemicals; UN TDG, United Nations Transport of Dangerous Goods.

TABLE 10.2 Human Health Hazard Classifications for a Variety of Regulatory Scenarios

Human Health Hazard Endpoint	UN GHS	UN TDG (ADR/ ADN/ RID)	IMO-IMDG	IMO-IMSBC	IMO-Marpol Annex V	EU SEVESO[a]
Acute oral	Category 1	DG Class 6.1	DG Class 6.1	Group B class 6.1	–	H1
	Category 2	DG Class 6.1	DG Class 6.1	Group B class 6.1	–	H2
	Category 3	DG Class 6.1	DG Class 6.1	Group B class 6.1	–	–
	Category 4	–	–	–	–	–
Acute dermal	Category 1	DG Class 6.1	DG Class 6.1	Group B class 6.1	–	H1
	Category 2	DG Class 6.1	DG Class 6.1	Group B class 6.1	–	H2
	Category 3	DG Class 6.1	DG Class 6.1	Group B class 6.1	–	–
	Category 4	–	–	Group B—MHB	–	–
Acute inhalation	Category 1	DG Class 6.1	DG Class 6.1	Group B class 6.1	–	H1
	Category 2	DG Class 6.1	DG Class 6.1	Group B class 6.1	–	H2
	Category 3	DG Class 6.1	DG Class 6.1	Group B class 6.1	–	H2
	Category 4	–	–	Group B—MHB[b]	–	–
Skin corrosion /irritation	Category 1	–	DG Class 8	Group B—Class 8	–	–
	Category 2	–	–	Group B—MHB	–	–
Serious eye Damage	Category 1	–	–	Group B—MHB	–	–
	Category 2A	–	–	Group B—MHB	–	–
Respiratory sensitisation	Category 1 (A,B)	–	–	Group B—MHB	–	–
Mutagenicity	Category 1 (A,B)	–	–	Group B—MHB[c]	HME[d]	–
	Category 2	–	–	–	–	–
Carcinogenicity	Category 1 (A,B)	–	–	Group B—MHB[c]	HME[d]	–
	Category 2	–	–	–	–	–
Reproductive	Category 1 (A,B)	–	–	Group B—MHB[c]	HME[d]	–
Toxicity	Category 2	–	–	–	–	–
STOT—single	Category 1	–	–	Group B—MHB[c]	HME[d]	H3
exposure	Category 2	–	–	–	–	–
STOT-repeated	Category 1	–	–	Group B—MHB[c]	HME[d]	–
exposure	Category 2	–	–	–	–	–

[a]Risk management measures depend on the tonnages handled
[b]Effects due to cargo dust and/or effects due to toxic gases that are formed when the cargo is wet.
[c]Inhalation and dermal exposure routes
[d]Product is considered HME if it also meets the criterion of 'high bioaccumulation' and is 'not rapidly degradable'
Classification for oral and dermal routes or without specification of the exposure route in the hazard statement: *ADR*, International Carriage of Dangerous Goods by Road; *ADN*, European Agreement concerning the International Carriage of Dangerous Goods by Inland Waterways; *IMO*, International Maritime Organisation; *IMDG*, International Maritime Dangerous Goods Code; *MARPOL*: International Convention for the Prevention of Pollution from Ships; *RID*, European Regulations concerning the International Transport of Dangerous Goods by Rail; *UN GHS*, United Nations Globally Harmonised System of Classifications and Labelling of chemicals; *UN TDG*, United Nations Transport of Dangerous Goods; Risk management measures depend on the tonnages handled.

It follows that international regulations on ores and concentrates are largely hazard driven. This chapter will therefore focus on hazard assessment. Further assessment aspects are discussed in Chapter 7.

10.2 SUBSTANCE IDENTITY

A correct identification of the ore or concentrate is a first and essential step for its hazard identification and classification (see Chapter 7 for a worked example). Ores and concentrates are more complex and usually less well-defined than most substances, making standard substance identification for hazard classification purposes difficult.

Largely in response to European requirements, the mining industry has globally adopted a three-step approach determining how to best identify ores and concentrates for the purposes of hazard identification and classification, posing the following questions about the substance under study:

1. Can its composition be fully defined with relevant and accessible data? (Answer: No. Their composition is to some extent unknown and variable.)
2. Is one constituent substance present in concentrations >80% w/w? (Answer: No. As per European requirements, ores and concentrates are multi-constituent substances with no single constituent exceeding 80% w/w)
3. Is the material the result of a chemical process and has its chemical structure changed as a result? (Answer: No. As per European requirements, ores and concentrates are naturally occurring substances resulting from processing by mechanical means or flotation.)

To assist with answering the third question, the most typical beneficiation processes can be distinguished according to whether removal of gangue minerals also entails a chemical modification of the retained minerals or not. Mineral processing by and large involves methods that exploit differences in the physical properties of naturally occurring minerals, for example, by sorting, magnetic separation, electrostatic separation, preferential crushing, grinding and milling, sieving and screening, hydrocycloning, filtration, and flotation. In these cases, the retained minerals are still the same naturally occurring minerals as were originally mined (ECHA, 2012, 2017a).

Following European conventions consistent with UN GHS, identifiers of substances for which the answer to all three questions is 'no', include generic information on their source, the most relevant process steps used to obtain them and the typical concentrations and concentration ranges of their known and classified constituents. With regard to the latter, two types of data can be available: elemental composition and/or mineral composition/speciation. For hazard assessment of naturally occurring sparingly soluble inorganics, mineral speciation is more relevant but chemical speciation data is often a cost-effective proxy and therefore more readily available.

A generic substance identity will describe the fullest range of known elemental constituents as well as the predominant minerals present.

If the answer to any of the above questions is 'yes', additional information may be required to properly identify the substance in a distinctive way. By this convention, for example, complex inorganic materials that have resulted from sintering, ion-exchange, solvent extraction, electrowinning, pressure digestion in acid or alkali solutions, roasting, smelting, calcination with release of carbon dioxide, precipitation, or gas precipitation would be more appropriately identified as a different kind of intermediate for further manufacturing of metallic substances (i.e., no longer an ore concentrate) to which additional or different distinctive identifiers may need to be applied (see also Chapter 11).

10.2.1 Substance Identity—General Approach to Characterisation

From the above, it follows that accurate characterisation of ores and concentrates, and existing data relevant to each UN GHS endpoint is required to initiate the identification and classification process. Accurate and precise analytical characterisation of ores and concentrates is critical to ensure accurate hazard classification for health and environmental endpoints. The key information to predict any liberation of an element from geological material is the mineralogical composition of the sample. However, from an analytical point of view, this information is difficult to obtain fully. Traditional techniques such as optical microscopy and X-ray diffraction (XRD) have strong limitations for quantification and indirect quantification methods are expensive and not always reliable (CEN 2012, Dold, 2016). Characterisation therefore requires collection of the best available analytical information on the (in situ) ore and the concentrate, as well as best professional judgement on the part of experienced geologists, geochemists, and mineralogists familiar with both the ore body and concentration process.

Specific elemental and speciation/mineralogical data help to ensure an accurate and realistic hazard identification and classification of ores and concentrates. Key properties that may drive classifications are the concentration, form (e.g., valence and particle size), and solubility of the minerals present, for example, as a carbonate mineral, an oxide mineral, a sulphide mineral, etc. Most minerals have discrete CAS numbers and, for certain endpoints, have existing toxicological data searchable under that CAS number. Constituents may be materials that are tightly complexed in a natural mineral matrix, only slightly soluble, and essentially nonbioavailable. General information on the properties of the mineral (stability, solubility, pH, particle size) can therefore often complement the toxicological data by dimensioning the likelihood of breakdown to free metal ions or exposure through various routes. Finally, this approach underscores that the constituents of ores and concentrates are naturally occurring minerals extracted from the ground. Recognition of their natural origins—typically associated with long-term weathering ('ageing') processes that maintain the majority of soluble inorganic components of rocks and soils immobilised in natural sinks—helps to explain why they are, by definition, in a stable chemical form and therefore less hazardous than a mixture of pure metal salts.

More recent advances in analytical chemistry (e.g., Microprobe assays and advanced X-ray crystallography, SEM-EDX, Mineral Liberation Analysis or QEMSCAN) have made it possible to define more precisely the structure and composition of ores and concentrates in mineralogical terms. Many modern mining companies now use these techniques for geometallurgical purposes and mineral speciation data can be expected to become more readily available in the future for more critical business purposes that would justify the expense (Dold 2016). Compositions that are defined mineralogically provide toxicologists with more relevant data with which to overrule the default application of total elemental concentrations and thus refine conservative first tier classifications.

10.2.2 Case Example on Copper Concentrates

In response to new reporting requirements in the EU, a database covering most copper concentrates transported globally, and containing both chemical and mineralogical speciation data, was compiled from 2010 onwards. The detailed elemental and mineral characterisation of more than 100 copper concentrate samples, collected world-wide, generated a chemical substance identity for copper concentrates. Table 10.3 demonstrates the presence of copper, typically around 30% and the co-occurrence of various minor metals. The mineralogical speciation data indicated that the dominant minerals (median values) are chalcopyrite (CuS 63%), pyrite (FeS 11%), and quartz (SiO_2 2%). Fig. 10.1 further demonstrates the dominance

TABLE 10.3   Elemental Composition of Worldwide Copper Concentrates

N=119	Cu	Sb	As	Zn	Pb	Ni	Ag	Cd	Co
Min	14.000	0.000	0.000	0.000	0.000	0.000	0.000	0.000	0.000
p50%	26.670	0.010	0.110	0.620	0.140	0.002	0.006	0.004	0.005
p60%	27.570	0.015	0.139	1.307	0.266	0.004	0.008	0.006	0.009
p70%	28.452	0.022	0.180	2.872	0.562	0.008	0.011	0.010	0.013
p80%	29.958	0.042	0.272	3.652	1.478	0.010	0.017	0.014	0.024
p90%	34.000	0.102	0.410	5.632	2.910	0.024	0.068	0.026	0.040
Max	51.050	7.250	7.500	9.280	12.710	1.030	1.907	0.072	0.250

Min, minimum; *Max*, maximum; *pXX%*, percentile.

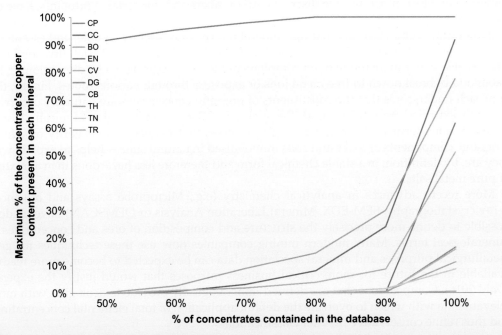

FIG. 10.1   The copper-bearing minerals present in copper concentrates (N=112). *CP, chalcopyrite; CC, chalcocite; BO, bornite; EN, enargite; CV, covellite; DG, digenite; CB, cubanite; TH, tetrahedrite; TN, tennantite; TR, tenorite.*

of chalcopyrite (CP) in copper sourcing globally, as well as the presence of chalcocite (CC), bornite, and covellite in the majority of transported copper concentrates at that time.

Particle size distributions, measured in a range of concentrates, usually by laser diffraction, indicated typical particles sizes <100 μm with some respirable fraction (<10 μm) (see Table 10.4).

The relative densities of copper concentrates typically range between 3.5 and $4.7 \text{ g}/\text{cm}^3$ at 20°C.

The elemental composition was analysed using several techniques and methodologies including, inductively coupled plasma atomic emission spectroscopy (ICP-AES), or inductively coupled plasma mass spectrometry (ICP-MS) after total dissolution. The amount of silicon oxide, SiO_2, was analysed colourimetrically. Sulphur and carbon contents were determined in automatic analysers (i.e., ELTRA CS2000) and the amount of sulphate, SO_4, by ion chromatography.

The mineralogy was determined using methodologies of scanning electron microscopy with energy dispersive X-ray spectrometer (SEM/EDS) and XRD, also using integrated methodologies such as QEMSCAN (Quantitative Evaluation of Minerals by SCANning electron microscopy) (Goldstein et al., 1992; Lipson and Steeple, 1970).

The comparison demonstrates a good correlation between copper content from elemental and mineral speciation analysis (Fig. 10.2).

Criteria for data acceptance considered as generally relevant are presented in Table 10.5.

TABLE 10.4 Summary of the Particle Size Distribution Measured in 23 Copper Concentrates

Particle Size Distribution	Range	10th Percentile	50th Percentile	90th Percentile
D50—μm	6–80	16	27	55
<10 μm—volume based %	5–71	18	24	33

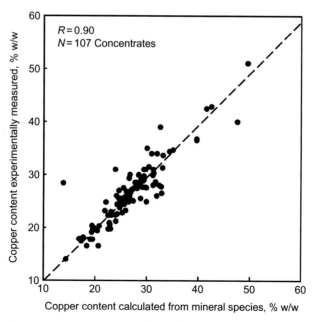

FIG. 10.2 Correlation between copper content experimentally measured and that calculated from copper mineral species.

TABLE 10.5   Criteria of Acceptance of the Data

Input Data	Control Parameter	Acceptance Criterion
Mineralogy	Total %	>95
Elemental	Total %	>90
Elemental measured/ calculated from mineralogy	% difference	95–105 if elemental >5%
Quartz	%	Reported
Particle size with respirable fraction < 10 μm	%	Reported
Density	kg/dm^3	Reported

10.3 ENVIRONMENTAL HAZARD IDENTIFICATION— CLASSIFICATION (UN GHS/EU CLP)

Given the natural variability of ores and concentrates even from the same ore body, full hazard profiles of ores and concentrates as a whole are not available. Classifications are therefore based on grouping and read-across (bridging) to ecotoxicological data for one or more 'source compounds' (in this case a representative source ore or concentrate) with similar physical/chemical properties and a known ecotoxicological profile. In this way, environmental classification can be based on data available for a representative ore or concentrate (e.g., another ore or concentrate that is used to produce the same commodity). The UN GHS offers guidance on the criteria for selecting or deselecting 'source compounds' in case of bridging.

Ores and concentrates, if released into the environment, will release a combination of metal ions based on the interaction between and solubility of the components contained. A range of elements occurs naturally in ores and ore concentrates, which depending on their mineral speciation have the potential to cause ecological risk (e.g., cadmium, copper, lead, mercury, thorium, thallium, selenium, zinc). During the transportation, handling, or storage of ores and concentrates with hazardous properties, environmental hazards may arise from released metal ions. Examples include propensity to release hazardous constituents as or within gases, dust, or aqueous solutions.

As such, the potential hazard of the resulting metal mixture needs to be assessed, even though the source material is considered a single substance. Guidance on the application of the EU CLP criteria (ECHA, 2017a,b) recognises that ores and concentrates are not simple mixtures of metals or metal compounds and that their solubility properties can differ substantially from what is observed for each individual constituent metal. The EU CLP guidance therefore recommends measuring the soluble fraction of each metal constituent in the ore or concentrate, according to the OECD's Transformation/Dissolution (TD) protocol (Skeaff et al., 2006, 2011; OECD 2001; UN GHS 2011). Finally, each soluble metal fraction is compared with the corresponding ecotoxicity reference value (ERV) of the soluble metal ion, for 7 days (acute endpoints) and 28 days (chronic endpoints) (see also Chapter 7). Based on these comparisons, the UN GHS and EU CLP criteria for acute and chronic hazard classification are applied.

Another relevant parameter for chronic environmental hazard classification is persistence. If a substance persists in the environment it is considered more hazardous than rapidly degradable substances. In such a case, a more severe hazard classification should therefore result (e.g., see Table 4.1.1. in UN GHS, 2015). Historically, degradability criteria were developed for organic substances and they do not apply to metals and metal-bearing minerals (Annex 9, A9.7.3.1 to UN GHS, 2015). It has been shown that released metal ions, being the toxic form of the metal, are removed from the water column by various physicochemical processes. For instance, some metals form compounds that precipitate out of solution (iron, aluminium, antimony, tin, and chromium), while others (copper, zinc, nickel, and lead) bind to acid volatile sulphides in aquatic sediments. This evidence of potentially rapid loss from the water column has been obtained from kinetic analysis of metals introduced in the water column and evaluation of sediment mineralisation and remobilisation of metal ions (Rader 2013a,b). This information combined with a modified TD protocol, field/mesocosm data) in a weight of evidence approach suggests that several metals fulfil the rapid removal criteria (equivalent to rapidly degradable for organics) and discussions are ongoing on the integration of rapid removal in UN GHS.

The bioaccumulation potential of a substance, measured as bioaccumulation factor (BAF) is also a general criteria used for hazard identification. An exhaustive analysis performed in European risk assessments for metals such as copper, zinc, and nickel concluded that these metals are not bioaccumulative (Adams et al., 2003; McGeer et al., 2003; ECI, 2008; JRC, 2010).

In the following sections, a methodology for the environmental hazard classification of ores and concentrates is presented, using chemical and mineralogical speciation data, together with TD results following the tiered approach shown in Fig. 10.3.

10.3.1 TD Test Protocols

Substance identity information should be used to identify the representative materials that will undergo T/D testing following the procedures described in Annex 10 of the GHS (UN GHS, 2015; OECD, 2008; OECD 2001) and Chapter 7.

The rate and extent of metal ion release from ores and concentrates to the aquatic media is evaluated through TD testing (see Chapter 7 for a description of the technique), with tests run for 7 days (acute endpoints) and 28 days (chronic endpoints). To strengthen and validate the grouping/read-across approach being taken, relevant pure minerals, ground to a similar particle size (see copper example) may be tested in parallel using the same protocol.

10.3.2 TD Tests Results

10.3.2.1 Zinc Example

Zinc concentrates are dominated by the mineral sphalerite (ZnS based; sulphidic, 35%–58%), which explains why they are often called zinc sphalerites. A total of 7- and 28-day TD results from testing 20 representative zinc concentrates were therefore obtained (Table 10.6). The environmental solubility of the other metals present in the concentrates (copper, arsenic, cadmium, nickel, lead), is characterised by large variations (see also Table 7.1 in Chapter 7). For all metals, conservative environmental solubility values were retained to define the hazard classification profiles of different zinc concentrates. Conservative values were defined as the 90% percentile of the measured ionic solubility from this set of 20 zinc concentrate results (see Chapter 7, Section 7.5.1).

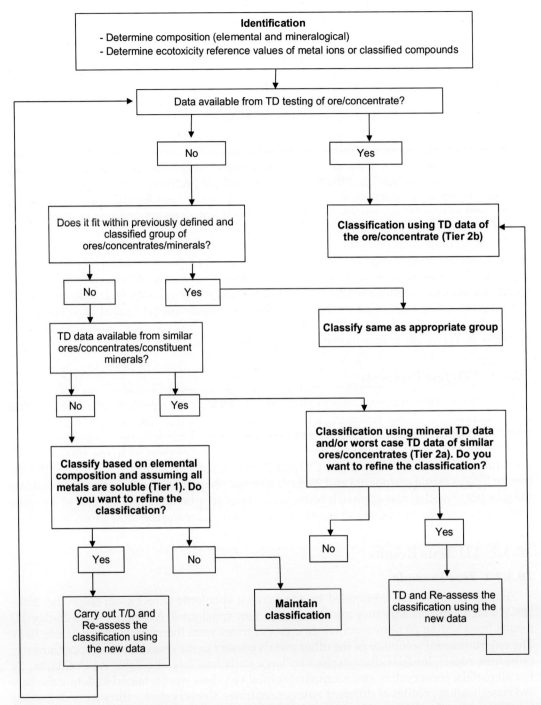

FIG. 10.3 General approach to assess the environmental classification of metal ores and concentrates (TD: Transformation/Dissolution).

TABLE 10.6 Acute and Chronic Environmental Solubility (%) for Cu, As, Cd, Pb, and Zn From Transformation/Dissolution (TD) Results of Representative Zinc Concentrates

| | % Environmental Solubility at 1 mg/L | |
| | 7-days | 28-days |
Metal	Percentile 90%	Percentile 90%
As	5.37	6.55
Cd	8.54	13.44
Cu	1.89	2.53
Pb	48.55	65.45
Zn	6.32	13.85

Metal releases from 20 zinc concentrates during short-term (7-days) and long-term (28-days) transformation/dissolution test results are expressed as: (μg metal ions released/μg total metal) × 100.

10.3.2.2 Copper Example

The results from TD tests, carried out on pure copper minerals (chalcopyrite, arsenopyrite, chalcocite, digenite, bornite, covellite, enargite, and tennantite) showed a broad range of copper releases. All TD time series for the copper-bearing minerals (except digenite) displayed a good fit with a second-order release model, indicating the participation of at least two metal species in copper dissolution. Results for digenite fit a first-order linear release model (not shown). A comparison between time series of the least soluble mineral (CP) and the most soluble one (CC) is shown in Fig. 10.4.

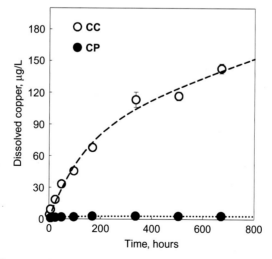

FIG. 10.4 Time course of copper release from pure ground minerals. Hollow symbols correspond to copper release from chalcocite (CC) mineral; filled symbols correspond to copper release from chalcopyrite (CP) mineral. *Dashed* and *dotted lines* correspond to the fit of a second order model to the experimental data.

TD testing of pure minerals enables calculation of the mineral-specific environmental solubility of copper after 7 and 28 days (Table 10.7).

As might be expected, the data show important differences in copper solubility across specific minerals. The environmental solubility percentage of copper from CP is about 10–20 times lower than that from CC, depending on the duration of the test. The TD results from testing of 12 representative copper concentrates with varying mineral composition [2%–80% chalcopyrite (CP); below detection limit (BDL) to 9.7% covellite (CV); BDL to 42.1% bornite (BN); BDL to 0.66% enargite (EN); BDL to 1.5% tennantite (TN); and BDL to 8.61% chalcocite (CC)] and elemental composition also resulted in a wide range of results: 0.5% and 7.5% copper release after 7 days and from 0.7% to 13.7% after 28 days (Table 10.7). Again, the observed range in the environmental solubility of copper from copper concentrates reflects the heterogeneity of their mineral composition. The mineral-specific results can be used for grouping/read-across as illustrated in Section 10.3.4. As in zinc concentrates, the environmental solubility of the other metals present in the copper concentrates (arsenic, cadmium, cobalt, nickel, lead, zinc), is characterised by large variations (Table 10.8).

For these metals, conservative environmental solubility values are retained to define the metal release of other copper concentrates, with a known elemental composition.

$$\sum_i ES_i^* [metal]_i = [metal_i - dissolved - Concentrate]_j \qquad (10.1)$$

where ES_i=environmental solubility (%) of the metal i after 7 or 28 days (from Table 10.8); [metal]=% metal i present in the concentrate j (from elemental composition) and [metal$_i$-dissolved Concentrate]$_j$=% metal i release from the concentrate j.

Values that are so defined can be considered as conservative because the maximum measured ionic solubility from the representative set of 12 copper concentrates (Table 10.8) is used.

10.3.3 Validation of Read-Across: Predicting Metal Releases From Concentrates

When comparing the environmental metal releases from zinc and copper concentrates, some similarities can be seen: zinc, arsenic, and cadmium containing constituents are less soluble than nickel containing constituents, which in turn are less soluble than constituents containing

TABLE 10.7 Environmental Solubility (%) of Copper From Copper Minerals During Short-Term (7-days) and Long-Term (28-days) Transformation/Dissolution Tests at 1 mg/L Loading and pH 6 are Expressed as: (µg Metal Ions Released/µg Total Metal) × 100

	% Environmental Solubility	
Mineral	**7-days**	**28-days**
Chalcopyrite (CP)	0.8 ± 0.080	0.9 ± 0.081
Digenite (DG)	0.8 ± 0.072	2.7 ± 0.243
Enargite (EN)	1.3 ± 0.195	2.2 ± 0.198
Covellite (CV)	0.9 ± 0.045	2.2 ± 0.264
Tennantite (TN)	3.4 ± 0.170	5.8 ± 0.522
Bornite (BN)	4.4 ± 0.176	7.0 ± 0.350
Chalcocite (CC)	9.9 ± 0.396	20.9 ± 0.627

TABLE 10.8 Acute (7 days) and Chronic (28 days) Environmental Solubility (%) at pH 6 for Cu, As, Cd, Co, Ni, Pb, and Zn From Transformation/Dissolution (TD) Results of Representative Copper Concentrates

Metal	% Environmental Solubility	
	7 days	28 days
Ag	ND	ND
As	2.7–**5.0**	3.2–**14.2**
Cd	ND–**9.8**	ND–**10.0**
Co	ND–**11.7**	ND–**30.0**
Cu	0.5–7.5	0.74–13.7
Ni	ND–**7.3**	ND–**29.2**
Pb	5.7–**50.3**	11.3–**53.3**
Zn	0.9–**9.1**	1–**11.6**

Four representative copper concentrates were tested at 100 mg/L loading for 7 days, 12 representative copper concentrates were tested at 1 mg/L loading for 7 and 28 days. Data are expressed as: (µg metal ions released/µg total metal)×100.

The highest values, indicated in bold and underlined, are retained for read-across. Measured concentration below limit of detection and/or coefficient of variation >20% were not considered as valuable for read-across.

lead. However, in the case of copper concentrates, the ranking of the metal-bearing minerals according to the release of metal ions will depend on the main copper-bearing minerals present.

The grouping/read-across procedure for copper release from copper concentrates was validated as follows (Fig. 10.5).

Using the copper environmental solubility (ES%) of different pure minerals, and the % of each mineral in each reference concentrate, the % copper release from each copper concentrate was estimated (see Eq. 10.2).

$$\sum_i ES_i^* [Cu - Mineral]_i = [Cu - dissolved - Concentrate]_j \qquad (10.2)$$

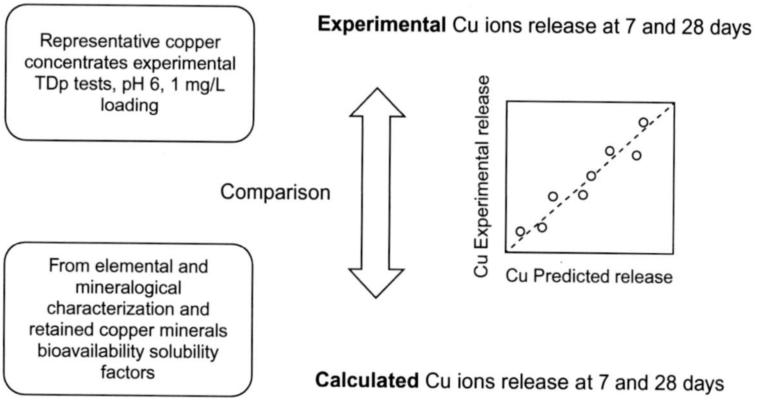

FIG. 10.5 Validation procedure comparing measured transformed/dissolution (TD) of copper from representative copper concentrates with calculated TD of copper using pure copper minerals.

where ES = environmental solubility (%) of the pure mineral i after 7 or 28 days (from Table 10.7) [Cu-Mineral] = % of the mineral i present in the concentrate and [Cu-dissolved-Concentrate] = % copper release from the concentrate j.

Fig. 10.6 shows the predicted vs estimated copper release rates, from the 7 and 28 day TD tests (at 1 mg/L loading), for the 12 representative concentrates. The correlation coefficient (R) is 0.75.

10.3.4 Environmental (Aquatic) Hazard Classification of Metal Concentrates

Metal release data from TD testing of pure metal-bearing minerals (e.g., for read-across to copper concentrates, Table 10.7) and/or conservative release data for companion metals from TD testing of the metal concentrates themselves (Tables 10.7 and 10.8) are retained for hazard classification purposes using the scheme shown in Fig. 10.7.

The classification is thus determined for each concentrate by determining the metal releases for each concentrate from (a) both the elemental and mineral composition and (b) the retained environmental solubility obtained from TD testing.

As several metals are typically present and released in the TD medium, combined toxicity is derived from the Toxic Units (TU), obtained by dividing the metal released after 7- or 28-days TD by the corresponding acute or chronic ERVs, and the application of the additivity formulas. This is further described in Chapter 7.

The classification approach is described in Section 3.2.2 of the UN GHS, and summation rules are thus applied to calculate the sum of acute and chronic TU with the dissolved concentrations and the appropriate ERVs (Rodriguez et al. 2013).

In the case of chronic classification, environmental degradation must also be considered. Among the potentially hazardous metals present in the concentrates, copper, cadmium, nickel, lead, antimony, and zinc are considered as equivalent to rapidly degradable. Discussions are

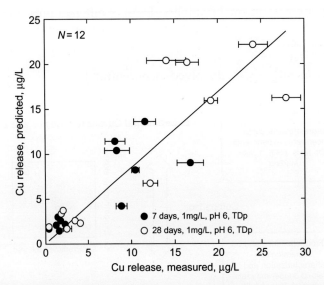

FIG. 10.6   Experimental validation results of predicted vs measured copper concentrate solubility using read-across from copper-bearing minerals to copper.

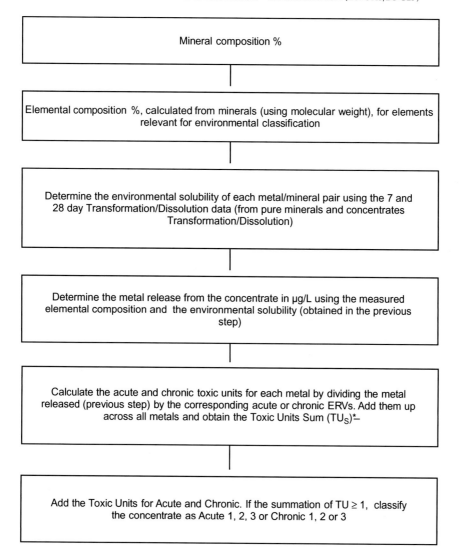

FIG. 10.7 Step-by-step environmental hazard classification procedure for copper concentrates. Concentrates classified as Acute 1, Chronic 1 and /or Chronic 2 also need to be classified as 'Harmful to the Marine Environment' under MARPOL Annex V.

ongoing at regulatory level on how to consider this in a classification scheme. The proposed stepwise approach is summarised in Fig. 10.8.

This hazard assessment is easily done with the assistance of appropriate software tools such as MeClas (http://www.meclas.eu) (Chapter 9).

The resulting UN GHS environmental hazard classification can then be used to comply with international regulations (Table 10.1), which may lead to certain risk management measures for transport of ores and concentrates (IMO IMDG; IMO MARPOL Annex V; UN TGD) and accident prevention (e.g., EU Seveso Directive).

Cu, Cd, Ni, Pb, Sb, and Zn as rapidly degradable

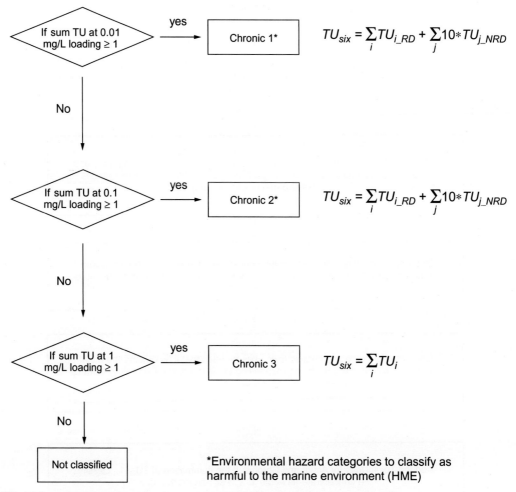

$$TU_{six} = \sum_i TU_{i\_RD} + \sum_j 10 * TU_{j\_NRD}$$

$$TU_{six} = \sum_i TU_{i\_RD} + \sum_j 10 * TU_{j\_NRD}$$

$$TU_{six} = \sum_i TU_i$$

*Environmental hazard categories to classify as harmful to the marine environment (HME)

FIG. 10.8   Decision tree for chronic environmental classification of copper concentrates, where TU_{i_RD} corresponds to the toxic units of the readily degradable and TU_{i_NRD} to the nondegradable components.

10.4 HUMAN HEALTH HAZARD IDENTIFICATION—
CLASSIFICATION (UN GHS)

Given the natural variability of ores and ore concentrates even from the same ore body, full hazard profiles of ores and concentrates as a whole are not available. As for the environmental hazard classifications, human health hazard identification therefore makes extensive use of grouping/read-across to use the toxicological data available for one or more 'source compounds' (in this case a representative source concentrate) with similar physical/chemical properties and a known toxicological profile. In this way, for each endpoint, a classification is based upon data available for a representative ore or concentrate (e.g., another ore or concentrate that is used to produce the same commodity) (bridging). The UN GHS offers guidance on the criteria for selecting or deselecting 'source compounds' for particular endpoints. Each toxicological endpoint has its own specific rules on bridging with respect to the concentration limits that impact the final hazard classification. The different exposure routes and endpoints should be carefully considered when applying grouping/read across for human health hazards, as well as key factors impacting on the possible toxicity (bioavailability, particle size, etc.).

Ores and concentrates, if in contact with body fluids, will release a combination of metal ions based on the interaction between and bioavailability of the contained constituents. Bioelution tests can be used to estimate how much of a constituent metal is bioaccessible following oral exposure to a metal ore or concentrate and this information can be used in a weight-of-evidence approach to read-across to reference materials (see Chapter 8). Bioelution test systems measure the metal ion releases from the material and allows it to determine whether bioavailability is influenced by, for example, the presence of a matrix, by comparing those releases with the releases from other ores and concentrates or from the reference metal compounds obtained in the same conditions (i.e., relative bioaccessibility). Discussions are ongoing in regulatory forums on how to incorporate bioaccessibility data in the classification schemes. Regulatory efforts currently focus on the inclusion of a possible refinement of the classification system for systemic effects from the oral route of exposure for materials characterised by a matrix. This refinement makes use of 'bioelution test results' to replace the 'concentration' in the UN GHS and EU CLP schemes by the bioaccessible concentration, reflecting the impact of the matrix on the bioavailability.

In the meantime, the classification may be determined by means of an appropriate software tool such as MeClas (http://www.meclas.eu) using the elemental concentrations in first tier (Tier 0), speciation data in Tier 1, and a limited bioelution correction (systemic effects, oral route) in Tier 2 (see also Chapter 9).

When assessing potential hazard from inhalation exposure of ores and concentrates during handling, transport and storage, information on particle size distribution and density will be critical to assess deposition in the lung and absorption. This is further discussed in Chapters 2 and 8. Nonsystemic local hazards to the lungs from nonsolubilising materials such as respirable crystalline silica (e.g., quartz) will additionally need to be assessed where relevant.

Considering the low solubility of ores and concentrates, skin absorption and related dermal contribution to systemic toxicity are often considered negligible. The skin sensitisation hazards may however be relevant when potential skin sensitisers such as nickel and cobalt are present.

All information needs to be integrated and the assessment based on a weight of evidence approach.

10.4.1 Copper Example

The dominant minerals (median values) in copper concentrates are chalcopyrite (CuS 63%), pyrite (FeS 11%), quartz (SiO_2 2%), and other gangue minerals. Their hazard profiles are summarised as follows:

As indicated in Table 10.9, copper concentrates contain minor amounts of arsenic, cadmium, cobalt, nickel, and lead. The occurrence and concentration of these elements depends on the particular mineralogy of the ore body in the mine of its source. If their elemental

TABLE 10.9 Hazard Profiles and Data Source of the Dominant Minerals in Copper Concentrates

Constituent	Hazard Profile	Source
CuS	No acute oral, dermal nor acute inhalation toxicity of copper sulphide No concern for repeated dose toxicity (STOT), genotoxicity, reprotoxicity, and carcinogenicity for copper-containing materials	OECD Screening Information Dataset (SIDs) Initial assessment Report summary (OECD, 2014) CuS REACH dossier
Pyrite	Iron containing mineral with no known hazards	
Others (elemental constituents (aluminium, calcium, magnesium, potassium, silicon, manganese) incorporated in gangue minerals, such as calcite, dolomite, hornblende, clinochlore, feldspar, kaolinite, and biotite	No hazards are attributed to these minerals.	

concentrations are well documented, the mineralogical speciation is sometimes not as regularly reported. The toxicity profiles of the copper concentrates were further assessed considering the hazard profiles of these metals/metal compounds (obtained, e.g., from their EU REACH dossiers). In the absence of information on the speciation, the worst-case speciation is considered (most severe hazard profile). Additional information on bioaccessibility and the particle size were used when applying the UN GHS/EU CLP mixture rules.

Metal release rates from 10 representative copper concentrates and 8 minerals typical of copper concentrates were assessed from in vitro bioelution tests in gastric fluids (pH 1.5) and results compared with reference materials. The reference material was selected to be as much as possible chemically similar to the form of the metal present in the tested material (ore or concentrate) and with a well-documented hazard profile. Considering that the hazards of the exact (geo-)chemical form (speciation and particles size) of the metals in the concentrate are often not known, pure soluble metal compounds were used as references. The results consistently demonstrate lower bioaccessibility of the metals from the concentrates compared to the soluble reference metal compounds (with measured bioaccessibility > 90% and known hazard profile). The systemic oral toxicity hazard profile of each copper concentrate was assessed from the total metal content (as obtained from the chemical composition) and the maximum bioaccessibility determined from the test results of representative copper concentrates (Table 10.10).

No correlation was observed between the bioaccessibility of copper from the copper minerals and the amount of copper mineral in the concentrates. Therefore, the maximum bioaccessibility was retained for all relevant elements, including copper.

The particle size distributions (Table 10.4) indicate that contemporary copper concentrates are inhalable and partially respirable: typically, 18%–33% of the copper concentrate is <10 µm. When also considering density, the application of SWeRF (Size Weighted Relative Fine Fraction) methodology demonstrates a probability for particle deposition in the lung (Pensis et al. 2013) ranging between 3% and 11%. To assess the systemic hazards after inhalation, the metal release rates observed for the representative copper concentrate samples in gastric fluids (pH 1.5)

TABLE 10.10 Highest Measured Metal Bioaccessibility (Amount of Metal Actually Released Compared to the Amount of Metal That Could Potentially be Released) From Copper Concentrates in Standard Bioelution Gastric Media (ASTM 5517-07)

Metal	Bioaccessibility (%) Max.
As	1.8
Cd	14.1
Co	4
Cu	6.5
Ni	11
Pb	56.8
Zn	8.1

The values were determined using only reliable measured releases from 11 representative concentrates (measurements above the quantification limit (QL) and with coefficient of variation <20%).

are relevant for the fraction of the inhaled particles that are swallowed and solubilised in the gastrointestinal tract. No extensive bioelution tests in inhalation fluids are currently available for the copper concentrates samples. Henderson et al. (2014) provides initial insight into solubilisation of metals from various matrices (including one copper concentrate) and suggests that 2-hours gastric bioelution data (at pH 1.5) can be considered as a conservative measure for metal release in biological fluid with the exception of long-term solubilisation (for 168 hours) in lysosomal fluids. Further information on bioelution in lung fluids would therefore reduce the uncertainty related to potential inhalation exposure and systemic hazard.

Nonsystemic local respirable hazards from nonsolubilising materials such as respirable crystalline silica (e.g., quartz) additionally need to be considered when concentrations are above the classification threshold value. The application of the SWeRF methodology (Pensis et al., 2013) may allow to refine the classification assessment relevant to the respirable fraction.

No dermal bioelution tests are currently available for copper concentrates. Skin absorption, as a possible contribution to systemic toxicity is considered negligible.

The dermal exposure assessment nevertheless needs to be considered with regards to local sensitisation hazards for concentrates with nickel and cobalt concentrations above classification threshold value.

Bioaccessibility is a key element in the hazard assessment of concentrates. The approach described above is expected to provide a conservative approximation of the bioavailable fraction of metals and other substances that may be released from materials under physiological conditions. Regulatory discussions are ongoing on the standardisation of the test protocol and the use of bioaccessible results for classification. In the meantime, such results are used in a weight of evidence approach and ideally verified using available animal toxicokinetics or toxicity data. For the oral exposure route, two validation sets are particularly relevant to concentrates: The US EPA has conducted a detailed validation programme for an in vitro method to predict the relative bioavailability of lead from soils (OSWER, 2007). The BARGE group have data from similar studies carried out with other metals (Denys et al., 2012).

KEY MESSAGES

A range of elements occurs naturally in ores and ore concentrates, which depending on their mineral speciation have the potential to become harmful to human health (e.g., cadmium, lead, mercury, thorium, thallium, selenium, etc.). During the transport, handling or storage of ores and concentrates with hazardous properties, environmental, and human health hazards may arise. The use of TD tests as described above demonstrates a direct relationship between mineralogical composition and potential environmental release rates of specific metals, where for example, chalcopyrite-rich concentrates release low levels of copper compared to chalcocite-rich concentrates. To assess human health hazards, information on particle size, density, elemental and mineral composition, and bioaccessibility in relevant human health fluids can be integrated into a weight of evidence approach to assess the hazard profiles.

The practice of materials stewardship—as adopted by the mining industry globally—provides guidance for identifying, and acting to prevent and manage unintended impacts on health and the environment from the storage and handling of ores and ore concentrates. Materials stewardship is thus a management technique that seeks to drastically reduce, if not eliminate, the pathways by which hazards associated with ores and ore concentrates can translate into risks to human health or the environment, such as skin irritation or water pollution. In practice, the actual degree of control or influence that an individual company has at different points in manufacturing processes will shape the materials stewardship activities they undertake. For example, an integrated company involved in exploration, extraction, smelting, and refining will have a greater opportunity to influence transportation distances and modes. Implementing materials stewardship therefore requires that mining companies develop relationships and, to varying degrees, encourage action outside of their direct control.

References

ADN: *(European agreement concerning the International Carriage of Dangerous Goods by Inland Waterways (ADN) including the annexed regulations, applicable as from)* 1 January 2013, 2012, ISBN: 978-92-1-139145-9. United Nations.

ADR: *European Agreement concerning the International Carriage of Dangerous Goods by Road,* 2016, ISBN: 978-92-1-139156-5. United Nations.

Adams WJ, DeForest DK, Brix KV: In *Evaluating copper bioaccumulation in aquatic risk assessments,* Proceedings Cobre 2003, 30-3 December, Santiago, Chile. Symposium. 2003, pp 1–11.

CEN: *(European Committee for Standardisation): Characterization of waste - Overall guidance document for characterization of wastes from extractive industries. CEN/TR 16376,* 2012.

Denys S, Caboche J, Tack K, et al: In vivo validation of the unified BARGE method to assess the bioaccessibility of arsenic, antimony, cadmium, and lead in soils, *Environ Sci Technol* 46:6252–6260, 2012.

Dold B: Acid rock drainage prediction: a critical review, *J Geochem Explor* 172:120–132, 2016.

ECHA: *Guidance for Annex V Exemptions from the obligation to register, version 1.1,* 2012. ECHA-10-G-02-EN.

ECHA: *(European Chemicals Agency): guidance for identification and naming of substances under REACH version 2.1,* 2017aISBN: 978-92-9495-711-5.

ECHA: *Guidance on the Application of the CLP Criteria. Guidance to Regulation (EC) No 1272/2008 on classification, labelling and packaging (CLP) of substances and mixtures Version 5.0. Annex IV Metals and inorganic metal compounds,* 2017b, ISBN: 978-92-9020-050-5. https://echa.europa.eu/documents/10162/13562/clp_en.pdf.

ECI (European Copper Institute): *Voluntary Risk Assessment—Copper and Copper Compounds. Submitted to the European Chemicals Agency (ECHA),* http://echa.europa.eu/web/guest/copper-voluntary-risk-assessment-reports, 2008.

European Parliament: *Recommendation for second reading on the Council common position for adopting a regulation of the European Parliament and of the Council concerning the Registration, Evaluation, Authorisation and Restriction of Chemicals (REACH), establishing a European Chemicals Agency, amending Directive 1999/45/EC of the European Parliament and of the Council, A6-0352/2006 FINAL, (Amendment 74)*, 2006.

EU CLP: *Guidance on Labelling and Packaging in Accordance with Regulation. EU CLP*, 2008. N° 1272/2008.

Goldstein J, Newbury D, Echlin P, et al: *Scanning electron microscopy and X-ray microanalysis, ed 2*, New York, 1992, Plenum press.

Henderson R, Verougstraete V, Anderson K, et al: Inter-laboratory validation of bioaccessibility testing for metals, *Regul Toxicol Pharmacol* 70:170–181, 2014.

IMDG (International Maritime Dangerous Goods Code): *Amendment 38-16*, edition 201, 2016, IMO Publishing.

IMO (International Maritime Organization), adopted resolution MEPC 219(63), 2012.

IMSBC: *International Maritime Solid Bulk Cargoes code resolution MSC 354(92)*, 2013.

JRC (Joint Research Centre): *Scientific and Technical Reports. European Union Risk Assessment Report—Zinc Metal*, JRC, European Commission CAS: 7440-66-6, EINECS No: 231-175-3 EUR 24587 EN—2010, 2010.

Lipson H, Steeple H: *Interpretation of x-ray powder diffraction patterns*, London, 1970, McMillan and co.

McGeer JC, Brix KV, Skeaff JM, et al: Inverse relationship between bioconcentration factor and exposure concentration for metals: implications for hazard assessment of metals in the aquatic environment, *Environ Toxicol Chem* 22(5):1017–1037, 2003.

OECD: *Screening Initial Assessment Profile for copper in its online Existing Chemicals Database*, Agreed Conclusions, 2014.

OECD (Organisation for Economic Co-operation and Development): *Series on Testing and Assessment N°29: Guidance Document on Transformation/Dissolution of metals and metal compounds in aqueous media*, ENV/JM/MONO (2001)9, 2001.

OECD (Organisation for Economic Cooperation and Development): *Report of the ring test and statistical analysis of performance of the guidance on transformation/dissolution of metals and metals compounds in aqueous media (transformation/dissolution protocol)*, Paris, 2008, OECD Environment Health and Safety Publications. Series on Testing and Assessment, No 87, ENV/JM/MONO(2008)8.

Oswer 9285.7-77: *Estimation of relative bioavailability of lead in soil and soil-like materials using in vivo and in vitro methods*, 2007.

Pensis I, Luetzenkirchen F, Friede B: SWeRF-A method for estimating the relevant fine particle fraction in bulk materials for classification and labelling purposes, *Ann Occup Hyp* 58(4):501–511, 2013.

Rader K: *Assessment of Time-Variable Solutions for Copper in the Unit World Model for Metals in Lakes*, Belgium, 2013a, European Copper Institute (ECI). Project EUCI.002; Report 001, submitted to the European Chemicals Agency (ECHA).

Rader K: *Metal Classification using a Unit World Model*, Belgium: Eurometaux report, submitted to the European Chemicals Agency (ECHA), 2013b.

RID: *Council Directive 96/49/EC of 23 July 1996 on the approximation of the laws of the Member States with regard to the transport of dangerous goods by rail*, 1996.

Rodriguez PH, Arbildua JJ, Urrestarazu PF, et al: In Copper International Conference, Santiago, Chile 2013, p 307. Environmental Hazard Classification of Copper Concentrates. Economics and Markets/Sustainable Development, Environment, Health and Safety, vol. 1.

Skeaff J, Adams WJ, Rodriguez P, Brouwers T, Waeterschoot H: Advances in Metals Classification under the United Nations Globally Harmonized System of Classification and Labeling, *Integr Environ Assess Manag* 7:559–576, 2011.

Skeaff JM, Ruymen V, Hardy DJ, et al: *The Standard Operating Procedure for the Transformation/Dissolution of Metals and Sparingly Soluble Metal Compounds*, Division Report MMSL 06-085 (TR), Ottawa, Canada, 2006, CANMET, Natural Resources Canada.

UN GHS (United Nations Globally Harmonized System of Classification and Labelling of Chemicals GHS): *Sixth revision ST/SG/AC10/30/Rev6*, 2015, New York and Geneva, United Nations. http://www.unece.org/fileadmin/DAM/trans/danger/publi/ghs/ghs_rev04/English/14e_annex10.pdf.

UN GHS (United Nations Globally Harmonized System of Classification and Labelling of Chemicals GHS): *Fourth revision, ST/SG/AC10/30/Rev4*, 2011, New York and Geneva, United Nations. https://www.unece.org/fileadmin/DAM/trans/danger/publi/ghs/ghs_rev04/English/ST-SG-AC10-30-Rev4e.pdf.

Further Reading

ICMM (International Council on Mining & Metals and Euromines): *Ores and Concentrates: an industry approach to EU Hazard Classification*, http://www.icmm.com/en-gb/publications/materials-stewardship/hazard-assessment-of-ores-and-concentrates-for-marine-transport, 2009.

Risk Assessment of Exposure to Inorganic Substances of UVCBs (Unknown or Variable Composition, Complex Reaction Products, or Biological Materials) During Manufacturing (Recycling) of Metals

Frederik Verdonck*, Federica Iaccino[†], Katia Lacasse[‡], Koen Oorts[†], Frank Van Assche[§], Violaine Verougstraete[¶], Daniel Vetter[‖]

*ARCHE Consulting, Ghent, Belgium [†]ARCHE Consulting, Leuven, Belgium [‡]European Copper Institute, Brussels, Belgium [§]International Zinc Association, Brussels, Belgium [¶]Eurometaux, Brussels, Belgium [‖]EBRC Consulting GmbH, Hannover, Germany

11.1 INTRODUCTION

Several types of complex inorganic materials are used and generated during the primary manufacturing or recycling of metals, such as minerals/ores or refined versions resulting from the use of these minerals, ores, or other raw materials. Most of these materials qualify as inorganic substances of Unknown and/or Variable composition, Complex reaction products, or of Biological origin (UVCB) (Rasmussen et al., 1999, US EPA, 2005a,b) due to the fact that they cannot be sufficiently identified by their chemical composition only.

The complex inorganic materials contain multiple (10 or more) and varying concentrations of constituents, that is, metals, metal compounds, and/or minerals with sometimes unknown speciation. Inorganic UVCBs are typically coming from the metal refining and recycling sectors. The manufacture of metals (e.g. copper, lead, zinc, nickel precious metals) requires several refinement steps to remove the so-called nontarget constituents and produce (single) enriched metal substances. During those refinement steps, UVCB-enriched intermediate substances, such as blister/anode copper, doré, and lead bullion are produced and further processed at the same site or transported to other companies. Because of the process conditions and for trading reasons (e.g. economic value of precious metals), the elemental composition of the enriched metal UVCBs are typically well known. Other examples of such materials are matte (material resulting from metallurgical process-ing of primary and/or secondary sources, mainly composed of one particular nonferrous metal sulphide with minor sulphides of other metals) or flue dust (dust obtained from the exhaust during refining of materials from primary or secondary sources containing various metals).

As the manufacture and the use of these UVCBs are associated with exposure of the envi-ronment and workers, a comprehensive assessment is required.

In the EU, several guidance documents (ECHA, 2011, 2016a,b, 2017) describe the specifics for the registration of UVCBs under REACH. These documents are largely based on the expe-riences gained on organic UVCBs.

However, for inorganic UVCBs, a different approach is needed, taking into account the unique specificities of these complex inorganic materials:

- usually of well-known elemental composition, but with high variability in concentrations;
- chemical speciation of constituents that can be known or unknown, but generally less variable;
- data richness of most constituents of the inorganic UVCBs (a number of metals have indeed been risk assessed in different chemical management systems, due to their hazards and wide uses and large datasets that are available);
- the hazard and risk will also depend on the physical form of the UVCB and on the effect of the three-dimensional (mineralogical/crystallographic) matrix in which the chemical constituents are embedded. This matrix may indeed impact the potential for the metal ions to become available to the surrounding environment;
- the difficulties in selecting a representative sample for (eco)toxicity testing: the intrinsic variability of the composition of the UVCB complicates the selection of a sample that would unambiguously be representative of the (eco)toxicological hazard profile of the inorganic UVCB and that could be used for testing. This is also discussed in Chapters 7 and 8. Therefore, instead of gathering data on the UVCB as a 'whole', it is proposed to treat it as a complex inorganic material containing a number of discrete constituents such as metals, metal compounds, and nonmetal inorganic compounds. It should be noted that selecting a representative sample is less critical—although still arduous—for physicochemical testing because the endpoint is either homogeneous (e.g. particle size distribution, physical state, specific surface area) or less variable (e.g. smelting point, relative density). The sample selection process for the physicochemical testing is often iterative and should however balance the need to obtain a faithfully representative sample with the need to select a reasonably worst-case 'version' of the UVCB. This can be done using knowledge on the chemical processes and sources in combination with proper knowledge on the chemical composition, particle size, and the predicted level of

solubility and associated bioavailability. It shall be noted that a sample representative for one physicochemical property may not necessarily be representative anymore for another one.

This chapter proposes an inorganic UVCB constituent-based risk assessment approach in which it is assumed that the overall fate and hazard properties of the UVCB are driven by the fate and hazard properties of its individual constituents.

11.2 OVERALL RISK ASSESSMENT FRAMEWORK

The risk assessment of the inorganic UVCB will be to identify undue risks and/or describe the conditions under which the manufacture and use of a substance is considered to be safe. It includes the major steps of a 'traditional risk assessment', such as substance identification, hazard assessment, exposure assessment, and risk characterisation followed by risk management, where appropriate. The focus will be on human populations and different environmental compartments (water, sediment, and soil) potentially exposed during metal production and recycling. The inorganic UVCB-specific aspect of this framework is that the constituent-based approach that is followed generates parallel risk assessments for each of the constituents of the UVCB. In the last step, combined exposure of constituents is considered. This is illustrated in Fig. 11.1.

The number and the variability of inorganic UVCBs occurring on an industrial site may make it challenging to assess each and every material specifically. It has been proposed in Chapter 7 to structure, as a first step, the environmental assessments to be carried out by collating the materials based, for example, on the main metal contained in the material (e.g. zinc concentrates). These groups can further be split into more specific groups, for example, by considering within the group of zinc concentrates, the materials originating from sulphidic zinc ore. Ultimately, grouping should result in a 'functional complex inorganic materials group' of materials with expected similar physicochemical and ecotoxicological properties. Within a group, a *generic* assessment can be made on the 'representative materials' of the group instead of analysing every single specific material of the group. This requires the representative materials to be relevant for all the materials of the group (see Chapter 7).

The type of grouping that can be performed has to be in line with the requirements of the chemical management systems. Some legislations, such as the EU REACH Regulation, impose a requirement for 'joint submission of data by multiple registrants'. These provisions require that when the same substance is intended to be manufactured in the EU Community by one or more manufacturers and/or imported by one or more importers, the information relating to properties of the substance and its classification shall be collected and submitted jointly. Registrants of the same substance have thus to comply with important data-sharing obligations (see also Chapter 6). EU REACH registrants of an inorganic UVCB substance, produced during the manufacture/recycling of metals (e.g. doré), will collect the information reported by the companies involved in its production (e.g. composition, processes) and from the gathered information describe and submit the identity of the material in a *generic* way. If testing (e.g. physicochemical testing or TD testing for refinement of the hazard) has to be performed, representative materials will be selected.

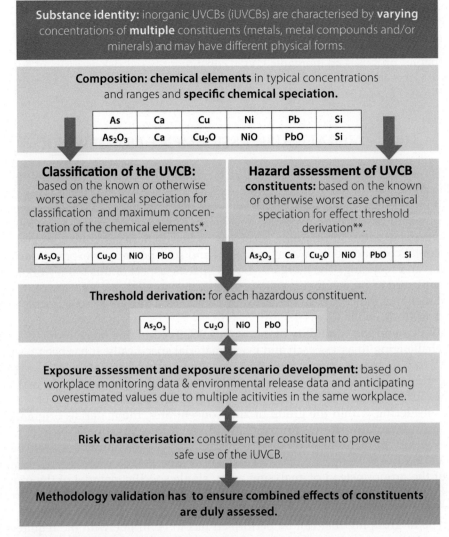

Substance identity: inorganic UVCBs (iUVCBs) are characterised by **varying** concentrations of **multiple** constituents (metals, metal compounds and/or minerals) and may have different physical forms.

Composition: chemical elements in typical concentrations and ranges and **specific chemical speciation.**

As	Ca	Cu	Ni	Pb	Si
As_2O_3	Ca	Cu_2O	NiO	PbO	Si

Classification of the UVCB: based on the known or otherwise worst case chemical speciation for classification and maximum concentration of the chemical elements*.

As_2O_3		Cu_2O	NiO	PbO	

Hazard assessment of UVCB constituents: based on the known or otherwise worst case chemical speciation for effect threshold derivation**.

As_2O_3	Ca	Cu_2O	NiO	PbO	Si

Threshold derivation: for each hazardous constituent.

As_2O_3		Cu_2O	NiO	PbO	

Exposure assessment and exposure scenario development: based on workplace monitoring data & environmental release data and anticipating overestimated values due to multiple acitivities in the same workplace.

Risk characterisation: constituent per constituent to prove safe use of the iUVCB.

Methodology validation has to ensure combined effects of constituents are duly assessed.

* MeCLas: tiered tool for classification of complex inorganics, based on CLP/GHS mixture rules (www.meclas.eu)
** MMD: multimetallic database reporting hazard data, as shared under EM umbrella

FIG. 11.1   Risk assessment process of an inorganic UVCB (iUVCB).

11.2.1 Identification and Selection of Representative Materials

The two main influencing factors for the observed variation in composition and speciation of the inorganic UVCB are the source of the material (e.g. varying ore characteristics) and the processes used. These two elements can significantly influence the substance identity, relevant to the hazard profile and need to be considered when grouping materials. For operational and commercial purposes, complex inorganic materials are usually characterised by their main constituents (e.g. 'iron silicates in slags') and by a detailed elemental composition. This information on the main constituents is essential, also when it comes to fulfilling data requirements. For example, under EU REACH information is required at least for the main constituents with a content >10%. Moreover, any other constituent at a lower concentration, which may influence the hazard classification, shall be identified by name and typical concentration (ECHA, 2016a).

Considering that it is the metal species of each element that will determine its solubility and as a consequence, its hazard, information on the speciation (e.g. metallic form, sulphide form, oxide form, minerals) of the different metals in the complex inorganic material is often useful. The speciation depends on the source material (e.g. sulphidic concentrates vs. metal scrap) and the process (e.g. reduction of metal sulphides to metal oxides). Besides the speciation, the granulometry of the material (e.g. powder form, particulate of different sizes) may also be relevant for the release of metal ions.

In practice, for a joint assessment the first step will be to collect the information on the complex inorganic material reported by the companies involved in its production. The key information as mentioned above will be the detailed elemental composition, the involved processes, and other physicochemical characteristics such as granulometry.

A typical information sheet to be filled in by the companies is presented in Table 11.1. From the gathered information, the complex inorganic material identity is described in a generic way, using the median and ranges reported across companies.

The elemental information is usually available from historical samplings within and across companies, since the metal content represents the potential value of the material. From the data on elemental composition follows the variability in content of the different constituting elements. This variability in elemental composition can be large because the source materials (e.g. ores, concentrates) can have a variable composition.

For generating information on composition (and physicochemical data, as explained above), companies need to take representative composite samples. Composite samples are by preference obtained according to standard sampling protocols. Some examples are provided below:

• For massive samples (e.g. blister), one composite sample is prepared by selecting massive forms from different cycles/baths. Sampling is subsequently done on those different massive forms, according to a defined template so that 4 drills per massive

TABLE 11.1 Information Sheet Used for the Gathering of Data on the Complex Inorganic Material From the Producing Industry

Site x—Yearly Characteristics (Median and Ranges)
• Source material
• Process description
• Generic description of the complex inorganic material
• Main constituents (e.g. >10% under REACH, ECHA, 2016b)
• Elemental composition
• Granulometry

form (1 drill in each quadrant) are made (ECHA, 2008). All drillings are collected and reduced to <5 mm by grinding or milling. All grinds are finally mixed and sampled as needed. During drilling, excessive speed and heating is to be avoided, cooling may be appropriate.

- For liquids and fine solids, representative samples are collected by taking different samples during different time slots, mixing, and dividing the samples (e.g. ISO 12743, 2006).
- For fine solids at stock piles, representative samples are collected by taking different samples in different areas of the stock pile (e.g. CEN TR 15310-3, 2006).

The sampling protocols need to be documented and compliant with international standards (e.g. CEN TR 15310-3, 2006 and ISO 12743, 2006, ISO 14488, 2007 and/or, ISO 3082, 2009).

The second step will be to define representative samples within the group that are typical for characteristics of the group as a whole. The main drivers for the selection of the representative samples will be process and source material: when the processes (e.g. type of furnace used) and/or source materials (e.g. concentrates and/or recycled materials) are comparable within a group, a representative sample can be selected in this group (e.g. see, Example 1: copper matte). When the materials are produced by different processes and/or from different source materials, a more extensive data collection is needed, data can be taken from each production site (e.g. see Example 2: copper slag).

The elemental composition of these representative samples is determined by means of ICP analysis. Metal speciation can then be analysed by means of, for example, XRD analysis or QEMSCAN. This composition will be considered representative of the complex inorganic materials group for this particular process and source.

In both the examples, the selected representative materials were used for TD testing, for refinement of the hazard assessment as explained under Sections 7.5.1.1 and 7.5.1.2.

EXAMPLE 1: COPPER MATTE

The main constituents of copper matte are copper (45%–75%) and sulphur (15%–30%). For this example, several samples of matte, corresponding to four different types of process and input materials were assessed:

- Type I: matte originating from a flash furnace
- Type II: matte originating from an electric furnace
- Type III: matte originating from a blast furnace
- Type IV: 'white' matte originating from a Converter furnace–material as imported

Each type was assessed for its composition (elemental and mineral composition) and physico-chemical properties (particle size distribution). The chemical and mineralogical data demonstrated that for all tested matte, a common mineralogical pattern existed (mainly sulphides). This confirmed the 'grouping' in copper mattes.

EXAMPLE 2: COPPER SLAGS

In this assessment, 12 copper slag samples were characterised. Slags were defined as substances produced from heterogeneous mixtures of materials containing copper formed during the copper production, by reduction at high temperature in molten state (i.e. melting and processing in a furnace) or by flotation processes. The main constituents were iron silicate and calcium–aluminium silicates, with the amount of nonferrous metals reduced to the lowest extent that are economically and technologically viable. Samples were chosen as representatives for the production processes depending on the origin of raw material (primary or secondary) and slag cooling rate (rapid cooling/granulation or slow cooling) and accordingly four types were distinguished:

Type I: slag, granules, primary
Type II: slag, granules, secondary
Type III: slag, slow cooled, primary
Type IV: slag, slow cooled, secondary

However, there was no evidence for a direct link between the grouping and initial hazard profiles, and thus all forms of copper slags (stones, granules, and powders) were assessed for composition (elemental and mineral composition) and physicochemical properties (particle size distribution) and subjected to further testing.

11.2.2 Hazard Assessment

11.2.2.1 *Hazard Identification and Classification of the Inorganic UVCB*

The identification and characterisation of hazards of a substance is a key step in the risk management of chemicals, and as such part of all the regulatory schemes aiming at ensuring safe use of chemicals.

The hazard assessment of the inorganic UVCB is assumed to be driven by the hazard(s) of its individual constituents (chemical elements and relevant speciation). The inorganic UVCB is therefore treated as a complex inorganic material with a number of discrete constituents (i.e. chemical element with discrete species). The hazard classifications of each constituent are then factored into a combined classification of the UVCB as a whole.

The MeClas classification tool has been developed for this purpose (see also Chapter 9). This web-based tool addresses the specific challenges associated with the human health and environmental hazard assessment and classification of complex inorganic materials. It can be freely used since 2010 and is accessible at www.meclas.eu. The tool is compliant with the UN GHS—(UN GHS, 2011), EU CLP—(EC, 2008), and US OSHA (US, 2012) requirements and rules, and relies on the data of the constituents (Verdonck et al., 2017), using the most up-to-date information available on UVCB constituents.

Although it has been broadly used by the industry since its development and awareness is further increasing, it has not been formally validated yet. However, the GHS mixture rules are intrinsically considered to be robust, predictive estimation algorithms, internationally accepted by regulators. In addition, comparisons between the MeClas predictions-based hazard classifications and the (eco)toxicity testing-based hazard classifications demonstrate that the MeClas predictions result in the same hazard category or overestimates it because of a precautionary tiered approach (Verdonck et al., 2017).

MeClas allows to generate classifications for inorganic UVCBs, while dealing consistently with the specificities and challenges of the inorganic UVCB:

- Known or unknown speciation: speciation analysis of the substance 'as manufactured' or 'as put on the market' is required for classification and labelling. When unknown, the speciation with the worst-case hazard classification is selected. The selection of the worst-case classification is done by the MeClas tool when no information on speciation is provided.
- Variable composition concentration ranges: as explained above for one UVCB substance, the variability in elemental composition can potentially lead to different hazard profiles. To address this, a worst-case composition (e.g. defined as the maximum of all companies' (across industry) typical concentrations for each constituent, which is leading to more than 100% composition) is selected. There can be a practical need (for the purpose of SDS and labelling) to differentiate more hazardous from less hazardous individual streams within the inorganic UVCB. Generic groups/grades/clusters within one UVCB—each group with a common worst-case classification profile—can be developed to increase general understanding of the variability of the hazard of the UVCB and to allow companies to easily derive a worst-case classification for possible new streams.
- Constituents-driven assessment: the hazard of the inorganic UVCB is predicted based on the hazard of the individual constituents. The combined exposure for hazard classification is taken into account in the same way as the GHS mixture rules internationally accepted by regulators. For example, for acute toxicity, additivity is assumed.

11.2.2.2 Hazard Assessment: Threshold Derivation/Selection

As outlined above, a constituent-based approach is proposed for the hazard assessment and threshold derivation. It is recommended to include all constituents (i.e. independently of their individual hazard profile) in the threshold derivation and in the environmental fate properties assessment: that is:

(1) PNEC (predicted no-effect concentration) values for all relevant environmental compartments.
(2) DNEL (derived no-effect levels under EU REACH) or N(L)OAEL values for any type of effect (local and systemic), exposure route (oral, inhalation, and dermal), and exposure duration (acute or chronic).
(3) Fate properties, that is, assessing bioconcentration/bioaccumulation and solid water partition coefficients (K_d) values are to be collected.

What are the inorganic UVCB's specific challenges for hazard assessment?

- Dealing with unknown speciation: whereas speciation analysis of the substance 'as manufactured' or as 'put on the market' is required for substance identification and classification and labelling, any human health and environmental risk assessment should be based on the chemical species a worker and the environment are actually exposed to. In case the speciation is unknown, a worst-case approach is followed, that is, by selecting the chemical species leading to the lowest DNEL/N(L)OAEL or PNEC for a constituent. Plausibility considerations or expert judgement (e.g. oxides may be assumed to be present/generated in oxidic conditions) may help to refine this worst-case approach. A further approach could be to measure the speciation in exposure monitoring. For the environment, most often, it is the metal ion that is the toxic driver (ECHA, 2008). Difference in speciation is therefore typically not or less relevant when selecting PNEC.

- Dealing with variable composition concentration ranges: Using the DNEL/N(L)OAEL and PNECs of all individual constituents indirectly circumvents the issue of varying composition of an UVCB as it implicitly assumes that each metal constituent is 100% bioavailable from the UVCB substance. In addition, the exposure of metals being measured and expressed as total metal, and thus compared with the corresponding metal DNEL (or PNEC) that is expressed in the same units, the actual concentration of the metal in the inorganic UVCB does not matter. The two practices together make that it can be considered as a conservative approach. A main outcome of the constituents' based assessment is the selection of all the constituents for which any human health or environmental hazard is identified.

11.2.3 Exposure Assessment

As for the hazard assessment, a constituent-based approach is proposed for the exposure assessment. Doing the exposure assessment for each constituent does not differ substantially from standard metal exposure assessment. For workers, hazardous constituents are generally monitored on a regular basis and measurements (e.g. personal, static measurements expressed as concentration/volume) are available. Such monitoring is normally conducted for inhalation and, less frequently, also for dermal exposure and/or is assessed with exposure modelling tools (see Chapter 9). In some cases, biomonitoring is conducted for specific constituents.

Emissions are also measured in the environmental compartments (air and wastewater) for most well-regulated metals (e.g. cadmium, copper, lead, zinc). When measured data are not available, default release fractions can be used for the constituents (see Chapter 9).

Information on the development and use of consensus standards and sampling and analytical methods that are key tools for chemical risk/hazard minimisation and management activities of these materials can be found in Chapter 2.

The hazard assessment conclusions delineate the scope of the subsequent exposure and risk assessment. Table 11.2 outlines the 'scope of assessment' rationale for environmental and occupational exposure assessments, respectively, based on the conclusions of hazard assessment. This scoping exercise should be done for each constituent and for each exposure route and environmental compartment (inhalation, dermal, oral, water, sediment, soil).

Note: Indirect exposure or exposure of man via the environment:

As explained in Chapter 2, for metals, the assessment of exposure of man via the environment is predominantly determined by the respective metal ion. The concentration thereof in environmental media is in most cases determined by measurement data. Intrinsic to such monitoring data is that they cover the potential exposure of the general population resulting from all emissions (including historical emissions) and natural background. Indirect exposure is generally assessed on a local and on a regional spatial scale. In this context, local

TABLE 11.2 Scope of Assessment Rationale to Be Considered for Each Constituent and for Each Exposure Route and Environmental Compartment

Hazard Threshold Conclusion	Explanation	Subsequent Exposure Assessment and Risk Characterisation
DNEL/PNEC	When DNEL/PNEC is available (and data sharing agreement) and exposure cannot be excluded	Quantitative
	When DNEL/PNEC is available but no or low exposure, hazard, and risk expected	'Quantitative' assessment of emission and hazard potential (benchmarking)
Other threshold (OEL[a],...)	When no DNEL/PNEC is available but hazard is identified, relevant exposure cannot be excluded and alternative threshold is available (e.g. OEL[a], ...)	Quantitative based on other threshold
Qualitative (no threshold)	When no DNEL/PNEC is available because of nonthreshold related effect (e.g. irritation) but hazard is identified (low–medium–high)	Qualitative
No hazard identified	When no hazard	Not needed

[a] *Occupational exposure limit.*

and regional environments are not to be considered as actual sites or regions, but instead as standardised environments. When it comes to UVCBs produced during manufacturing or recycling of metals, it should be acknowledged that their life cycle is usually limited to the production sites, and without wide-dispersive uses or any downstream uses that would lead to a significant contribution to regional emissions. The regional assessment is usually based on monitoring data. Thus, exposure of the general population via the environment is already covered in a precautionary way when assessing the environmental exposure of the metal constituents since monitoring data cover all emission sources. On the local scale, the contribution of specific sites has to be calculated or measured, based on the measured or modelled contributions of the constituents.

What are the inorganic UVCB's specific challenges in the exposure assessment?

- Dealing with unknown speciation: whereas speciation analysis of the substance 'as manufactured' or 'as put on the market' is required for substance identification and classification and labelling, any human health and environmental risk assessment should be based on the chemical species a worker and the environment are actually exposed to. The same approach as for the hazard assessment is used for exposure assessment. In case the speciation is unknown, a worst-case approach is followed, that is, selecting the chemical species leading to the lowest DNEL or PNEC. Plausibility considerations or expert judgement (e.g. oxides may be assumed to be present/generated in oxidic conditions) may help to refine this worst-case approach. A further approach could be to measure the speciation in exposure monitoring. For the environment, most often, it is the metal ion that is the toxic driver (ECHA, 2008). The difference in speciation is therefore typically not considered/less relevant when selecting PNEC.

- Dealing with variable composition concentration ranges: The monitoring practice has the following implications. The variability of the elemental constituent concentration on the inhalation or emission exposure levels of that constituent is automatically reflected. The impact of varying process conditions and risk management measures are also directly reflected in the measurements. Chapter 2 provides more information on monitoring practice. The contribution of other substances (including UVCB substances) handled in parallel and having the same constituents as the inorganic UVCB under investigation to overall exposure is also automatically reflected in the measurements. In addition, a worst-case percentile (or the maximum) of the measurements is further considered (hereby covering the variation in the monitoring data). This is a conservative approach in case multiple inorganic UVCBs are handled on the same site or in the same workplace, in particular for substance-driven chemical legislations (e.g. REACH), where combined exposure from multiple substances is not required.

11.2.4 Risk Characterisation

The constituent-based approach for inorganic UVCBs has several consequences for risk characterisation.

First, multidimensional risk characterisation ratio (RCR) tables need to be developed (constituent × exposure route × local/systemic effects, short-term/long-term). For each constituent, RCRs are given for all environmental compartments and occupational exposure routes and type of effect (local/systemic, acute/chronic). In case of qualitative risk characterisation, this should be duly justified and all relevant information reported.

Second, the potential additional effects of constituents due to combined toxicity mechanisms need to be further considered. Information on combined toxicity is, however, not readily available for complex inorganic materials, including UVCBs, except for a few metals where interactions have been documented. Regulatory guidance to assess and include combined toxicity in the assessment of a material is under development in several jurisdictions.

11.2.4.1 Dealing With Combined Exposure for Human Health

Several guidance documents already provide a framework that can be used in developing robust argumentation (IGHRC, 2009, SCHER-SCCS-SCENIRH, 2012, ECHA, 2014). The methods currently proposed to assess 'combined toxicity' usually consider additive effects (such as dose/concentration addition or response/effect addition) as a first tier, in particular if the mode of action is unknown. It is expected that more data on the modes of action and interactions shall become available in the coming years due to the regulatory focus on combined toxicity as well as from regular literature screening.

An agreed inventory of mode of actions/target organs for metals and a defined set of criteria on how to characterise or predict a mode of action for data-poor chemicals would facilitate the consideration of combined toxicity for complex inorganic materials. An additional knowledge gap to fulfil relates to the lack of exposure information. Recently, a metals 'health initiative' has been launched (i.e. Health Toxicology Advisory Panel, HeTAP) that aims at supporting an advisory panel consisting of experts from the academia, institutes, etc. and industry experts, who will reflect on scientific issues of common interest to the mining and metals industry and provide scientific advice. Combined toxicity is one of the identified major topics on the agenda and actions have been launched to collect data and experience.

For UVCBs, an approach is proposed under EU REACH—as outlined below. It refines the assessment of effects by using knowledge on target organs as a surrogate for information on mechanism of action, of exposure (based on measurements in representative workplaces), and risk management. Fig. 11.2 presents the tiered framework currently used:

At Tier 1, *by default*, it is proposed to add the RCRs of the different constituents and to compare the sum to 1 as a default approach for each type of derived no-effect level. This is done as a first filter, to analyse whether there may be a concern.

To arrive at a more realistic risk assessment, a next Tier 2 of refinement is proposed. The refinement can be done on the hazard or the exposure part of the risk characterisation—whatever is practically more feasible/robust.

REFINEMENT OF HAZARD

The refinement is based on the knowledge of the mode of action and/or information on the target organ of the constituents:

If information is available showing that two or more constituents of the UVCB have the same mode of action, the corresponding constituent-specific RCRs are to be added and this 'combined RCR' is to be used to implement risk management measures. In case of a different mode of action, RCRs are not to be added.

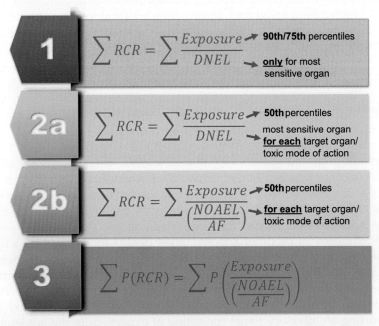

FIG. 11.2   Tiered approach for combined occupational exposure.

If information on the mode of action is not available but the target organs of the UVCB constituents are known (by endpoint) it is proposed to work as follows.

- For repeated dose, systemic effects: if two or more constituents have a same target organ, RCRs shall be added. Otherwise no combined toxicity is assumed.
- For repeated dose, local effects: if two or more constituents have the same target organ, RCRs shall be added. Otherwise no combined toxicity is assumed.
- For acute, systemic effects: add RCRs (if thresholds available).
- For acute, local effects: add RCRs (if thresholds available) or address via RMMs if peak exposures can be expected.
- For reproductive and carcinogenicity effects (at this stage no consideration has been given to a threshold vs a nonthreshold MOA), RCRs shall be added.
- Mutagenicity: as for the individual constituents, only a qualitative risk assessment is possible since this is a nonthreshold effect.

A further distinction can be made between Tiers 2A and 2B depending on whether a toxicological threshold is derived for a specific target organ ('target-organ-specific DNEL' as in ECHA, 2015). In this case, where specific target organ is not the most primary target organ, a specific DNEL can be calculated.

REFINEMENT OF EXPOSURE

This refinement is based on the knowledge of the workplace exposure patterns. In the calculation of the RCR for the risk assessment of an individual UVCB constituent, reasonable worst-case estimates should be based on upper percentiles (such as 90th or 75th percentiles) are often proposed. However, if RCRs are to be summed up for individual constituents, such summation could easily lead to an unrealistic overestimation of risk as worst-case estimates of exposure would be combined. Thus, for the assessment of combined effects, the contribution to risk of individual constituents should be calculated on the basis of estimates of typical exposure. It is noted that such a refinement is only justified if no (significant) positive correlation was discovered between individual constituents. Consequently, any such analysis requires comprehensive monitoring databases including information on contextual information such as the content of the specific constituents in the materials handled/processed during actual generation of the monitoring data.

The most complex Tier 3 assessment would address the aforementioned (lacking) correlation of exposure levels of individual constituents and would also include the refinements proposed for the hazard side (i.e. mode of action and/or target organ). Since various constituents are commonly analysed from the same dust sample the correlation matrix for exposure levels to constituents is intrinsically included in monitoring data of such type.

If RCRs were to be calculated for each exposure data point for all constituents measured, a summation of all relevant RCRs according to the refinements—as proposed above—for the hazard side at the measurement level would be a realistic measure of the risk of combined effects. Since such calculations would be repeated for all measurements, a distribution of RCRs would be the result from which, in turn, a realistic worst-case estimate could be calculated, for example, in the form of 75th percentile and/or its upper confidence limit.

Such an approach is naturally very data hungry since all relevant constituents need to be analysed in each measurement. Thus, when some data are missing, extrapolation could be conducted by taking into account the correlation matrix between individual constituents.

11.2.4.2 *Dealing With Combined Exposure for Environment*

The reader is referred to Chapter 7 for the assessment of combined exposure for environment.

KEY MESSAGES

Inorganic UVCBs typically originate from metal refining and recycling sectors. Their intrinsic variability in known chemical composition and their sometimes unknown speciation raise challenges during hazard and exposure assessment and risk characterisation.

A strategic approach has been developed to cope with both uncertainty and variability. The hazard profile of the inorganic UVCB is dependent on the individual constituents and their chemical speciation.

Depending on the level of knowledge, the following situations can be distinguished:

- when the chemical speciation of the constituent in the inorganic UVCB is known, this speciation is used for classification;
- when the chemical speciation of the constituent as present in the workplace is known, this speciation is used for risk characterisation;
- when information on chemical speciation of the constituent is not complete, the worst-case speciation (in UVCB for the purpose of classification and in workplace/environment for the purpose of risk characterisation) is assumed, that is, the speciation that would lead to the most severe classification or to the lowest DNEL/PNEC. It is noted that different chemical species could be relevant.

All relevant identified constituents, workplace monitoring and environmental release data and metal-specific models are considered in the exposure assessment ensuring that the unknown and variability are addressed in a precautionary way. Potential combined exposures of the inorganic UVCB constituents are assessed in a tiered way: the first—very conservative—tier is based on the summation of the associated RCRs and can be refined by considering toxicity and mechanistic data (e.g. adding RCRs by target organ for human health). RCRs can also be refined on the basis of statistical considerations before summing them up.

References

CEN TR 15310-3: *Characterization of waste—Sampling of waste materials—Part 3: Guidance on procedures for sub-sampling in the field*, 2006.

EC (European Commission), Regulation (EC) No. 1272/2008 on Classification, Labelling and Packaging (CLP) of Substances and Mixtures, 2008.

ECHA (European Chemicals Agency): *Guidance on Information Requirements and Chemical Safety Assessment.: Appendix R.7.13-2 Guidance on Environmental Risk Assessment for Metal and Metal Compounds*, 2008.

ECHA (European Chemicals Agency): *Guidance on Information Requirements and Chemical Safety Assessment—Part B: Hazard Assessment*, 2011.

ECHA (European Chemicals Agency): *Transitional Guidance on the Biocidal Products Regulation Transitional Guidance on Mixture Toxicity Assessment for Biocidal Products for the Environment*, https://echa.europa.eu/documents/10162/15623299/biocides_transitional_guidance_mixture_toxicity_en.pdf, 2014.

ECHA (European Chemicals Agency): *Guidance on the Application of the CLP Criteria*, Version 4.1, 2015, ISBN: 978-92-9247-413-3.

ECHA (European Chemicals Agency): *Guidance for Identification and Naming of Substances Under REACH and CLP*, Version 2.0, 2016a, ISBN: 978-92-9495-711-5.

ECHA (European Chemicals Agency): *Guidance on Registration*, Version 3.0, 2016b, ISBN: 978-92-9495-072-7.

ECHA (European Chemicals Agency): *Guidance in a Nutshell: Identification and Naming of Substances Under REACH and CLP*, Version 2.0, 2017, ISBN: 978-92-9495-792-4.

IGHRC: *Chemical Mixtures: A Framework for Assessing Risk to Human Health (CR14)*, Cranfield, UK, 2009, Institute of Environment and Health, Cranfield University.

ISO 12743: *Copper, Lead, Zinc and Nickel Concentrates—Sampling Procedures for Determination of Metal and Moisture Content*, 2006.

ISO 14488: *Particulate Materials—Sampling and Sample Splitting for the Determination of Particulate Properties*, 2007.

ISO 3082: *Iron Ores—Sampling and Sample Preparation Procedure*, 2009.

Rasmussen K, Pettauer D, Vollmer G, Davis J: Compilation of EINECS: descriptions and definitions used for UVCB substances: complex reaction products, plant products, (post-reacted) naturally occurring substances, micro-organisms, petroleum products, soaps and detergents, and metallic compounds, *Toxical Environ Chem* 69(3/4):403–416, 1999.

SCHER, SCCS, SCENIHR: *Opinion on the Toxicity and Assessment of Chemical Mixtures*, http://ec.europa.eu/health/scientific_committees/environmental_risks/docs/scher_o_155.pdf, 2012.

UN GHS: *Globally harmonized system of classification and labelling of chemicals (GHS)*, fourth revised edition, 2011. United Nations, ST/SG/AC.10/30/Rev.4, New York and Geneva, https://www.unece.org/fileadmin/DAM/trans/danger/publi/ghs/ghs_rev04/English/ST-SG-AC10-30-Rev4e.pdf.

US (United States). 2012. OSHA GHS HCS 2012, March 26, 2012, FR 77: 17574 [Based on 2009 GHS/Rev3].

US EPA: *Toxic Substances Control Act Inventory Registration for Products Containing Two or More Substances: Formulated and Statutory Mixtures*, http://www.epa.gov/opptintr/newchems/mixtures.txt, 2005a.

US EPA: *Toxic Substances Control Act Inventory Registration for Combinations of two or more substances: complex reaction products*, http://www.epa.gov/opptintr/newchems/rxnprods.txt, 2005b.

Verdonck F, Waeterschoot H, Van Sprang P, et al: MeClas: an online tool for hazard identification and classification of complex inorganic metal-containing materials, *Regul Toxicol Pharmacol* 89:232–239, 2017.

Further Reading

ASTM E 1833 07a: *e 1. Standard Practice for Sampling of Blister Copper in Cast Form for Determination of Chemical Composition*, 2013.

EC (European Council): *Regulation (EC) No. 1907/2006 of the European Parliament and of the Council of 18 December 2006 Concerning the Registration, Evaluation, Authorisation and Restriction of Chemicals (REACH)*, Official Journal of the European Union L396: L136/133-L136/280, 2006.

Risk Assessment for the Manufacture and Formulation of Inorganic Pigments

Daniel Vetter, Tina Bodenschatz, Rüdiger Vincent Battersby

EBRC Consulting GmbH, Hannover, Germany

12.1 INTRODUCTION

Inorganic pigments are commonly produced by high-temperature calcination of a mixture of metal oxides (MeOx) in varying amounts to form a crystalline matrix. Owing to their chromophoric properties, inorganic pigments are widely used to colourise other substances, mixtures (e.g., paints, articles, plastic articles), or in solid form as dry powder. They are insoluble in the carrier material meaning that they are incorporated into the matrix of the carrier material as a dispersed compound and not as a solute (ECHA, 2015).

Owing to their stable crystalline structure, most inorganic pigments are perceived as toxicologically neutral substances. However, the increasing toxicological knowledge about the individual constituents of the inorganic pigments and the related—sometimes very low—effect thresholds for those constituents required the development of an approach to assess whether this perception is justified. While such an evaluation is essential for occupational hygienists to demonstrate safe handling of the respective substances in the workplace, this type of information could also be used, for example, to justify waiving toxicological testing, as would be required for a standard substance under EU REACH.

The inorganic pigments manufacturing and formulation sector has proposed the approach explained below, which evaluates the risk of their inorganic pigments based on their constituents, similar to the generic approach applied for other inorganic materials (see Chapters 7, 8, and 11).

12.2 THE PROPOSED APPROACH

In the inorganic pigment sector, inhalation exposure is commonly monitored by measurements of airborne dust in the breathing zone of workers. Such measurements are gravimetrically analysed and generally gives an indication of exposure to inhalable generic dust. Consequently, such gravimetric results could be compared with occupational exposure limits (OELs) given for inhalable generic dust. However, when specific constituents are of concern (e.g., MeOx) a subsequent chemical (elemental) analysis of the dust should be used for comparison with an OEL specific to the constituent. For inorganic pigments, for which it can safely be assumed that all constituents are tightly bound to the inorganic pigment matrix, such a simple approach would most likely lead to significant overestimation of health risks for workers by the inhalation route because the approach does not account for the very low bioavailability of the constituents (see also Chapter 8).

An alternative approach is therefore proposed, estimating exposure to the inorganic pigment on the basis of measurements of an element (by chemical quantification) that can be used as a first step as an 'indicator substance of exposure,' refining the assessment subsequently by considering the bioaccessibility data for individual constituents. The contribution by the dermal route could be assessed by measuring dermal exposure, but such data are often not available. An alternative is to use an appropriate exposure modelling tool such as MEASE, 2010 (see also Chapter 9). Still, one could also consider that the approach outlined above for the inhalation route—as it is very conservative—will also control effects mediated by the dermal route as long as the associated effect is threshold related. For the oral route, good occupational hygiene practices may be adopted to effectively minimise exposure.

The approach is exemplified below by applying it to a generic inorganic pigment containing three metal oxide constituents ($MeOx_1$, $MeOx_2$, and $MeOx_3$). It is demonstrated in the example that the approach integrates a number of aspects of the inorganic pigment: its composition, its particle size distribution (PSD), its hazard profile and one of its constituents, the constituents' bioaccessibility, and the identification of an indicator substance for exposure to the inorganic pigment.

12.3 DESCRIPTION OF THE GENERIC INORGANIC PIGMENT SELECTED AS EXAMPLE

12.3.1 Composition

The composition of the inorganic pigment at or above a concentration of 0.1% (w/w) is given in Table 12.1.

12.3.2 Hazard Profile of the Inorganic Pigment and of Its Constituents

The generic inorganic pigment is not classified according to EU CLP, 2008 or UN GHS, 2015 and based on the information available, hazards for human health are not expected to occur. One of the consequences is that a NOAEL or a DNEL or a comparable toxicological threshold value is not available.

TABLE 12.1 Composition of the Inorganic Pigment

Constituent	Concentration (%)	Molecular Weight (g/mol)	Metal Content in Pigment (wt%)
$MeOx_1$	53.7	101.96	28.4
$MeOx_2$	25.5	81.38	20.5
$MeOx_3$	19.4	74.93	15.3
Total	98.6	–	–

MeOx, metal oxide.

It is therefore proposed to identify a reference value to which the assessed exposure levels can be compared with. From this comparison, the large margin of safety inherently given by the (very) low bioaccessibility of the individual constituents (see Section 12.3.3) can be quantified.

This means that one or multiple intended constituents of the inorganic pigment need to be selected, which could be used for both: (i) as an indicator for assessing exposure to the inorganic pigment and (ii) to derive a reference value.

To identify the appropriate constituent for the aforementioned purposes, the hazard conclusions for the MeOx, which are the constituents of and/or used as raw materials for the inorganic pigment are evaluated (see Table 12.2). The hazard conclusions for those MeOx should however not be extrapolated to the inorganic pigment as such without considering the fact that the constituents are tightly bound in the inorganic pigment matrix. During the manufacturing process that leads to the generation of the inorganic pigment, all of the raw materials are transformed into the newly created substance (i.e., the inorganic pigment), and the different MeOx are homogeneously interdiffused within the crystalline structure, where they remain tightly bound, rendering the pigment inert and the metals (or their oxides) biologically unavailable.

TABLE 12.2 Toxicological Information of Individual Constituents of the Inorganic Pigment

	Constituent	$MeOx_1$	$MeOx_2$	$MeOx_3$
Route	Type of Effect	Toxicological Information Relevant for Workers (EU REACH DNELs)		
Inhalation	Systemic long-term	$15.63\,mg/m^3$	$5\,mg/m^3$	[a]
	Systemic acute	[a]	[a]	[a]
	Local long-term	$15.63\,mg/m^3$	$0.5\,mg/m^3$	$54.5\,\mu g\ MeOx_3/m^3\ (=40\,\mu g\ Me_3/m^3)$
	Local acute	N/A	[a]	[a]
Dermal	Systemic long-term	N/A	$83\,mg\ Me_2/kg\ bw/day$	[a]
	Systemic acute	N/A	[a]	[a]
	Local long-term	N/A	[a]	[a]
	Local acute	N/A	[a]	[a]
Eye	Local effects	[a]	[a]	[a]

N/A, no information available (hazard conclusion is missing).
[a] No hazard identified.

12.3.3 Bioaccessibility of the Inorganic Pigment

As mentioned above, for a correct interpretation of the collected exposure monitoring data, the bioavailability of each constituent of the inorganic pigment must be taken into account. Bioavailability can be estimated by measuring the bioaccessibility (i.e., release) of the metal ion from the constituents in bioelution test systems. In practice, the metal ion releases (Me_1, Me_2, Me_3 releases) from the constituents of the inorganic pigment are measured in five different synthetic biological fluids at two different time periods, 2 and 24 h, to determine the total released amount. The five different synthetic biological fluids used in the bioaccessibility testing are listed below:

- Gamble's solution (GMB): Mimics the interstitial fluid within the deep lung under normal health conditions.
- Phosphate-buffered saline (PBS): Standard physiological solution that mimics the ionic strength of human blood serum.
- Artificial sweat (ASW): Simulates the hypoosmolar fluid, linked to hyponatraemia (loss of Na+ from blood), which is excreted from the body upon sweating.
- Artificial lysosomal fluid (ALF): Simulates intracellular conditions in lung cells occurring in conjunction with phagocytosis and represents relatively harsh conditions.
- Artificial gastric fluid (GST): Mimics the very harsh digestion milieu of high acidity in the stomach.

For the test, a loading of 0.1 g/L was used to reflect a physiologically relevant concentration, that is, to be comparable to realistic exposure conditions of the lung, on the skin, or in the gastrointestinal tract. When comparing the released fraction of each element (metal ion) with the total loaded mass,<0.3%, <0.03%, and <0.02% of the Me_1, Me_2, and Me_3 contents, respectively, were transformed during 24 h of exposure at these acidic conditions in the GST (see Table 12.3).

By considering the hazard profile, the bioaccessibility of Me_3 may be of particular interest for the inhalation route, whereas Me_2 may be most relevant for the dermal route. On an elemental basis, Me_3 reaches its maximum in the physiologically relevant surrogate media ALF of 0.014% for inhalation and Me_2 in ASW of 0.021% for the dermal route, respectively (Table 12.3, values indicated in bold numbers).

TABLE 12.3 Elements Transformed (mass%), Equivalent to Their Percentage of the Elemental Content of the Total Amount of Particles Loaded for the Three Constituents of the Inorganic Pigment

Test Item	Exposure Period	GMB pH 7.4	PBS pH 7.2	ASW pH 6.5	ALF pH 4.5	GST pH 1.7
Me_1 release	2 h	0.015 ± 0.0050	0.014 ± 0.0042	0.041 ± 0.014	0.059 ± 0.0093	0.077 ± 0.0082
	24 h	0.024 ± 0.0053	0.018 ± 0.0030	0.017 ± 0.0051	0.071 ± 0.0082	0.27 ± 0.0073
Me_2 release	2 h	0.0066 ± 0.0037	0.010 ± 0.0040	**0.021 ± 0.0030**	0.019 ± 0.0021	0.040 ± 0.038
	24 h	0.0052 ± 0.0068	0.0096 ± 0.0032	0.020 ± 0.0079	0.021 ± 0.0010	0.028 ± 0.0025
Me_3 release	2 h	0.010 ± 0.011	0.0028 ± 0.0024	0.0073 ± 0.0026	**0.014 ± 0.0043**	0.0076 ± 0.0005
	24 h	0.0007 ± 0.0007	0.0016 ± 0.0013	0.0043 ± 0.0009	0.0054 ± 0.0002	0.011 ± 0.0003

12.3.4 PSD of Airborne Dust Generated From Inorganic Pigments

Since the PSD of the inorganic pigment will not be modified into smaller particles during any downstream use (confirmed by an extensive life-cycle research conducted by the industry), laboratory testing of the particle size of the material as placed on the market could be considered as being relevant during handling at the manufacturing stage and in any downstream use. The particle size investigation is relevant for airborne dust generated through light agitation of the inorganic pigment that may also occur during packaging and mixing activities. From the result displayed in Table 12.4, it can be seen that the inorganic pigment yield coarser airborne particles, with the majority of their particles having a 'mass median aerodynamic diameter' (MMAD) of more than 40 µm (reference anonymised).

The derived PSD enables the estimation of fractions of the airborne dust that may be deposited in the human respiratory tract. The 'multiple-path particle dosimetry model' (MPPD; CIIT, 2006) is used to predict this fractional deposition behaviour for workers. Using the morphological data of the human respiratory tract and the aerodynamic diameter of a specific particle as input, this model predicts the fraction of inhaled material that would get deposited in the extrathoracic (Head), tracheobronchial (TB), and alveolar (PU) regions. The model calculations are displayed in Table 12.5.

The total deposited fractions of the material that would become airborne under simulated workplace conditions are less than 44%. The rest of the airborne material would not get inhaled due to the physical phenomena related to air streams and turbulences close to the mouth or would be exhaled (i.e., not deposited). When assessing the exposure of specific regions of the respiratory tract, the data indicate that almost all the deposited material would actually have an impact on the pharynx (as reflected in the 'Head' column) and is subsequently most likely to be swallowed. Thus, almost all the deposited material would contribute to oral and not to inhalation exposure. As can be seen from the table below, only a marginal fraction of the airborne material will be deposited in the alveolar region of the human lung (PU = 0.7%).

TABLE 12.4   Particle Size Distribution of the Inorganic Pigment

Test Material	MMAD (µm)	GSD
Inorganic pigment	49.04	7.72

GSD, Geometric standard deviation; MMAD, mass median aerodynamic diameter.

TABLE 12.5   MPPD Model Outputs for Workers (%)

Test Material	Head	TB	PU	Total
Inorganic pigment	42.5	0.6	0.7	43.8

Head, extrathoracic region; MPPD, multiple-path particle dosimetry model; PU, pulmonary region; TB, tracheobronchial region.

12.3.5 Identification of an Indicator Substance for Inhalation Exposure

As several substances may contribute to generic dust exposure, an indicator substance needs to be identified and monitored to assess exposure to the inorganic pigment as a whole. In this example, considering the concentration of the constituent $MeOx_3$ in the inorganic pigment (roughly 20%; see Table 12.1) and the routine basis upon which the element Me_3 is analysed from dust in the workplace atmosphere (because of its severe health classification and corresponding low DNEL), Me_3 appears to be the most suitable indicator element to reflect occupational exposure to this inorganic pigment. It needs to be highlighted that the Me_3 content reported after analysis of the dust sampled during exposure monitoring has been chemically quantified after dissolution of the samples in sulphuric acid. It is therefore not representative of exposure to Me_3 but instead reflective of exposure to the inorganic pigment, in which Me_3 is tightly bound. Thus, after taking into account the Me_3 content in the inorganic pigment, exposure to inorganic pigment can be calculated. Selecting Me_3 as an indicator substance provides a good compromise between availability of exposure data and (highest) severity of effects of the individual constituents (if they were completely bioaccessible), and therefore representing the most robust basis. To calculate exposure to the inorganic pigment, when using Me_3 as an indicator substance an extrapolation factor of approximately 6.5 (100%/15.3% Me_3 content, see Table 12.1) should be used.

In a second step, one then needs to account for the bioaccessible fraction of Me_3 contained in the inorganic pigment of 0.014% (see Table 12.3) to be compared with the currently available toxicological threshold value. Since the effects in the pulmonary region of the lung are relevant for Me_3, such estimation of bioaccessible Me_3 would also need to account for the fact that only parts of the airborne particles (<1% (PU), see Table 12.5) are able to penetrate in that region. However, because of the intrinsic uncertainties related to such particle size considerations and in the absence of robust particle size information from the relevant workplaces, further refinement of the 'accessibility factor' was not applied. Thus, to estimate exposure to bioaccessible Me_3 ($Me_{3access}$), a factor of 1.4×10^{-4} needs to be applied to inorganic pigment exposure. Such derived concentration is expressed in µg $Me_{3access}/m^3$.

As indicated before, the proposed bioaccessibility factor is assumed to be highly conservative and to cover uncertainties of up to two orders of magnitude, deposition in the human lung is not considered. The margin of safety is calculated by comparing the estimated exposure concentration for bioaccessible Me_3 ($Me_{3access}$) with the relevant threshold value.

12.4 COLLECTION OF INHALATION EXPOSURE DATA

As indicated above, monitoring data were available at the industry level. A survey was therefore conducted to collect recent inhalation monitoring data and contextual information from companies in the inorganic pigment industry sector. A submission form was developed focussing exclusively on workplaces at manufacturing and formulating sites of inorganic pigments. The submitted data were screened for quality and consistency (see Section 12.4.2).

12.4.1 An Outline of the Data Submission Form

For reporting inhalation monitoring data, common workplaces that were assumed to exist at inorganic pigment manufacturing and formulation sites were defined together with industry experts. A questionnaire in which monitoring data were to be assigned to these workplaces was circulated. A surrogate workplace called 'job-rotation' was additionally inserted to, for example, reflect measurements that were taken on workers conducting multiple tasks. Qualifying information on sampling methodology (e.g., sampling head used, measured fraction according to EN 481, sampling duration and date) was requested because such information is required for quality screening (see Section 12.4.2). In addition, surveyed contextual information included, for example, the physical form of the materials handled or a brief description of tasks performed during sampling. Together with the total dust results, results were reported per chemical element as chemically quantified from the obtained samples.

12.4.2 Quality Screening

All data entered into the inhalation exposure database and to be used for the exposure assessment had to fulfil strict quality criteria (Chapter 2, Section 2.2.2.1). For the sake of brevity, only the most important criteria are listed below.

- In general, only measurements of personal inhalation exposure have been used.
- Depending on the exposure duration, these values have to be either full-shift-representative (with a minimum of 120 min measurement duration) or must have been obtained during the entire task duration.
- Time-weighted averages were not used as such but instead back-calculated to measure the exposure levels (if possible) to reflect full-shift exposure.
- The measured fraction of dust must be 'inhalable' according to EN 481.
- All measurements have to be assigned to a specific workplace, process, or task.
- Additional information such as measurement date, sampling equipment and method of analysis had to be provided for individual data sets.

Data not meeting the above-mentioned criteria could only be used as supportive information for the exposure levels to be derived.

12.4.3 Summary of Submitted Exposure Data

The manufacturing process of the inorganic pigment can be divided into two parts depending on the type of materials being handled. These parts are namely (i) 'precalcination', that is, handling of materials that have not yet been chemically transformed into an inorganic pigment and (ii) subsequent 'postcalcination' activities, consisting of the tasks and processes that include handling of the inorganic pigment and which are hence subject for this exposure assessment.

Table 12.6 presents the data situation for the constituents of the inorganic pigment (for measured constituents having concentrations of $\geq 0.1\%$ in the inorganic pigment) after conducting the quality screening.

TABLE 12.6 Number of Monitoring Data Points (Inhalable, Personal) Relevant for Specific Constituents (Given as Metal) of the Inorganic Pigment

Workplace	Me_1	Me_2	Me_3
Precalcination (not relevant for this exposure assessment)			
Raw material handling	2	17	35
Postcalcination (relevant for this exposure assessment)			
Calcination	2	7	14
Furnace unloading	N/A	13	13
Washing/drying	N/A	N/A	2
Milling/mixing	1	4	25
Pigment handling	N/A	15	28
Packaging of formulated solid product	4	1	3
Sampling	1	1	3
Cleaning and Maintenance	N/A	N/A	5
All postcalcination workplaces	8	41	93

N/A, not data available in the database.

12.5 INHALATION EXPOSURE ASSESSMENT AND RISK CHARACTERISATION

For the derivation of inhalation exposure estimates for the individual workplaces, data analyses were performed by using Me_3 as an indicator substance (i.e., by multiplying the analysed Me_3 levels by 6.5 as described in Section 12.3.5). According to current REACH Guidance R.14 (ECHA, 2016), the percentile to reflect the exposure level for workers has to be determined according to the specificity of the exposure situation's data to be assessed and the variability of the data as reflected by the geometric standard deviation (GSD). For each workplace, reasonable worst-case (RWC) estimates of inhalation exposure levels to the inorganic pigment were derived as presented in Table 12.7.

In the absence of any identified human health hazard for the inorganic pigment, an OEL or similar threshold has not been derived so far. Thus, the exposure estimates as derived above may be compared with generic dust limits set in many European countries to $10 \, mg/m^3$ (e.g., BMAS, 2015 or HSE, 2013), resulting in a safety factor of larger than 16 when dividing $10 \, mg/m^3$ by the maximum exposure estimate of $0.609 \, mg/m^3$.

When considering individual constituents, exposure to the Me_3 content (that is tightly bound in the inorganic pigment matrix) may be seen as a worst case, further justifying its selection as an indicator substance. Table 12.8 extrapolates the exposure to accessible Me_3 ($Me_{3access}$) by considering the Me_3 content and bioaccessibility. A safety factor was calculated by considering the current DNEL for Me_3 of $40 \, \mu g/m^3$ for effects after long-term inhalation.

As can be seen from Table 12.8, maximum RWC $Me_{3access}$ exposure levels were estimated for milling/mixing of inorganic pigments at less than $0.1 \, \mu g \, Me_{3access}/m^3$. Even for 'Cleaning

TABLE 12.7 Analysis of Monitoring Data of Inhalation Exposure to the Inorganic Pigment for Individual Workplaces (Results Provided in µg Inorganic Pigment/m^3, Inhalable, Personal)[a]

Workplace	Counts	Min	Median	GM	GSD	P75	P90	P95	Max
Precalcination (not relevant for this exposure assessment)									
Raw material handling	N/R	N/R	N/R	N/R	N/R	N/R	N/R	N/R	N/R
Postcalcination (relevant for this exposure assessment)									
Calcination	14	1.6	22.9	21.1	4.2	70.4	140.2	166.1	170.4
Furnace unloading	13	6.6	19.7	25.8	3.2	26.2	150.7	211.0	281.8
Washing/drying	2	1.6	1.6	1.6	1.0	1.6	1.6	1.6	1.6
Milling/mixing	25	1.6	45.9	33.4	4.9	59.0	344.7	452.2	530.8
Pigment handling	28	0.3	19.7	19.8	6.0	47.5	260.2	313.2	747.0
Packaging of formulated solid product	3	6.6	6.6	8.3	1.5	9.8	11.8	12.5	13.1
Sampling	3	13.1	19.7	19.5	1.5	24.2	27.0	27.9	28.8
Cleaning and Maintenance	5	1.6	39.3	23.9	13	183.5	317.2	361.7	406.3 (609.4)[a]
All postcalcination workplaces	**93**	**0.3**	**19.7**	**22.2**	**5**	**52.4**	**237.2**	**343.4**	**747.0**

GM, geometric mean; *GSD*, geometric standard deviation; *Max*, maximum; *Min*, minimum; *N/R*, not relevant; *P75*, 75th percentile; *P90*, 90th percentile; *P95*, 95th percentile; Shaded cells, selected percentile.
[a] Estimated exposure level (maximum multiplied by 1.5 (as reported in brackets)

TABLE 12.8 RWC Estimates of Personal Exposure for Postcalcination Workplaces and Calculated Me$_{3access}$ (Results Provided in µg/m^3, Inhalable Fraction)

Workplace	RWC Exposure Estimates			
	Inorganic Pigment	EF$_{access}$	Me$_{3access}$	Safety Factor
Calcination	140.2	1.4×10^{-4}	0.0196	2037
Furnace unloading	150.7		0.0211	1896
Washing/drying	1.6		0.0002	174,402
Milling/mixing	344.7		0.0483	829
Pigment handling	260.2		0.0364	1098
Packaging of formulated solid product	12.5		0.0017	22,948
Sampling	27.9		0.0039	10,235
Cleaning and maintenance	609.4		0.0853	469

EF$_{access}$, extrapolation factor for calculation of bioaccessible Me$_3$ content in inorganic pigment dust (see Section 12.3.5); *RWC*, reasonable worst-case.

and Maintenance', the RWC exposure levels of $Me_{3\,access}$ exposure was calculated at 0.085 µg $Me_{3access}/m^3$ still providing a safety factor of 469 even under the applied worst-case assumptions (e.g., particle deposition not considered). As mentioned before, these results do not consider any respiratory protective equipment.

12.6 DERMAL EXPOSURE ASSESSMENT AND RISK CHARACTERISATION

As mentioned already, the inorganic pigment is not classified according to EU CLP/UN GHS and hazards for the dermal route are not expected based on available information. In contrast to the assessment of the inhalation route, dermal exposure data representing actual workplace measurements were not available.

A dermal DNEL that can be used to compare is only available for the constituent $MeOx_2$ for dermal systemic, long-term effects (83 mg Me_2/kg bw/day) (see Table 12.2), whereas for the other constituents of the inorganic pigment a hazard is not identified for the dermal route.

Using the MEASE tool (see Chapter 9 for further details on MEASE), the highest achievable estimate for dermal exposure may be used as a very conservative estimate for potential dermal exposure to the inorganic pigment (assuming the highest dermal exposure potential for all determinants considered by the tool with an exposed skin area of 1980 cm² and not considering the use of gloves). The maximum dermal exposure estimate of 990 mg inorganic pigment/day includes approximately 20.5% Me_2 (see Table 12.1). In addition, for the calculation of systemic, dermal exposure a body weight of 70 kg for workers should be taken into account, which results in an exposure estimate of 2.9 mg Me_2/kg bw/day (=990 mg/day/70 kg × 20.5%). Furthermore, the very low bioaccessibility of Me_2 from the inorganic pigment as derived in the leaching test for ASW of 0.021% (see Table 12.3) needs to be taken into account—resulting in an exposure estimate of approximately 6×10^{-4} mg $Me_{2access}$/kg bw/day.

According to the HERAG, 2007 fact sheet on dermal exposure, a default dermal absorption factor of 1% has been proposed for dermal absorption of inorganic substances from liquid aqueous media that is reflected in the final exposure estimate of approximately 6×10^{-6} mg $Me_{2access}$/kg bw/day.

When comparing this estimate with the currently available DNEL of 83 mg/kg bw/day, a safety factor of larger than 1.36×10^7 is derived.

12.7 SUMMARY AND CONCLUSIONS

It has been shown that the constituent of the highest inhalation hazard potential of the inorganic pigment is $MeOx_3$ (assessed as Me_3). It is nevertheless noted that workers are exclusively exposed to the inorganic pigment as such and not to its individual constituents. By considering the deposition behaviour of the inorganic pigment in the human lung and by considering the very low bioaccessibility of Me_3 from the inorganic pigment, a correction factor of 10^{-4} has been established on a highly conservative basis (it has been shown above that a hypothetical correction factor could even be calculated at 10^{-6}). This factor should be used as a minimum when transforming any exposure measurements expressed as mass of Me_3 dispersed in workplace air into 'accessible Me_3 content' ($Me_{3access}$), the latter of which could be considered as being the fraction of exposure that may be of potential toxicological relevance for the inhalation route. It has been shown by quality screened exposure monitoring data and derived exposure estimates based on conservative assumptions that RWC exposure levels of $Me_{3access}$ are well below $0.1\,\mu g/m^3$. Considering a DNEL of $40\,\mu g\ Me_3/m^3$, the absence of significant inhalation exposure resulting from the handling of the inorganic pigment has been demonstrated for the processes and tasks conducted by workers at the assessed workplaces, with a corresponding minimum margin of safety of 469.

For dermal exposure, Me_2 has been identified as a toxicity driver. The highly conservative assessment resulted in a safety factor of approximately 1.36×10^7, indicating that the dermal route does not lead to any significant exposure either.

KEY MESSAGES

Inorganic pigments, used for the colourisation of other substances or mixtures, are usually incorporated into the matrix of the carrier material as a dispersed compound.

Owing to their stable crystalline structure, they are perceived as toxicologically neutral substances. However, as they can contain constituents with low effect thresholds, the pigment industry has developed an approach to check 'safe use'. This approach makes use of information on the inorganic pigment such as its composition, its constituents' hazard profile, and the particle size distribution. Exposure to the inorganic pigment is estimated on the basis of measurements of an element that can be used as an 'indicator substance of exposure' and the assessment is subsequently refined by considering the bioaccessibility data for individual constituents.

A combination of the consideration of the deposition behaviour of the pigment in the human lung and the very low bioaccessibility of the 'indicator' has allowed to establish a correction factor of 10^{-4} that can be used to transform actual exposure measurements (expressed as mass dispersed in workplace air) into 'accessible metal content', which could be considered as being the fraction of exposure that may be of potential toxicological relevance for the inhalation route.

References

BMAS (Bundesministerium für Arbeit und Soziales): *Technische Regel für Gefahrstoffe 900—Arbeitsplatzgrenzwerte (TRGS 900)*, 2015.

CIIT: *MPPD model (v2.0)*, CIIT Centers for Health Research, P.O. Box 12137, Research Triangle Park, NC 27709, USA, 2006.

ECHA (European CHemicals Agency): Guidance on Information Requirements and Chemical Safety Assessment Guidance, Chapter R.12. In *Use description. Version 3.0*, 2015, ISBN: 978-92-9247-685-4.

ECHA (European CHemicals Agency): Guidance on Information Requirements and Chemical Safety Assessment Guidance, Chapter R.14. In *Occupational exposure estimation (Version 3.0)*, 2016, ISBN: 978-92-9495-081-9.

EU CLP: *(Classification, Labelling and Packaging (CLP) of Substances and Mixtures) EC No.1272/2008*, 2008.

HERAG: *Assessment of occupational dermal exposure and dermal absorption for metals and inorganic metal compounds*, https://www.icmm.com/website/publications/pdfs/chemicals-management/herag/herag-fs1-2007.pdf, 2007.

HSE (Health Safety Executive): *Workplace exposure limits EH40/2005*, 2013.

MEASE: *Exposure assessment tool for metals and inorganic substances (Version 1.02.01)*, http://www.ebrc.de/mease.html, 2010.

UN GHS: *Globally harmonized system of classification and labelling of chemicals (GHS)*, Sixth revised edition ST/SG/AC.10/30/Rev.6, 2015, United Nations, New York and Geneva, https://www.unece.org/fileadmin/DAM/trans/danger/publi/ghs/ghs_rev06/English/ST-SG-AC10-30-Rev6e.pdf.

13

Risk Assessment of Alloys

Eirik Nordheim*, *Tony Newson*[†]

*Aluminium REACH Consortium, Brussels, Belgium †Consultant, Rotherham, United Kingdom

13.1 INTRODUCTION

While some metallic elements are used as engineering materials in their pure elemental state, the majority of metals are used in the form of alloys that have widespread applications in the technological society of today. The definition internationally used for an alloy is *a metallic material, homogeneous on a macroscopic scale, consisting of two or more elements so combined that they cannot be readily separated by mechanical means* (UN GHS, 2015). As such, alloys should not be considered as simple mixtures but as 'special mixtures' in which the properties of the ingredient substances are modulated by their inclusion within the chemical matrix of the mixture. Each alloy is unique with physical, mechanical, and chemical properties that differ significantly from those of their ingredients.

Consequently, the description of hazardous properties and human health and environmental risks of alloys based on the hazard and risk profiles of its individual ingredients may be incorrect. Alloys should be evaluated based on their own intrinsic properties rather than those of their alloying elements. The following key issues require specific recognition when classifying or performing a risk assessment for alloys:

- Alloys exhibit unique properties with their own intrinsic hazard profile that differs from the hazard profile of its ingredients.
- Central to the issue of hazard identification and the determination of the alloys' toxicity is the rate of transformation/dissolution (TD) of the alloy to its (bio)available form.
- Similar to metals and metal compounds, speciation is of paramount importance. The released metal ingredients can occur in different valences, associated with different anions or cations. In the environment, released ions can be associated to adsorptive agents, such as the dissolved organic matter (DOM) in water or bound to minerals in sediment or soil resulting in differences in (bio)availability. Speciation and valence are also important factors, determinants of toxicity, as explained in Chapter 3 and the speciation of released metal ions should be checked. Since a vast number of alloys exist and are available in different physical forms, grouping alloys and read-across of human health and environmental endpoints within an alloy group merits special attention.

13.2 PRODUCTION AND USE OF ALLOYS

In general, a metallic alloy consists of a metal or a metalloid base element, constituting the largest percentage of the material and one or several intentionally added elements to achieve specific and improved mechanical, physical, or chemical properties compared with its individual alloy ingredients. One of the earliest examples of the use of alloys is in the production of bronze from copper and tin around 3000 BC.

The most widely used alloy is steel, which consists primarily of iron and carbon, with varying amounts of other elements added to it. Nearly all aluminium products are alloys based on aluminium and with small amounts of metals such as iron, silicon, magnesium, copper, zinc, and manganese, depending on the properties needed for the designed use of the alloy. Hence, most of the metallic materials we use in our daily life are alloys.

Most alloys are produced by melting two or more metals or metalloids together. Generally, the major ingredient is melted first and the other ingredients are added to it as molten metals or in some cases in solid form (e.g. solid zinc is added to molten copper to prevent large losses due to volatilisation). Alloys may also be produced by a process known as smelting, where metal ores are heated in an electric arc furnace along with a suitable reducing agent (e.g. carbon or charcoal). Similarly, metallothermal processes use electric arc furnaces in combination with aluminium or silicon as reducing agents to transform mixtures of the more stable metal oxides directly into alloys of their constituent metals. Diffusion alloying of compacted metal powders is another example of an alloy production technique used extensively in powder metallurgy.

This chapter discusses successively hazard identification and classification for human health and environmental hazards.

13.3 HUMAN HEALTH HAZARD IDENTIFICATION AND CLASSIFICATION

The classification of mixtures under UN GHS/EU CLP is the same for hazards as for substances (UN GHS, 2015; EU CLP, 2008). Table 13.1 lists the endpoints for classification.

TABLE 13.1   List of Endpoints for Classification

Acute toxicity
Skin corrosion/irritation
Serious eye damage/irritation
Respiratory or skin sensitisation
Germ cell mutagenicity
Carcinogenicity
Reproductive toxicity
STOT-single exposure
STOT-repeated exposure

13.3.1 Hazard Identification and Classification of Mixtures: General Reasoning

According to the UN GHS and the EU CLP, as a general rule, when a mixture has to be classified, all available information on the mixture should be gathered in a first step. The next steps (evaluation of the available data and use for classification) are outlined in Fig. 13.1.

For health hazards, the classification of the mixture should preferably be based on available information (including toxicity test data) on the mixture as a whole, except when classifying for carcinogenicity–mutagenicity–reproductive toxicity (CMR) effects under EU CLP. In Europe, according to EU CLP Article 6 (3), the classification of mixtures for CMR effects must be based on the toxicological information on the ingredient substances.

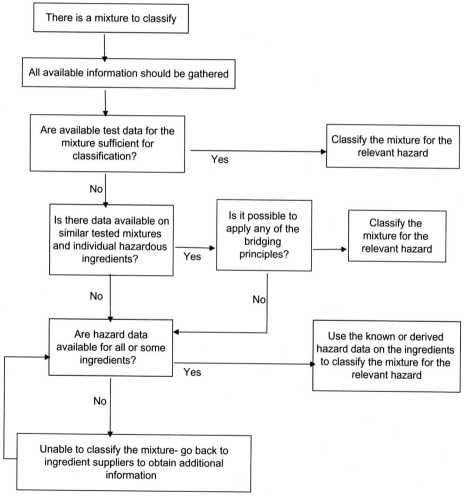

FIG. 13.1 How to classify a mixture. *Based on Fig. 1.6.1A, ECHA: Guidance on the application of the CLP criteria. Guidance to Regulation (EC) No 1272/2008 on Classification, Labelling and Packaging (EU CLP) of substances and mixtures Version 5.0., 2017. ISBN: 978-92-9020-050-5. Available from: https://echa.europa.eu/documents/10162/13562/clp_en.pdf.*

The UN GHS specifies that in most cases, it is not anticipated that reliable data for complete mixtures will be available for germ cell mutagenicity, carcinogenicity, and reproductive toxicity hazard classes and that therefore the mixtures will generally be classified based on the available information for the individual ingredients of the mixtures. There is however a possibility to have the classification modified case by case based on available test data for the complete mixture if such data are conclusive.

In practice, health data that directly apply to a mixture are rarely available. In those cases, further approaches to mixture classification may be applied: 'bridging principles' can be applied for some health hazards, using data on similar tested mixtures and information on individual hazardous ingredient substances (see also Chapter 8).

If you cannot apply bridging principles, the key to the classification of the mixture could be sufficient information on its ingredients. Part 3 of the UN GHS (UN GHS, 2015) and the Guidance on the application of the EU CLP criteria (ECHA, 2017) provide specific rules for the classification of mixtures based on the classification of the individual substances in the mixture, their content and the concentration of hazardous substances, referring either to an acute toxicity estimate (acute toxicity endpoint) or to cut-off/concentration limits, as proposed in the UN GHS (2015) and EU CLP (2008) for germ cell mutagenicity, carcinogenicity, reproductive toxicity, and specific target organ toxicity. Examples of such cut-off values/ concentrations are provided in Tables 13.2 and 13.3 for mutagenicity and STOT-RE, respectively.

It shall be noted that some regions set specific concentration limits (SCL) for specific substances, to take into account their potency as well as classification of other substances containing these substances as impurities, additives, or individual constituents. In the EU, the SCLs can be found in Part 3 of Annex VI of the EU CLP (EU CLP, 2008).

TABLE 13.2 Cut-Off Values/Concentration Limits of Ingredients of a Mixture Classified as Germ Cell Mutagens That Would Trigger Classification of the Mixture (UN GHS, 2015)

	Cut-Off/Concentration Limits Triggering Classification of a Mixture as:		
	Category 1 Mutagen		
Ingredient Classified as:	Category 1A	Category 1B	Category 2 Mutagen
Category 1A mutagen	≥0.1%	–	–
Category 1B mutagen	–	≥0.1%	
Category 2 mutagen	–	–	≥1%

TABLE 13.3 Generic Concentration Limits (GCL) of Ingredients of a Mixture Classified as a Specific Target Organ Toxicant That Triggers Classification of the Mixture (EU CLP, 2008)

	Generic Concentration Limits Triggering Classification of a Mixture as:	
Ingredient Classified as:	Category 1	Category 2
Category 1 Specific target organ toxicant	Concentration ≥10%	1.0% ≤ concentration <10%
Category 2 Specific target organ toxicant		Concentration ≥10%

In case a specific concentration limit (SCL) has been established for one or more ingredients these SCLs have precedence over the respective generic concentration limit.

13.3.2 Classification of Alloys: General Reasoning Under GHS/CLP

Alloys are considered as mixtures under EU CLP and UN GHS. The proposed classification approach follows the same general reasoning as outlined in Fig. 13.1 (Are the available test data for the mixture(s) sufficient for classification? Is there a possibility to apply bridging principles? Are hazard data available for the ingredients?).

A discussion has started at the EU level on whether the EU CLP approach could possibly be refined to consider the specificities of alloys, in particular the presence of a matrix that can impact the release of the metal ion (and hence the bioavailability). The debates focus on systemic effects, following exposure by the oral route. This refinement is embedded in the general classification scheme, at the level where ingredients' data are used to classify the mixtures (Fig. 13.2). This refinement (alloy-specific approach) uses the 'bioaccessible concentration' (BC, %) in oral fluid instead of the bulk concentration of the ingredient in the alloy. It does not modify the classification of the individual ingredients or possible cut-off limits that have been established (the generic cut-off limits and the EU SCL). This alloy-specific approach is presented in Figs. 13.3–13.6 and is described further in the text.

For all other endpoints, and at this stage of the regulatory discussions, it is proposed to follow the normal classification approach based on the ingredients of the alloy (expressed as concentrations) and to compare those to cut-off values/concentration limits).

13.3.3 In Practice

The first step in the hazard identification and classification process for all mixtures—including alloys—involves the collection and evaluation of all available data on the mixture, its composition, and its ingredients.

The data on the mixture and its ingredients can be listed using, for example, the following template (Table 13.4):

Depending on the sufficiency and quality of the data collected for the alloys, the following approaches can be used for each systemic health effect endpoint as deemed appropriate:

- When test data on systemic toxicity specific to the alloy (as a whole) are available, these data can be directly compared with the criteria for hazard classification. This applies to all systemic endpoints except in the EU CLP, where for CMR endpoints "the manufacturer, importer or downstream user of a mixture shall only use the relevant available information referred to […] for the substances in the mixture" (EU CLP Article 6(3), 2008). There are some instances where alloy-specific toxicity data exist for at least one health endpoint. In this case, these data would be applied to that endpoint, while for the other endpoints, bridging principles or data on ingredients would be used. For example, nickel-containing alloys in massive forms can be tested using EN1811 and based on the results from this test (i.e. rate of nickel ion release) their classification as dermal sensitisers can be determined, independent of their nickel content (EN1811:2011+A1:2015, 1998). There is one example where the toxicity of an alloy powder (SS316L) was tested in rats in a 28-day inhalation study. The data generated in this study can be used to decide whether a classification as STOT-RE is needed for SS316L (Stockmann-Juvala et al., 2013).
- In accordance with UN GHS and EU CLP guidelines for classification of mixtures, the use of bridging principles (i.e. read-across between mixtures) should be considered for the classification of alloys for which toxicology data are not available. This approach can

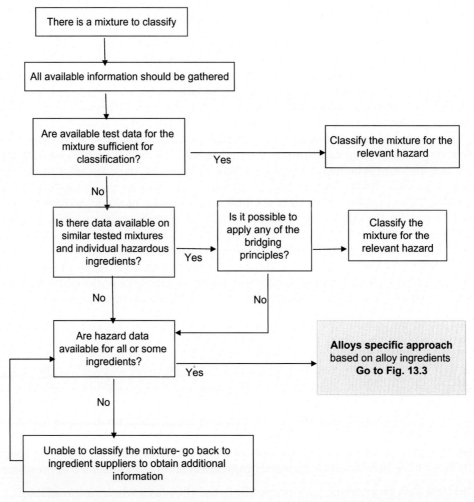

FIG. 13.2 How to classify an alloy. *Based on Fig. 1.6.1A, ECHA: Guidance on the application of the CLP criteria. Guidance to Regulation (EC) No 1272/2008 on classification, labelling and packaging (CLP) of substances and mixtures Version 5.0. 2017. ISBN: 978-92-9020-050-5. Available from: https://echa.europa.eu/documents/10162/13562/clp_en.pdf.*

be used to group target alloys with other similar alloys for classification where sufficient data on alloy characteristics (e.g. metal bioaccessibility and physicochemical properties, chemical composition, and technical performance, etc.) are available. This ensures that the classification process uses the available alloy data to the greatest extent possible without relying on additional, unnecessary animal testing. As explained in Chapter 8, bioaccessibility data can be used as part of the evidence that allow read-across and grouping to support the bridging approach for hazard classification of alloys. The main principle driving this approach is that bridging can be used if hazard classifications exist for a 'source' alloy and sufficient data exist to demonstrate that a 'target' alloy has similar exposure behaviour (e.g. release rate of metals) relative to the source alloy. It shall be

TABLE 13.4 Matrix of Data on the Mixture and Its Ingredients (Name, %, Classification, GCL, or SCL* (*under EU CLP), Classification Specified for a Specific Exposure Route)

Alloy Name:			
Source:			
Process:			
Designation: For Example, ASTM C36000, DIN CuZn36Pb3			
Physical Properties **Specific Gravity: For Example, 8.5 kg/L** **Melting Point: For Example, 1650°C**			
Ingredient name (% content up to the 0.01% level depending on ingredient SCL or GCL for the endpoint)	Ingredient 1	Ingredient 2	Ingredient 3
Acute oral toxicity			
Germ cell mutagenicity (incl. GCL/SCL*)			
Carcinogenicity systemic organ (incl. GCL/SCL*)			
Reproductive toxicity (incl. GCL/SCL*)			
STOT-single exposure systemic organ (incl. GCL/SCL*)			
STOT-repeated exposure systemic organ (incl. GCL/SCL*)			

noted that the EU CLP does not allow the use of bridging for CMR endpoints (EU CLP article 6.3), where only information on the ingredients in the mixture shall be used.

- If bridging is not possible, classification of the alloys needs to be based on the classification of the individual ingredients (e.g. metals) and their content, in line with the principles outlined in the EU CLP and the UN GHS (calculation of an ATE mixture for acute toxicity, comparison with 'concentration limits' for the other endpoints). The concentration limits to be considered are the generic concentration limits (GCLs) and the SCLs, as proposed in the UN GHS (2015) and EU CLP (2008). This is where the proposed alloy-specific refinement will come in, replacing the use of bulk concentration of alloy ingredients that are classified for systemic effects with the use of BC (gastric fluid) of classified ingredients in alloys. As explained above, at this stage, this refinement is currently discussed only for systemic effects after oral exposure. The different steps of this refinement are explained in Figs. 13.3–13.6.

The starting point is the collection of data on the alloy, and, in particular, on the composition of its ingredients, at or above 0.01%.

Then the assessor should move to **Step 1:** *Does the alloy contain an ingredient classified for systemic effects at concentrations above the generic concentration limit (GCL)/specific concentration limit (SCL)?*

This question can be answered using Table 13.4 with regard to systemic effects. For example, we could have an alloy XYZ (Table 13.5):

The GCL to be considered in this alloy-specific approach are those relevant to systemic effects after oral route. Also check if some of the ingredients have SCL listed in Annex VI to the EU CLP.

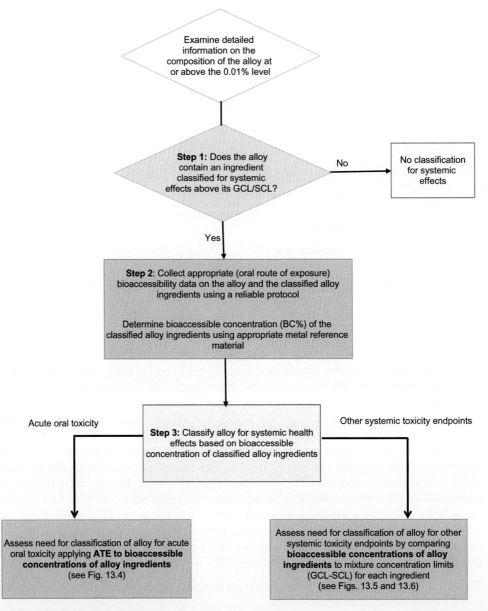

FIG. 13.3   Alloys specific approach.

Possible answers to **Step 1**:

⇨ **yes**: there is one or more classified ingredient above the SCL/GCL: go to **Step 2** of Fig. 13.3
⇨ **no**: no classification as per current EU CLP/UN GHS is warranted

Step 2: *Collect appropriate (oral route of exposure) bioaccessibility data on the alloy and its classi-fied ingredients using a reliable protocol.*

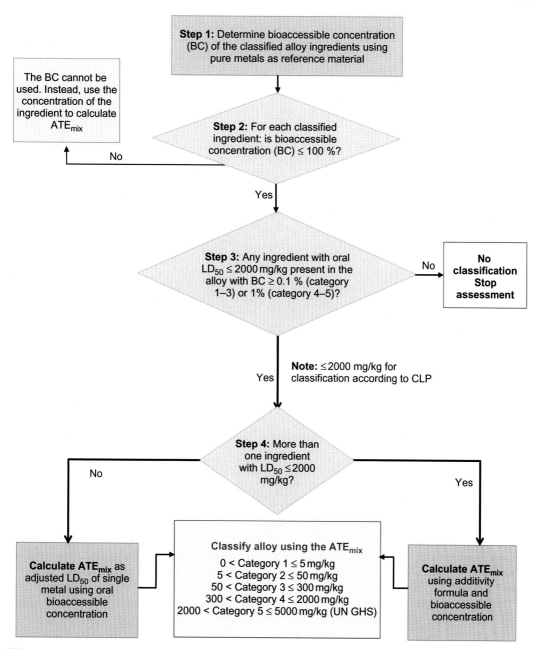

FIG. 13.4 Alloy specific approach, **acute oral toxicity**, with reference to the EU CLP for the relevant ingredients' concentrations to consider.

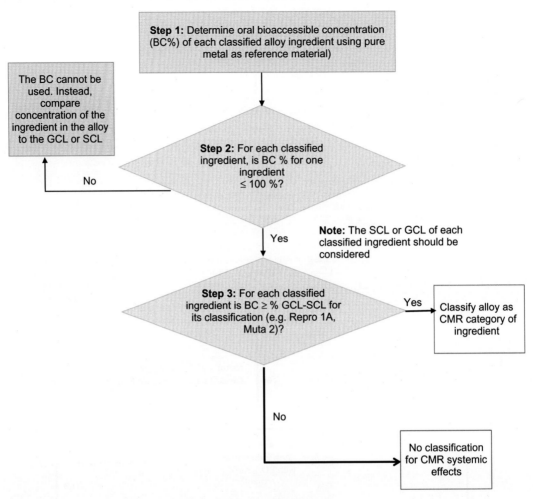

FIG. 13.5   Alloy-specific approach, **Other systemic endpoints (CMR).**

The BC, which is the concentration of the bioaccessible metal (i.e. the actual concentration that could be released from the alloy based on results from bioelution tests and potentially absorbed), should be determined using an appropriate metal reference material. In generic terms, the reference material should be as chemically (and physically, e.g. particle size distribution) similar as possible to the form of the metal present in the material and have a well-documented hazard profile. For alloys, it is suggested to use a pure metal as the reference material.

The BC is calculated as follows:

$$BC = \frac{\text{Release from alloy}}{\text{Release from pure metal}} \times 100\% \tag{13.1}$$

Once data are collected and BC% is determined, move to **Step 3**.

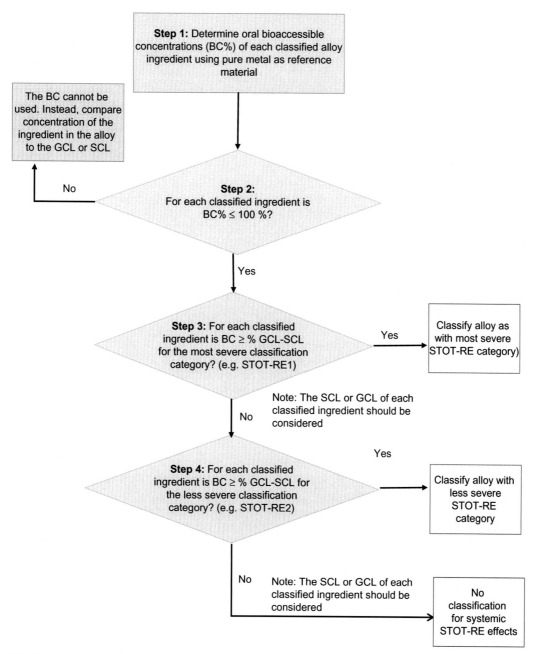

FIG. 13.6   Alloy-specific approach, **STOT-RE** effects.

TABLE 13.5   Matrix of Data on Alloy Ingredients Identifying the Ingredients Above the Concentration Limit

Alloy XYZ	Ingredient X	Ingredient Y	Ingredient Z
Ingredient name (% concentration)	Copper metal (65%)	Zinc metal (34%)	Lead metal (0.8%)
Reproductive toxicity *Oral route*	No classification	No classification	Cat 1A SCL: 0.3% massive SCL: 0.03% powder

SCL, specific concentration limit.

Step 3: *Classify using the BC*

To classify alloys for systemic health effects based on bioelution-derived bioaccessibility concentration of classified alloy ingredients move to Fig. 13.4 (acute toxicity), Fig. 13.5 (other systemic endpoints), and Fig. 13.6 (STOT-RE).

13.3.3.1 Acute Oral Toxicity

For acute oral toxicity, classification is based on a comparison between acute toxicity estimate of the mixture (ATE_{mix}) and acute toxicity category ranges defined under the EU CLP and the UN GHS.

The ATE_{mix} is derived through the use of an additivity formula and the ATEs for the individual ingredients present in the alloy at $\geq 1\%$ (nominal content). Please note that in the EU CLP, the generic cut-off will depend on the hazard class of the ingredient for acute toxicity (Table 13.6).

The UN GHS defines the relevant ingredients of a mixture as those present in concentrations $\geq 1\%$ (w/w for solids, liquids, dusts, mists, and vapours and v/v for gases), unless there is a reason to suspect that an ingredient present at a concentration <1% is still relevant for classifying the mixture for acute toxicity. This point is particularly relevant when classifying untested mixtures that contain ingredients that are classified under categories 1 and 2.

The approach to follow to classify alloys using the bioaccessibility concentration is explained in Fig. 13.4.

Fig. 13.4, **Step 1**: Determine the oral bioaccessible concentration (BC%) of the classified alloy ingredients using pure metals as reference materials.

Fig. 13.4, **Step 2**: Check if for each classified ingredient the BC% is less or equal to 100% (this means that the ion release from the ingredient in the alloy is lower or equals the release of the ion from the reference material).

⇨ **yes:** go to **Step 3**
⇨ **no:** the refinement cannot be used and the concentration of the ingredient in the alloy should be used to calculate the ATE of the mixture.

TABLE 13.6   Annex I EU CLP: Generic Cut-Off Values (EU CLP, 2008)

Hazard Class	Generic Cut-Off Values to be Taken Into Account (%)
ACUTE TOXICITY	
- Categories 1–3	0.1
- Category 4	1

Fig. 13.4, **Step 3**: Is there any ingredient with an $LD_{50} \leq 2000\,mg/kg$ (in the EU) or $5000\,mg/kg$ (under the UN GHS) and a $BC \geq 0.1\%$ (if the ingredient's ATE places it in categories 1–3 under the EU CLP) or $\geq 1\%$ (if the ingredient's ATE places it in category 4, under the UN GHS and EU CLP)?

Note: the LD_{50} of the reference material for the calculation of BC should be used in **Step 3**.
  ⇨ **no**: no classification for acute oral toxicity
  ⇨ **yes**: go to **Step 4**
Step 4: Is there more than one ingredient with an $LD_{50} \leq 2000\,mg/kg$ (EU CLP) or $\leq 5000\,mg/kg$ (UN GHS)?
  ⇨ **no**: if only one ingredient has an estimated toxicity value below the upper limit of classification, classify based on that ingredient. Calculate ATE_{mix} as adjusted LD_{50} of the single metal using the BC and compare to the categories below
  ⇨ **yes**: if there is more than one ingredient, calculate ATE_{mix} using the additivity formula and BC:

$$\frac{100}{ATE_{mix}} = \sum_n \frac{C_i}{ATE_i} \tag{13.2}$$

where Ci = bioaccessible concentration of ingredient i in the alloy (%w/w) and $AREi$ = acute toxicity estimate for ingredient i.
Classify the alloy based on the acute oral toxicity category in which the ATE_{mix} falls. For oral toxicity the following category cut-offs apply (mg/kg bw):

- $0 < Category\ 1 \leq 5$
- $5 < Category\ 2 \leq 50$
- $50 < Category\ 3 \leq 300$
- $300 < Category\ 4 \leq 2000$
- $2000 < Category\ 5 \leq 5000$

Hypothetical example using the BC for determining the hazard classification for acute toxicity via the oral route:
The oral LD_{50} values of the three constituent metals of alloy XYZ, to be used as the ATEs, are $3000\,mg/kg$, $1000\,mg/kg$, and $40\,mg/kg$ for X, Y, and Z, respectively. The content (i.e. nominal concentration) of the respective metals in the alloy is X: 60%, Y: 36%, and Z: 4%. Bioelution testing gives the following ion releases from the metals in the alloy, as compared with the ion releases from the reference materials (BC): X: 45%, Y: 18%, and Z: 2%.

Metal in Alloy	ATE (mg/kg)	Bioaccessible Concentration (%)
X	7000	45
Y	1000	18
Z	40	2

Compared with the classification scheme:

One constituent (X) has an $LD_{50} > 5000 \, mg/kg$ and it is not included in the ATE_{mix} calculation. Calculation of the ATE_{mix} proceeds, therefore, using the corresponding ATE and BC data for Y and Z as follows:

$$100 / ATE_{mix} = BC\%_Y / ATE_Y + BC\%_Z / ATE_Z = 18 / 1000 + 2 / 40 = 0.068;$$
$$ATE_{mix} = 1470 \, mg / kg \tag{13.3}$$

Based on this result, alloy XYZ would be classified as category 4 for acute oral toxicity when following the refined approach using the bioaccessibility data.

13.3.3.2 CMR Endpoints

By default, the classification for these endpoints compares the concentration of the ingredient in a mixture to generic cut-off values/concentration limits or SCL (EU CLP). The same reasoning is applied here in the refinement for systemic effects, oral exposure but the concentration is substituted by the bioaccessible concentration (BC) (Fig. 13.5).

Fig. 13.5, **Step 1**: Determine the oral BC% of each classified alloy ingredient using a pure metal as the reference material

Fig. 13.5, **Step 2**: Check if for each classified ingredient the BC% is less or equal to 100%?
  ⇨ **yes**: go to **Step 3**
  ⇨ **no**: the refinement cannot be used: compare the concentration of the classified ingredients with the available cut-offs (GCL or SCL)

Fig. 13.5, **Step 3**: Check for every classified ingredient, the BC ≥ is above the GCL, SCL (%) for its classification?
  ⇨ **yes**: classify the alloy with the CMR category of the ingredient
  ⇨ **no**: no classification for CMR systemic effects

Hypothetical example using the BC approach for determining the hazard classification for germ cell mutagenicity.

The alloy has three constituent metals: V, W, and X. V is not a mutagen; W is a category 2 mutagen; and X is a category 1 mutagen. The content (i.e. nominal concentration) of the respective metals in the alloy is V: 52%; W: 44%; and X: 4%. Bioelution testing gives the following ion releases from the metals in the alloy, as compared with the ion releases from the reference materials (BC): V: 42%, W: 22%, and X: 0.01%.

Metal in Alloy	Mutagen Classification	Bioaccessible Concentration (%)
V	None	42
W	Category 2	22
X	Category 1	0.01

Following the alloy-specific approach, alloy VWX would be classified as category 2 Muta, because W is classified as cat 2 and it has a BC% > 1% and X is classified as cat 1 but it has a BC% < 0.1%.

13.3.3.3 STOT-RE

For STOT-RE, the GCL triggering the classification of the mixture and defining the categories shall be considered (see Table 13.3). If there is an SCL established for one or more of the alloy's ingredients, the SCLs take precedence over the GCLs.

Fig. 13.6, **Step 1**: Determine the oral BC% of each classified ingredient using a pure metal as the reference material

Fig. 13.6, **Step 2**: Check if for all the classified ingredients of the alloy, the BC% is less or equal to 100%?

> ⇨ **yes**: go to **Step 3**
> ⇨ **no**: the scheme cannot be used: the concentration of the ingredient in the alloy should be compared with the GCL–SCL

Fig. 13.6, **Step 3**: Check for each of the classified ingredients of the alloy, whether the BC% is ≥GCL–SCL for the most severe classification category (%) (e.g. STOT-RE1)?

> ⇨ **yes**: classify the alloy with the most severe category
> ⇨ **no**: go to **Step 4**. Note: The SCL of the reference material used to calculate the BC should be considered

Fig. 13.6, **Step 4**: Check for each of the alloy's classified ingredients, whether the BC% is ≥GCL–SCL for the less severe classification category (%) (e.g. STOT-RE2)?

> ⇨ **yes**: classify the alloy as less severe category
> ⇨ **no**: no classification for systemic STOT-RE effects

Note: In Figs. 13.4–13.6 a provision has been included in case the BC% is higher than 100%. There are indeed some alloys (e.g. some sulphur-containing alloys) that may release more metals than predicted based on release from pure ingredients (i.e. exceptional alloys). This means that the matrix effect will result in increased releases of the classified metal ion from the alloy compared with those from the pure metal. For such alloys, where increased metal release is suspected (e.g. based on their composition or function), the refinement based on the BC% should not be used. It is recommended to use the concentration (%, content) in the classification schemes. In general, knowledge on basic physicochemical properties of the alloy such as corrosion characteristics, surface properties, data on the structure, Pourbaix diagrams, as well as information on technical performance of the alloy, and TD data can provide an indication about the potential enhanced metal releases.

13.4 ENVIRONMENTAL CLASSIFICATION SCHEME FOR ALLOYS

The typical physicochemical properties of solid and nonsolid (e.g. brazing pastes) alloys clearly suggest that a simple classification based on the hazardous properties of their constituents may be incorrect: alloys should be classified on the basis of their intrinsic properties rather than those of their alloying elements.

A proper characterisation of an alloy should contain the following components:

- Concentration of the metal constituents
- Physicochemical properties of the alloy and its constituents
- Crystal structure of the alloy
- Relevant ecotoxicological data on the metal(s) and alloy
- Speciation and bioavailability parameters
- Surface property of the alloy
- Corrosion data, metal release, and run-off studies
- Data from simple solubility tests or preferably transformation/dissolution tests (abbreviated as TD or T/Dp) (OECD, 2001)
- Electrochemical evidence
- Available information (including classification) of comparable alloys.

Specific attention should also be given to determine whether differences in minor substance content in the alloy could significantly affect corrosion or other physical properties that could lead to a different environmental classification category.

This characterisation and the further classification combines the collection of the best available analytical information on the alloy with the best expert judgement on the part of experienced metallurgists, chemists, and (eco)toxicologists (Fig. 13.7). However, depending on the amount of information that is available, some expert judgement-based assumptions may be required for several aspects of the alloy properties; this may lead to a worst-case approach in which the different alloying constituents are treated as highly soluble and highly bioavailable.

The collected information on the alloy can be summarised in a template as shown in Table 13.7.

Owing to a large number of alloys that exist and are currently produced, the individual ecotoxicological assessment of an alloy as a substance is deemed impractical for read-across purposes. In many cases, the ecotoxicity data of the individual constituents will form the basis for the classification of a specific alloy, taking into account the rate and extent to which metallic ions can be generated from the alloy.

TABLE 13.7   Matrix of Data on the Mixture and Its Ingredients (Name, %, Classification, M Factor)

Alloy Name:			
Source: **Process:**			
Designation: For Example, ASTM C36000, DIN CuZn36Pb3			
Physical Properties **Specific Gravity: For Example, 8.5 kg/L** **Melting Point: For Example, 1650°C**			
Ingredient name (% content)	**Ingredient 1 (%)**	**Ingredient 2 (%)**	**Ingredient 3 (%)**
Aquatic toxicity acute (including M Factor)			
Aquatic toxicity chronic (including M Factor)			

M Factors are used in the application of the summation method for classification of mixtures containing substances that are classified as very toxic. The concept of M Factors has been established to give an increased weight to very toxic substances when classifying mixtures. M Factors are only applicable to the concentration of a substance classified as hazardous to the aquatic environment (categories Acute 1 and Chronic 1) and are used to derive by the summation method the classification of a mixture in which the substance is present. They are, substance-specific and it is important that they are being established already when classifying substances (ECHA, 2017).

Alloys are grouped under numerous national and international standards (AFNOR, AISI, ASTM, DIN, JISI, etc.) mainly based on their chemical composition. Metal alloys, by virtue of their composition, are often grouped into two classes: ferrous and nonferrous alloys. Ferrous alloys (including steels and cast irons) are those in which iron is the primary constituent. Nonferrous alloys, on the other hand, are not iron based, and this group is further subdivided according to base metal or some distinctive characteristic that is shared by a group of alloys (e.g. copper, aluminium, magnesium and titanium alloys, refractory metals, superalloys, etc.). For classification purposes, this grouping may be less useful and it should be treated with care because it is unlikely that the chemical composition is the most important factor in determining alloy behaviour when exposed to biological systems or environmental media.

It could be worthwhile, however, to consider grouping alloys based on chemical compositions, microstructure, and properties (e.g. nickel-containing stainless steels (austenitic) based on 16%–30% Cr and 6%–22% Ni; low-alloy steels containing a few percent of elements such as Cr, Ni, Mo, and V; copper–nickel alloys). Within an alloy group, it can be assumed that these alloys display similar behaviour (e.g. release rate of metals assessed by dissolution tests) that can be related to the potential biological impact.

Grouping for environmental classification has value to facilitate read-across of the data available for a well-defined representative alloy to those alloys for which data are limited, to bridge data gaps and avoid unnecessary testing. Grouping should be applied in a very careful way recognising the 'mechanism of action' (biological and chemical factors) for the endpoint under consideration (e.g. for aquatic toxicity the free metal ion interacting with the gills). In this regard, the bioavailable fraction would be the ideal comparative level but this type of information is not always available. In this respect, the release under a given pH condition may in those cases be a valid alternative.

In practice, when assessing an alloy for its environmental hazards, it can first be assessed whether structural relationships can be used to classify the alloy based on comparison with alloys already environmentally classified. When an adequate review of the relevant information that was identified during the characterisation step shows that there is no evidence that the alloy under consideration would behave differently than the one(s) already classified, it is suggested to classify accordingly. This bridging principle could include similar composition, surface, and dissolution properties. For example: an alloy (I) ('target alloy') containing the same alloying elements as alloy (II), for which hazard data are available for the alloying elements A and B and that has already been classified ('source' alloy), could be assigned to the same hazard category if only the concentration of alloying metals A and B in alloy (I) are lower and the TD test or corrosion test results show lower or equivalent metal A, B release from alloy (II).

If there is no clear data of sufficient validity to show that the alloy can be allocated to a source group, a simple solubility test such as the 24-hour screening TD test (24-hour TD)—as described in Annex 10 of the UN GHS Document (UN GHS, 2011) could be conducted on both the alloy under evaluation and the peer reference alloy. The 24-hour TD test is chosen because of its high dose, rigid test conditions resulting in a high solubility and hence, better discriminatory power. As the equivalent of metals, the test should be conducted under conditions that avoid abrasion.

On the basis of the results of the 24-hour TD test, the likelihood that the alloy under consideration has the potential of releasing metal ions to a larger, smaller, or similar extent than its alloying constituents can be evaluated to ensure that the existing classification data on a similar alloy can be exploited for as many alloys as possible. The 24-hour screening test results could also help in determining if the summation method can be used (Fig. 13.8).

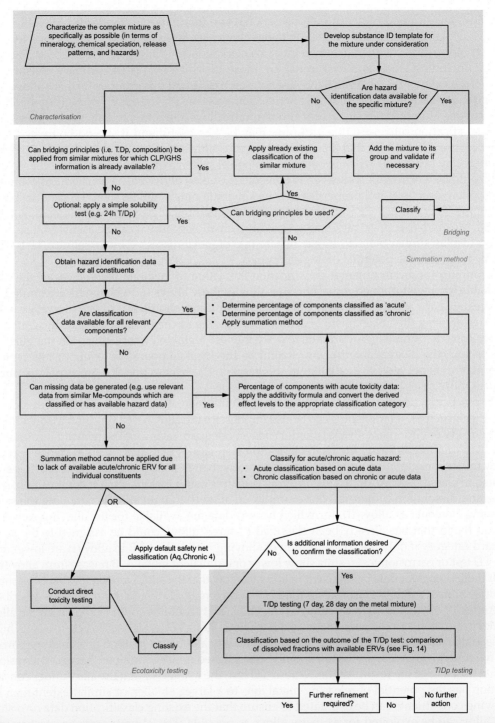

FIG. 13.7 The conceptual outline of the classification strategy for complex metal mixtures/materials (MERAG, in press).

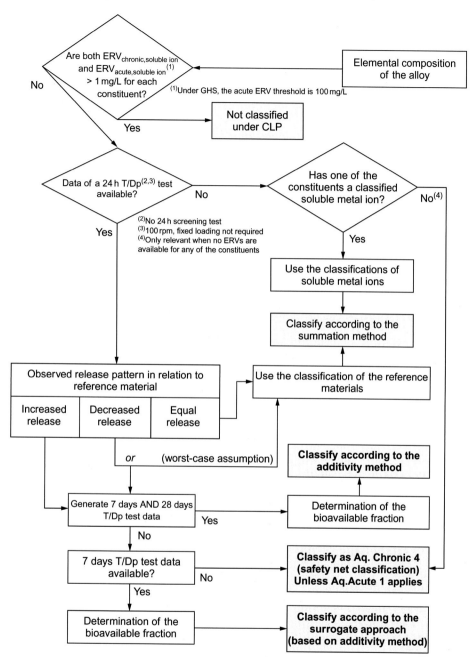

FIG. 13.8 The general classification scheme for alloys.

For a number of alloys tested up to now, available 24-hour TD data indicate that soluble metal ion concentrations after 24 hours are a reliable indicator for assessing whether the long-term release pattern of the alloy is similar to that of the pure metal, or whether an increased release or decreased release (alloy effect) can be expected. Therefore, it is strongly recommended to conduct a 24-hour TD test when deriving a reliable environmental classification of an alloy that contains hazardous constituents at concentration levels that would trigger an environmental classification (assumptions: 100% bioavailability is assumed, and classification is based on the summation method), and to share data/experience.

13.4.1 If 24-Hour TD Data Are Unavailable

When no 24-hour TD test data are available, the summation method can be applied when all metal constituents have a known classification. When the alloy contains a constituent for which no classification could be derived (i.e. no reliable ecotoxicity data are available), then the alloy has to be classified as Aquatic Chronic 4 (the default safety net classification), unless the summation method that is based on the other constituents already results in an Aquatic Acute 1 classification. It is expected that ion-based ecotoxicity data will be available for more metals by 2018 (EU REACH registration deadline for lower tonnage substances), making the safety net classification redundant.

13.4.2 If 24-Hour TD Test Data Are Available

- Measured soluble metal concentrations of the alloy are equivalent to those of the reference material (massive metal/metal powder): classification of the reference materials can be used, and classification for the alloy is based on the summation method;
- Measured soluble metal concentrations of the alloy are lower than released concentration from the reference material (massive metal/metal powder): classification of the reference materials can be used as a worst case, and classification for the alloy is based on the summation method. Refinement of the classification, however, is possible by conducting 7-day/28-day TD tests;
- Measured soluble metal concentrations of the alloy are higher than released concentration from the reference material (massive metal/metal powder): classification of the alloy has to be derived by using data from a 7-day and 28-day TD test (additivity method, taking into account the bioavailable fraction). If only 7-day TD data are available, the chronic classification will be based on the surrogate approach. As long as TD data are not available, the Aquatic Chronic 4 safety net classification is applied (unless Aquatic Acute 1 applies, based on the hazard of the soluble ion).

For some alloys, it has been demonstrated that the ratios between different constituents in the (thin) surface layer of an alloy can be significantly different from the overall composition; an alloy containing 2% of metal X may actually have a significant higher or lower percentage in the surface layer. The available information on such differences within an alloy particle may help to understand and explain enhanced/decreased alloy-effect rates.

It is recognised that conducting TD tests with nonsolid alloy forms (e.g. brazing pastes) may not be straightforward due to their fluidity and viscosity. When a water-based flux is used for dispersion of the powder particles during application of the paste, the bridging of the classification of the equivalent powder alloy (i.e. pure metallic powder form without addition of a flux medium) can be applied. Other types of fluxes, however, may affect the release of metal ions in a different way. Under these circumstances, the material has to be considered as 'difficult to test', and the proposed general classification method for alloys may not be valid anymore. Deriving a classification for such materials may require direct ecotoxicity testing that is conducted in line with the CLP guidance on 'Substances that are difficult to test' (ECHA, 2017).

A stepwise example of the classification of an alloy—taking into account more information on the alloy in each step—is presented below:

An alloy (massive) has three different constituents (Me_A, Me_B, and Me_C) and the following properties are known for its constituents:

- Composition of the alloy and acute/chronic ERV for the soluble ion of each metal in the alloy;
- Me_A-ion is Aquatic Acute1/Chronic1, M-factor 1;
- Me_B-ion is Aquatic Acute1/Chronic1, acute M-factor = 1; chronic M-factor = 10;
- Me_C-ion is not classified;
- Pure metal forms are not classified (based on 28-day TD testing with the massive pure metals);
- 24-hour release rate (in %) for each metal in the massive form;
- 24-hour release rate (in %) for each metal in an alloy configuration (outcome of 24-hour TD test);
- 28-day release rate (in %) for each metal in an alloy configuration (outcome of 28-day TD test).

The physicochemical and ecotoxicological properties of the alloy ingredients are as given in Table 13.8:

TABLE 13.8 Properties of the Alloy Ingredients

	Me_A	Me_B	Me_C
Percentage in alloy	12%	8%	80%
Classification of pure massive metal[a]	Not classified	Not classified	Not classified
Ion-based ERV$_{acute}$	400 µg Me_A/L Aq.Acute 1, $M=1$	110 µg/Me_B/L Aq.Acute 1, $M=1$	10,500 µg Me_C/L Not classified in CLP
Ion-based ERV$_{chronic}$	32 µg Me_A/L Aq.Chronic 1, $M=1$	3 µg/Me_B/L Aq.Chronic 1, $M=10$	2200 µg Me_C/L Not classified
Release rate of the pure massive (24 h T/Dp)	0.042%	0.028%	0.02%
Release rate (normalised) of metal in alloy (24 h T/Dp)	0.041% No alloy effect	0.21% Negative alloy effect	0.02% No alloy effect
Release rate (normalised) of metal in alloy (28 days T/Dp)	0.89%	3.9%	0.06%

[a]Based on 28-day T/Dp data (not reported).

Step 1: Classification based on elemental composition (summation method)

Assumption: all metal in the alloy is bioavailable
Is the alloy Aquatic Acute 1?
Summation of all Aquatic Acute 1 fractions, corrected with M-factor:

$$= 12\% + 8\% = 20\% < 25\% \left(\text{cut} - \text{off for Aquatic Acute1} \right) \tag{13.4}$$

=> Alloy is not classified for the acute hazard under EU CLP

Is the alloy Aquatic Chronic 1?
Summation of all Aquatic Chronic 1 fractions, corrected for M-factor:

$$= 12\% + \left(10 \times 8\% \right) = 92\% > 25\% \left(\text{cut} - \text{off for Aquatic Chronic1} \right) \tag{13.5}$$

=> Alloy is classified as Aquatic Chronic 1

Step 2: TD-based refinement of the chronic classification, using 24-hour data

Evaluation of an alloy effect requires the comparison between the 24-hour TD results of pure metals with those of metals in alloy configurations. If a comparable (no alloy effect) or a decreased release is observed for each hazardous metal ion, then read-across of the pure metal classification is applicable, using the summation method. As the massive form of each of the three metals is not classified, the alloy would not be classified either:

- 24-hour TD for Me_A and Me_C: metal release from massive metal and metal in alloy configuration are similar: no alloy effect;
- 24-hour TD for Me_B: metal release in alloy configuration is almost one order of magnitude higher than the release from the massive form: increased release.

Owing to the increased release for Me_B, read-across from the massive forms is not allowed.

Step 3: Classification based on 28-day TD test (critical surface area—toxic unit approach)

At loading of 1 mg/L (classification cut-off for Aquatic Chronic 2):

$$Me_{released} = \text{loading} \times \text{fraction of } Me_{A,alloy} \times 28 - \text{day release rate}$$
$$Me_{A,released} = 1000 \mu g / L \times 0.12 \times 0.0089 = 1.07 \mu g / L$$
$$Me_{B,released} = 1000 \mu g / L \times 0.08 \times 0.039 = 3.12 \mu g / L \tag{13.6}$$
$$Me_{C,released} = 1000 \mu g / L \times 0.80 \times 0.0006 = 0.52 \mu g / L$$

The alloy will be classified as Aquatic Chronic 2 if the sum of toxic contributions (i.e. concentration at 1 mg/L loading/$ERV_{chronic}$) exceeds 1 toxic unit (TU).

$$\Sigma TU = \left(\text{Conc.Me}_A / ERV_{chronic;MeA} \right) + \left(\text{Conc.Me}_B / ERV_{chronic;MeB} \right) + \left(\text{Conc.Me}_C / ERV_{chronic;MeC} \right)$$
$$\Sigma TU = \left(1.07 \mu g / L / 32 \right) + \left(3.12 \mu g / L / 3 \mu g / L \right) + \left(0.52 \mu g / L / 2200 \mu g / L \right)$$
$$= 0.03 + 1.04 + 0.0002 = 1.07$$
$$\tag{13.7}$$

This value is higher than 1.0, and therefore the alloy is classified as Aquatic Chronic 2.

A comparable analysis with a loading of 0.1 mg/L (classification cut-off for Aquatic Chronic 2) results in a TU of 0.10; this is well below 1.0, and therefore there is no need to classify the alloy as Aquatic Chronic 1.

Outcomes:

- Classification of the alloy, *based on elemental composition*: Aquatic Chronic 1
- Refinement: Classification of the alloy, *based on 24-hour and 28-day TD data*: Aquatic Chronic 2 (Fig. 13.9)

Conducting an ecotoxicity validation step with alloys is recommended especially when the existing information suggests potential antagonistic or synergistic effects, and when the composition of the alloy does not allow TD testing or bridging (e.g. several types of alloy pastes). As stipulated before, these tests should be in line with the available guidance (e.g. ECHA, 2017).

When direct aquatic testing with an alloy is conducted for reasons of validation/ refinement of a calculated classification, then the test should be carried out with the most sensitive aquatic species at dissolved ion concentrations equivalent to those measured in the TD medium. These tests can also be initiated for alloys classified as Aquatic Acute 1/Aquatic Chronic 1, 2, or 3 when antagonistic effects are expected. When toxicity is found, the alloy can then be classified directly following the outcome of the performed ecotoxicity tests.

13.5 EXPOSURE ASSESSMENT

The exposure assessment for alloys is, in principle, no different from exposure assessment for metals, but it is necessary to consider the different metal ingredients present. It will describe as accurately as possible, using measured data and/or estimates generated by modelling tools and/or analogous data, the pattern of exposures human populations and environmental compartments are subject to, investigating the different routes of exposure. For the consumer, exposure via food that has been in contact with the alloy (cooking pot) may be considered through data on food intake for specific population groups at risk and data on concentrations in different foods. Exposure via drinking water may be assessed using the same methods as for environmental exposures. For the environment, it is important to remember that metal concentrations in the environment are the result of the natural background, historical contamination and local diffuse emissions associated with the use pattern and the complete life cycle of the alloy.

Analytical methods may vary depending on the alloy, but in general ICP-MS is the common procedure and will measure the metal ingredients. Further information can be found in Chapter 2 and Vetter et al. (2016).

13.6 RISK ASSESSMENT AND MANAGEMENT

Risk characterisation for alloys is, in principle, no different from the general case of risk characterisation of metals; however, the presence of several ingredients in the alloy, and hence simultaneous exposures, poses the question of combined toxicity. Some guidance is provided in Chapters 7 and 8, but further research is required. Risk management measures will be driven by the exposure to hazardous ingredients in the alloy.

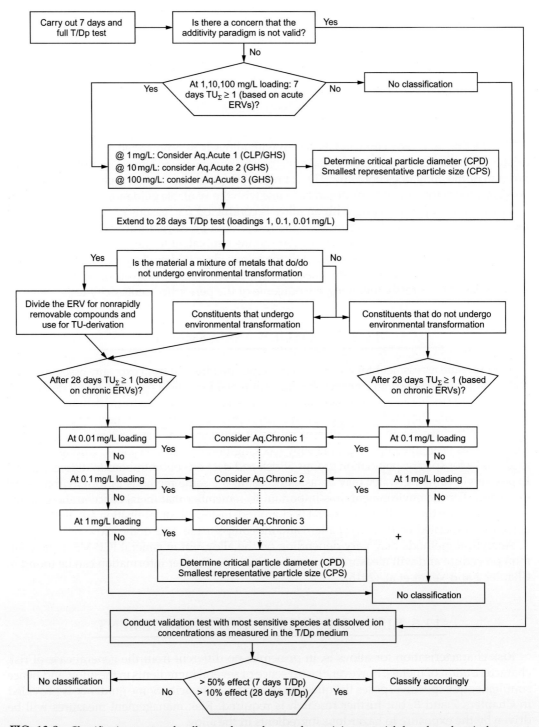

FIG. 13.9 Classification strategy for alloys and complex metal-containing materials based on chronic data.

KEY MESSAGES

Alloys are mixtures under EU CLP and UN GHS. The proposed classification approach follows the general reasoning outlined in the UN GHS and the EU CLP, making use in a tiered way of available test data for the mixture, bridging principles and/or data for the ingredients of the alloy. For human health classification, an alloy-specific approach is currently being discussed with the EU authorities, making use of 'bioaccessible concentration' (BC, %) instead of the bulk concentration. The scope of this refinement is currently limited to systemic effects, oral exposure.

For the environmental classification, the TD protocol allows to consider to their full extent the alloy's properties, including the effect of an alloy matrix on the release of the metal ion. Schemes are proposed for the classification for aquatic toxicity with strong recommendations to start with the 24-hour screening transformation/dissolution test (24-hour TD).

References

ECHA: *Guidance on the application of the CLP criteria. Guidance to Regulation (EC) No 1272/2008 on classification, labelling and packaging (CLP) of substances and mixtures Version 5.0.*, 2017, ISBN: 978-92-9020-050-5. Available from: https://echa.europa.eu/documents/10162/13562/clp_en.pdf.

EN1811:2011+A1:2015 (CEN European Committee for Standardization): *Reference test method for release of nickel from products intended to come into direct and prolonged contact with the skin*, Brussels: EC Directive 94/27/EC, 1998.

EU CLP (Classification, Labelling and Packaging of Substances and Mixtures): *EC No 1272/2008 of the European Parliament and of the Council of 16 December 2008 on classification, labelling and packaging of, substances and mixtures amending*, 2008.

MERAG (Metals Environmental Risk Assessment Guidance) *classification fact sheet, Application of classification criteria to mixtures of multi-metallic compounds, complex metal-containing substances (UVCBs) and alloys*, in press (chapter 3)

OECD (Organisation for Economic Co-operation and Development): *Series on Testing and Assessment N°29: guidance document on transformation/dissolution of metals and metal compounds in aqueous media*, ENV/JM/MONO(2001)9, 2001.

Stockmann-Juvala H, Hedberg Y, Dhinsa NK, et al: Inhalation toxicity of 316L stainless steel powder in relation to bioaccessibility, *Hum Exp Toxicol* 32(11):1137–1154, 2013.

UN GHS (United Nations): Globally harmonized system of classification and labelling of chemicals (GHS), Fourth revision ST/SG/AC.10/30/Rev.4, 2011, New York and Geneva, https://www.unece.org/fileadmin/DAM/trans/danger/publi/ghs/ghs_rev04/English/ST-SG-AC10-30-Rev4e.pdf

UN GHS (United Nations): Globally harmonized system of classification and labelling of chemicals (GHS), Sixth revision ST/SG/AC.10/30/Rev.6, 2015, New York and Geneva, http://www.unece.org/fileadmin/DAM/trans/danger/publi/ghs/ghs_rev04/English/14e_annex10.pdf

Vetter D, Schade J, Lippert K: *Guidance on the assessment of occupational exposure to metals based on monitoring data*, Final report, Available from: http://www.reach-metals.eu/index.php?option=com_content&task=view&id=216&Itemid=3242016, 2016.

Emerging Tools in the Assessment of Metals: Current Applicability

Koen Oorts*, Jelle Mertens[†], Vincent Dunon[‡],
Patrick Van Sprang[‡], Frederik Verdonck[‡]

*ARCHE Consulting, Leuven, Belgium [†]European Precious Metals Federation, Brussels,
Belgium [‡]ARCHE Consulting, Ghent, Belgium

14.1 INTRODUCTION

Apart from the tools and approaches currently used for the risk management of complex inorganic materials, as discussed in the preceding chapters, new tools with potential application in this field are (further) being developed. In silico modelling, such as quantitative ion–character activity relationships (QICARs) or combined toxicity models, may decrease the need for in vivo testing and further target resources towards most critical materials and endpoints. Similarly, adverse outcome pathway (AOP) analysis may help to combine the current knowledge from in vivo testing with data from biochemical and cellular assay systems and computational predictive methods, and utilise that knowledge to support risk-based decision-making. As such, it can be used to address knowledge gaps in the mechanisms of actions for contaminants and strengthen the justification for the extrapolation of (bioavailability) models across species. Finally, life-cycle assessment (LCA) approaches are a valuable tool to assess and compare impacts from substances or processes on a range of endpoints. However, metal-specific aspects are generally not yet sufficiently reflected in current methodologies. This chapter discusses the current status of these tools and their potential applicability for the risk management of complex inorganic materials.

14.2 IN SILICO MODELLING

14.2.1 Introduction

The assessment or screening of chemicals as required by today's regulatory frameworks is generally based on the available data. This inherently implies that the robustness of such assessment largely depends on the amount and quality of the available data. Some metals and inorganic metal compounds are associated with an extended dataset of (eco)toxicity data. This is typically related to the global industrial volume of the metal and the consequent (historical) regulatory interest to assess and control all potential risks. Well-known examples of these metals are cadmium, cobalt, copper, lead, nickel, and zinc. Depending on the individual toxicological profiles, data generation has been focused on selected environmental and/or mammalian endpoints. The hazard profiles, for example, for copper and zinc (compounds) are driven by ecotoxicological concerns, whereas the profiles for lead and cobalt are driven by mammalian toxicity. Such data-rich metals typically have experimental data for multiple test species per environmental compartment (water, sediment, and soil), or multiple studies for the most critical mammalian toxicity endpoint(s) (carcinogenicity, genotoxicity, reproductive toxicity...).

However, the majority of metals are much less data-rich due to, for example, low production volumes or limited regulatory interest. The available data are often focused on a few endpoints of concern (such as aquatic toxicity or carcinogenicity), and the amount of data for these endpoints is limited to minimal standard data requirements. For the other (eco)toxicological endpoints, little to no data are available. As an example, metals such as palladium, rhenium, gold, indium, or germanium presently have hardly any reliable (eco) toxicity data available. When checking the EU REACH registration dossiers for these metals, they are mostly associated with a few acute ecotoxicity test data with common test species (fish, daphnia, and/or algae), and usually lack chronic ecotoxicity data and data for compartments other than freshwater (https://www.echa.europa.eu/information- on-chemicals/registered-substances). Consequently, the assessments or prioritisation of these data-poor metals is difficult (if possible at all due to an absence of data), and conclusions are associated with a significantly larger amount of uncertainty compared to data-rich metals.

An even more complicated situation appears for complex inorganic materials (such as ores, slags, or alloys), which are often a mixture of data-rich and data-poor metals. The difference in data availability for the individual constituent metals complicates the comparability of the results. This severely complicates the constituent-based risk assessment (RA) often followed for these materials.

As mentioned in previous chapters, experimentally developing new data for each metal or complex inorganic material to fill data gaps is resource intensive and often problematic because of the variable composition of complex inorganic materials. As an alternative, scientists and regulators have investigated possible ways to model and predict the (eco)toxicological profile of individual metals or complex metal mixtures. Such models can help in filling data gaps for individual metals and in the assessment of combined toxicity of the individual constituents in complex inorganic materials.

14.2.2 Quantitative Ion–Character Activity Relationships (QICARs)

One option to predict (eco)toxicity data for data-poor metals is the quantitative structure–activity relationship (QSAR) concept, originally developed for organic substances. The QSAR concept is used to predict the properties of a target substance based on its molecular structure and on the properties of structurally similar source substances. The QSAR concept has successfully been applied by regulators and assessors worldwide. Well-known examples are the OECD QSAR Toolbox (http://www.oecd.org/chemicalsafety/risk-assessment/theoecdqsartoolbox.htm), the US EPA EPISuite (https://www.epa.gov/tsca-screening-tools/epi-suitetm-estimation-program-interface), or ECOSAR (https://www.epa.gov/tsca-screening-tools/ecological-structure-activity-relationships-ecosar-predictive-model).

The QSAR concept is, however, not directly applicable to metals and their inorganic compounds. Firstly, the generation of industrially relevant 'new' inorganic metal compounds is not frequent (in comparison to the organic industry where new compounds are continuously being developed). Secondly, the source database that can be used to predict properties of 'new' or 'unknown' inorganic metal compounds is often limited in the number of studies and endpoint coverage, or even completely lacking (Le Faucheur et al., 2011; Wang et al., 2016). Therefore, a different approach is required for metals and inorganic materials. For ecotoxicological endpoints, it is generally recognised that effects are related to metal-ion activity in solution. Further, metal ions vary widely in physicochemical properties (atomic weight, valence, electronegativity, etc.), but there is some predictability for many of these properties based on their position in the periodic table of Mendeleev. Consequently, there have been many attempts to develop the so-called 'quantitative cation–activity relationship' (QCAR) or 'quantitative ion–character activity relationship' (QICAR) concept for metals since mid-1950s, as summarised by, for example, Walker et al. (2003). An evident basis for the development of QICARs is the available metal toxicity data, and relevant physicochemical properties of both target and source metals such as molecular weight, electronegativity, or standard electrode potential. An overview of the most frequently used physicochemical characteristics is given by, for example, Walker et al. (2003) and Le Faucheur et al. (2011), and is summarised in Table 14.1.

Initial binding is a prerequisite for the interaction of metal ions with biological systems. Afterwards, the metal ion can potentially cause effects in the organism. In some cases, binding can be followed by subsequent events such as metal internalisation or disruption of cellular mechanisms. Pearson (1963) developed a theory which links metals to their predicted binding behaviour with ligands (of either organic or inorganic nature). According to this author, metals can be divided in two groups (plus an intermediate group): hard and soft Lewis acids and bases (abbreviated as 'HSAB'). The 'hard' acids (chromium, manganese, and aluminium, etc.) are non-polarisable and tend to form ionic bounds with ligands, whereas the 'soft' acids (palladium,

TABLE 14.1 Overview of Some Metal Properties That Have Historically Been Used as Predictor of (Eco) Toxicity in QICAR Models

Metal Property	Examples
Physical properties	Atomic weight, atomic volume, density, melting point, polarisability, and molar refractivity
Electronic structure	Atomic number, electronic configuration, ionisation potential, and electron affinity
Redox capacities	Oxidation number, standard electrode potential, and electrochemical potential
Binding properties	Ionic radius, atomic radium, covalent radius, electropositivity, and negativity
Indices	Ionic potential, ionic index, covalent index, covalent bond stability, first hydrolysis constant, equilibrium, or formation constant of metal–ATP complex, stability constant metal–NH_3, metal–EDTA or metal–AMP, and softness parameter

Based on Walker JP, Enache M, Dearden JC: Quantitative cationic-activity relationships for predicting toxicity of metals, Environ Toxicol Chem 22:1916–1935, 2003 and Le Faucheur S, Campbell PG, Fortin C: Quantitative ion character activity relationships (QICARs). Research report No R-1262, Final project report presented to Environment Canada, 2011, Science and Technology Branch, Ecological Assessment Division, Inorganics Unit.

platinum, and silver, etc.) are polarisable and tend to form covalent bounds with ligands. The hard acids tend to bind preferentially with O-donor ligands, followed by N- and S-donor ligands, whereas the soft acids tend to bind with S-donor ligands, followed by N- and O-donor ligands. In biological systems, metals binding to oxygen, nitrogen, or sulphur atoms are expected to be of high concern in causing toxic effects, as these elements are ubiquitously present in biological macromolecules. Since its development in the early 1960s, the HSAB theory has frequently been used to predict the biological effects of metal ions via the use of a 'softness index' σ_p, either or not in combination with other physicochemical properties (e.g. Williams and Turner, 1981).

Promising QICAR models are summarised in Table 14.2.

TABLE 14.2   Summary of Selected QICAR Models

Authors	Model	n	R^2	p	Toxicity Parameter
Wolterbeek and Verburg (2001)	$pT = a_0 + a_1 \Delta E_0 + a_2 \log \mathrm{IP} + a_3 X_m + a_4 \log(\mathrm{AR/AW})$ With: T, toxicity; E_0, the electrochemical potential; IP, the ionisation potential; X_m, electronegativity; and AR/AW, ratio of atomic radius to atomic weight	9–31	0.51–0.97	0.27–<0.0001	30 literature datasets
Kinraide (2009)	$T_{con} = a\sigma_{con} + b\sigma_{con}Z + cZ$ With: σ_{con}, $aE^0I_p + \beta\rho_{metal}$; E^0, electrode standard potential; I_p, first ionisation potential; ρ_{metal}, bulk metal density, and Z, ion charge	NR	0.923	NR	T_{con} derived from 10 studies
Wu et al. (2013)	LC_{50} or EC_{50} via two-variable regressions using combinations of softness index (σ_p), maximum complex stability constants ($\log(-\beta n)$), covalent index ($X_m^2 r$), atomic ionisation potential (AN/ΔIP), polarisation force parameters (Z/r), first hydrolysis constants ($|\log K_{OH}|$) and electrochemical potential (ΔE_0)	>6 (to max. 10)	0.77–0.96	0.07–0.004	Acute toxicity for eight species (covering five phyla)
Mu et al. (2014)	$\mathrm{Log(NOEC)} = a\sigma_p + b$ with σ_p = softness index	6–9	0.73–0.94	0.02–0.0001	Chronic toxicity for eight aquatic species (covering six phyla)
Chen et al. (2015)	$\mathrm{Log(LC_{50})} = a\sigma_p + b$ with σ_p = softness index	5–8	0.66–0.89	0.094–0.006	Acute toxicity predicted for eight marine species (covering five phyla)
Wang et al. (2016)	$\mathrm{Ln(CMC)} = -8.75X1 + 13.34$ With $X1 = 0.567CR + 0.568r - 0.597 \mathrm{AR/AW}$, CR, covalent radius; r, Pauling ionic radius, and AR/AW, electron density	7	0.63	0.019	US EPA Criteria Maximum Concentrations (CMC) values

n, *number of metals as training dataset*; NR, *not reported*; p, *level of significance*; R^2, *correlation coefficient*.

Le Faucheur et al. (2011) reviewed various QICAR models for their usefulness to predict the toxicity of data-poor metals for regulatory prioritisation purposes. These authors concluded that two models were most useful: the models of Wolterbeek and Verburg (2001) and Kinraide (2009).

The Wolterbeek and Verburg (2001) model was based on the work of Kaiser (1980), who successfully modelled metal toxicity to *Daphnia magna* reproduction using a linear combination of $\log(AN/\Delta IP)$ and ΔE_0 (where AN is the metal's atomic number, IP the ionisation potential, and E_0 the electrochemical potential). The model was later successfully applied to predict, for example, acute metal toxicity estimates with the *Microtox* bioluminescence assay (Newman and McCloskey, 1996). The main drawback of the Kaiser model is that the model parameters α_0, α_1, and α_2 depend on the metal softness (hard vs soft), test organism, and endpoint and, therefore, the applicability of the model was limited. Wolterbeek and Verburg (2001) applied a similar model on 30 existing literature datasets (covering different test species and endpoints). These authors used as model parameters E_0, IP, Xm (electronegativity), and $\log(AR/AW)$ (ratio of atomic radius to atomic weight) This model was promising for predicting the toxicity of a wide range of metals to many test species and endpoints.

The second model selected by Le Faucheur et al. (2011) was Kinraide's (2009) more recent one. This author used a 'consensus scale of softness' (based on 10 different scales), depending on the electrode standard potential, first ionisation potential and the bulk metal density, and a 'consensus scale of toxicity' (based on 10 different studies) (Table 14.2). This author successfully predicted toxicity by including ion charge in the model, although it might be doubtful that ion charge directly affects toxicity (Fig. 14.1).

Le Faucheur et al. (2011) concluded that the prioritisation of metals—by using the estimated toxic concentrations—differed largely between test species, exposure time, and toxic endpoints using the Wolterbeek and Verburg model. The Kinraide model inherently gives only one priority ranking. Nevertheless, when comparing the 'average' ranking of metals using the Wolterbeek and Verburg model with the ranking of the Kinraide model, the top seven priority metals were equal whereas the 'lower priority' metals differed between both models. It was concluded that both models were promising for prioritisation purposes. However, it needs to be added that the underlying training database is rather limited considering today's continuous development of new (eco)toxicological data under regulatory regimes such as EU REACH. This might be especially relevant for the data-poor metals that were ranked as highest priority elements. Also, it is unclear to what extent data quality has been assessed. As an example, the predictions using the Wolterbeek and Verburg (2001) model were largely based on total (or even nominal) metal concentrations, whereas it is well recognised that metal solubility and speciation are important considerations when deriving toxic threshold values.

In a more recent model, Wu et al. (2013) used a similar approach to Wolterbeek and Verburg (2001) to predict acute freshwater aquatic threshold values for 25 metals. The ecotoxicological threshold data of minimum six metals (and maximally 10 metals) for eight freshwater test species (covering five phyla) were used as the training dataset. Starting from a dataset of 14 physicochemical properties, the two parameters with highest correlation per test species were used to predict acute toxicity (per test species) using a two-variable linear regression model (Table 14.2). Softness index was significantly correlated with ecotoxicological thresholds for six of the eight test species ($R^2 = 0.68$–0.90; $P = .001$–$.024$). A similar strong correlation between softness and toxicity has been shown earlier, for example,

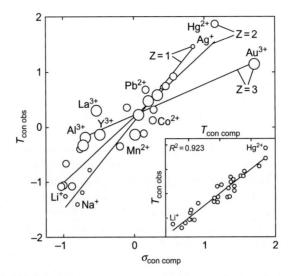

FIG. 14.1   Relationship between consensus scale of toxicity ($T_{con,obs}$) and consensus scale of softness ($\sigma_{con,comp}$). *Lines* are fitted for metals with the same oxidation state (+1, +2, and +3). The insert *(bottom right)* shows observed vs computed toxicity. *From Kinraide TB: Improved scales for metal ion softness and toxicity, Environ Toxicol Chem 28:525–533, 2009.*

in *D. magna* and several fish species (Newman et al., 1998; Ownby and Newman, 2003; Zhou et al., 2011). Subsequently, the predicted toxicity thresholds for each of the eight test species were used to derive individual species sensitivity distribution (SSD) curves for the 25 metals (i.e. 10 training metals and 15 'unknown' metals). The corresponding criteria maximum concentrations (CMC) were predicted as $HC_5/2$, with HC_5 = hazardous concentrations at which 5% of the species in the SSD exhibit an effect. The CMC values reported by the US EPA for the 10 training metals was predicted within 1 order of magnitude (except for arsenic and chromium VI that were predicted within 1.5 orders of magnitude). For the 15 'unknown' metals, predicted $logHC_5$ values strongly correlated with softness index, and the authors suggested that soft metals were more toxic than hard metals. Important to note here is that the training dataset only included 'softer' metals, and that this conclusion needs to be considered with care.

Mu et al. (2014) successfully applied the approach used by Wu et al. (2013) to predict chronic freshwater toxic thresholds for 34 metals or metalloids (of which nine were used as training metals). These authors used the QICAR-SSD model and the final acute-chronic ratio (FACR) method. Eight species were used (covering six phyla), and predictions were made using a one-variable linear regression model using the softness index as a predictive variable (Table 14.2). Predicted values were within 1 order of magnitude of US EPA's criteria continuous concentrations (CCC), except for arsenic and lead for which predictions were within 1.5–2 orders of magnitude. The data for the 'unknown' metals suggest better performance of the QICAR-SSD model compared to the FACR method.

The approach of Wu et al. (2013) was also successfully applied by Chen et al. (2015) to predict acute marine threshold values for 34 metals (of which eight were used as training metals). Whereas Mendes et al. (2013) only used red seaweed (*Gracilaria domingensis*) as test species, Chen and colleagues used data for eight different test species. The LC50 values of all eight test species correlated best with softness index (Table 14.2). The predicted values for five metals were within 0.5 orders of magnitude of the recommended water quality criteria by US EPA, and within 1 order of magnitude for three other metals. As mentioned by the authors, ionic conditions in marine media are different from freshwater systems, which might have significant implications on metal speciation and, thus, metal toxicity. Also, responses of marine species to metals might differ from those of freshwater species. Therefore, combining freshwater and marine species in a single model seems inappropriate.

Recently, Wang et al. (2016) published a study where they started from single parameter linear regressions to identify the most correlated physicochemical parameter out of a set of 26. These authors tried to directly predict the US EPA CMC values based on a training dataset consisting of seven metals. Also in this study, the softness index was best correlated with the US EPA CMCs ($R^2 = 0.88$; $P = .004$). In a next step, a reduced set of integrated variables was created using principal component analysis, resulting in a new single parameter linear relationship, and using 'integrated radius' as a variable to predict ln(CMC). The model was successful in predicting ln(CMC) for the seven training metals within the 95% confidence interval, except for chromium (VI) that was slightly outside the interval. The second part of this research was the prediction of CMCs for 56 'unknown' transition metals. The predicted CMCs were then correlated with median acute lethal LC50 concentrations for 31 of these metals in 1-week experiments with *Hyalella azteca*. The level of correlation between predicted and observed toxicity values differed between metals. Most likely, predictions can be improved if metal speciation and variations in the physicochemical properties (with e.g. metal oxidation state) are taken into account. Also, the toxicity prediction was compared to the experimental toxicity threshold from a single test species, which might explain part of the variation. Nevertheless, the work of Wang et al. further confirms that QICAR models are promising tools for screening and prioritisation purposes.

As mentioned earlier, metal speciation is considered an important toxicity determinant for metals. As an example, Newman et al. (1998) observed that one- and two-variable linear models perform, in general, quite well for prediction toxicity of 19 published datasets. However, predictions deviated most for metals where speciation in solution was significant (e.g. mercury, lead, and manganese). Via two test systems with bacterial bioluminescence and using MINTEQA2 modelling, Newman and colleagues were able to improve their predictions. Mendes et al. (2013) recently used the same MINTEQA2 software to calculate free median inhibitory concentrations (IC50F). These data were successfully used to model the toxicity of 14 metals to red seaweed using ion characteristics. The predicted log(IC50) was within 1 order of magnitude of the observed value for 13 metals (and within 2 orders of magnitude for Li). However, at the same time, Tatara et al. (1998) showed that using free metal ions rather than total metal concentration in solution does not always improve model performance. Obviously, factors such as medium composition, test species, and test metal will largely determine whether the model performance will improve.

The above studies have shown that QICAR is a promising tool to predict metal toxicity for single metal systems. However, it is less clear whether QICAR is useful for (complex) inorganic materials. Using the *Microtox* assay, interactions have been observed between some binary metal mixtures such as $Cu^{2+} + Pb^{2+}$ (Newman and McCloskey, 1996). In an attempt to predict the toxicity of binary metal mixtures using a *Microtox* assay, Ownby and Newman (2003) concluded that QICAR was a useful tool. The difference in softness ($\Delta\sigma_p$) was identified as the best correlated parameter to predict metal interaction (metal interaction coefficient $= 68.5 \times \Delta\sigma_p - 2.63$; $R^2 = 0.69$; Fig. 14.2), with a tendency to move from independent action to nonadditivity if $\Delta\sigma_p$ decreases.

A lot of additional scientific work has been published in the past decades on the toxicity of single metals vs complex inorganic mixtures in an attempt to describe or explain the (potential) interactions between metals. Metal interactions vary between test systems, and go from more than additive to less than additive (synergistic vs antagonistic). Despite this wealth of mixture data being generated, there are to date only very few publications available investigating the potential application of QICAR to predict the toxicity of (complex) inorganic materials. In reality, metal mixtures also often exceed binary or ternary in level of complexity, with some constituents being present in minor amounts, but for which the individual level of toxicity might be severe and for which potential interactions with other metals have not yet been tested. Nevertheless, as QICAR is a powerful tool to prioritise individual metals for expected toxicity, the obtained information gives a good indication of the potential toxicity of a metal mixture when combined with the mixture composition or mixture solubility data.

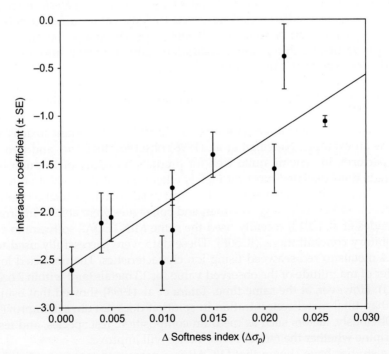

FIG. 14.2   Difference in softness index in binary metal mixtures versus metal interaction coefficient as observed using the Microtox assay. *From Ownby DR, Newman MC: Advances in quantitative ion character-activity relationships (QICARs): using metal-ligand binding characteristics to predict metal toxicity, QSAR Comb Sci 22:241–246, 2003.*

14.2.3 Combined Toxicity Models

In a constituent-based RA of complex inorganic materials, there is, next to the assessment of the individual constituents, a clear need for a realistic prediction of metal mixture effects on toxicity. In regulatory applications, (metal) mixture toxicity has generally been modelled by toxic unit or other additive approaches (Meyer et al., 2015). Data reviews (Norwood et al., 2003; Vijver et al., 2011) have shown that additive approaches based on dissolved-metal concentrations are not always sufficient in predicting mixture toxicity. Rather, metal-mixture toxicity tests have shown a wide range of organism responses, with no clear patterns in additive and nonadditive behaviour. Understanding the behaviour and toxicity of metal mixtures poses unique challenges for incorporating metal-specific concepts and approaches, such as bioavailability and metal speciation, in multiple-metal exposures. RAs of individual metals in freshwater and soil have successfully applied chemical speciation models to establish bioavailability-based regulatory benchmarks and conduct single-metal RAs (OECD, 2016). These bioavailability-based models and tools for assessing environmental risk to single metals are now being evaluated for their applicability to real-world mixture scenarios, such as mining or industrial effluents. Recently, several available higher tiered chemical speciation-based metal mixture modelling approaches have been evaluated for their prediction of toxicity of metal mixtures to freshwater organisms (Van Genderen et al., 2015; Farley et al., 2015). Two main modelling frameworks have been considered: the biotic ligand model (BLM) as first presented by Di Toro et al. (2001) and the Windermere humic aqueous model using the toxicity function (WHAM-FTOX) (Stockdale et al., 2010; Tipping and Lofts, 2015).

In the BLM framework, metal bioavailability is evaluated by considering competitive interactions of metals and cations for binding to dissolved organic matter and inorganic ligands using WHAM. Competitive binding of metals and cations is also assumed to occur at binding sites on or in biological organisms, which are referred to as the 'biotic ligand(s)'. The accumulation of metal on the biotic ligand is then correlated to the toxic response of the organism.

The WHAM-FTOX model was developed specifically to address the effects of metal mixtures on aquatic organisms. This approach also uses WHAM to evaluate the competitive interactions of metals and cations on DOM and inorganic ligands. In contrast to the BLM, WHAM-FTOX does not explicitly consider competitive binding of metals and cations to a biotic ligand. Rather, the model assumes that the nonspecific accumulation of metabolically active metals by the organism is proportional to metal concentrations predicted to accumulate on humic acid when exposed to the same exposure water (as calculated by WHAM). Accumulated metal is related to toxicity using the FTOX function, which is obtained by multiplying the calculated humic-bound metal and proton concentrations by cation-specific potency factors and then summing up the results over all cations. The resulting FTOX value is correlated to the toxic response of the organism.

Both modelling approaches perform consistently well for various metal combinations, organisms, and endpoints. Both illustrate the importance of bioavailability-based methods as a component of metal-mixture assessment and bioavailability models calibrated for single-metal exposures can be integrated to assess mixture scenarios. The models provided a reasonable description of additive (or nearly additive) toxicity for several individual toxicity test results. Less-than-additive toxicity was more difficult to describe with the available models. The WHAM-FTOX model seems very promising as it integrates both interactions in the environment and at the biotic ligand level and can be applied to water, sediment, and soils. BLM-based models may also provide reasonable predictions for water-exposure scenarios. The models can allow site-specific assessment when using site-specific parameters. The calibration strategy is critical for each of the models. WHAM-FTOX has an advantage in that it has been fitted to field data (Stockdale et al., 2010), but this was also limited to two or three metals.

These modelling approaches will allow us to further refine the assessment of the combined effects of the constituents of complex inorganic materials and will avoid the default application of additive approaches. Moving forward, efforts should focus on reducing uncertainties in model calibration, including the development of better methods to characterise metal binding to toxicologically active binding sites, conducting targeted exposure studies to advance the understanding of metal-mixture toxicity, and further developing tools to help constrain the model calibration (Farley and Meyer, 2015).

14.3 ADVERSE OUTCOME PATHWAY

Due to the costs and time involved, it is not practical or feasible to test exhaustively all chemicals that could adversely affect humans and ecosystems. Recognising these limitations of current in vivo testing approaches for toxicological assessment and the rapid development of new biochemical and cellular assay systems and computational predictive methods, regulators, and other stakeholders have been exploring ways to integrate existing knowledge from in vivo tests with the results of alternative methods and other sources of information. The purpose of this integration is to identify better schemes for making regulatory decisions (OECD, 2017). The AOP methodology is an approach which provides a framework to collect, organise, and evaluate relevant information on chemical, biological, and toxicological effect of chemicals. This approach supports the use of a mode (and/or mechanism) of action basis for understanding the adverse effects of chemicals (OECD, 2017). AOP analysis provides a framework for evaluating pathways by which a toxicant may interact with an organism to cause an adverse outcome. The analysis begins at the molecular level with molecular initiating events (MIEs) and provides a plausible set of linkages to increasingly higher levels of biological organisation, from macromolecular interactions to cellular responses to organs and then whole-animal responses and ultimately population-level and community-level responses that are the focus of environmental RA and water-quality management (OECD, 2016, 2017; Villeneuve et al., 2014).

An AOP analysis helps us to organise what we know and utilise that knowledge to support risk-based decision-making; it can be used to address knowledge gaps in mechanisms of actions for contaminants and remove concerns on the extrapolation of (bioavailability) models across species. The outcome of an AOP framework can contribute to directing testing resources by focusing on chemicals of highest concern and limiting testing to the most probable hazards and the most vulnerable species (Tollefsen et al., 2014). As such, AOP may be used to target resources to fill critical data gaps for data-poor metal constituents of complex inorganic materials, yield additional justification for cross-species extrapolation of bioavailability models, and provide evidence for the mode of action for the selection of the most appropriate combined toxicity models.

Brix et al. (2017) applied an AOP analysis to study the mechanisms of nickel toxicity in aquatic environments, taking advantage of genomic/transcriptomic techniques and extant mechanistic knowledge from other environmental compartments (e.g. terrestrial plants and mammals) to identify potential mechanisms not previously considered in aquatic environments. Five potential mechanisms for chronic Ni toxicity to aquatic organisms were evaluated within the AOP framework: disruption of Ca^{2+} homeostasis, disruption of Mg^{2+} homeostasis, disruption of $Fe^{2+/3+}$ homeostasis, reactive oxygen species-induced oxidative damage, and an allergic-type response of respiratory epithelia (Fig. 14.3).

At the organ level of biological organisation, these five potential MIEs collapse into three potential pathways: reduced Ca^{2+} availability to support the formation of exoskeleton, shell, and bone for growth; impaired respiration; and cytotoxicity and tumour formation. At the level of the whole organism, the organ-level responses contribute to potential reductions in

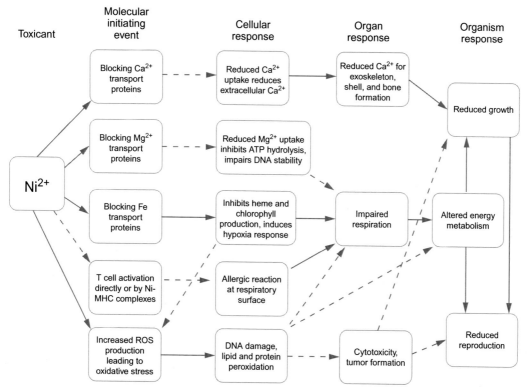

FIG. 14.3 Adverse outcome pathway (AOP) analysis for nickel (Ni) in aquatic systems. *Solid arrows* represent pathways with supporting data; *dashed arrows* are hypothesised pathways. *ATP*, denosine triphosphate; *DNA*, deoxyribonucleic acid; *MHC*, major histocompatibility complex; *ROS*, reactive oxygen species. *From Brix KV, Schlekat CE, Garman ER: The mechanisms of nickel toxicity in aquatic environments: an adverse outcome pathway analysis,* Environ Toxicol Chem *36:1128–1137, 2017.*

growth and reproduction and/or alterations in energy metabolism, with several potential feedback loops between each of the pathways. Each of the potential pathways was found to have some support in the literature, although the allergic response pathway is not considered likely based on our analysis. Further, data supporting one or more links between MIEs and whole animal-level or population-level effects were unavailable for each pathway. Overall, the AOP analysis was judged a useful tool for conceptualising interactions between potential mechanisms of action and identifying knowledge gaps that should be the focus of future research efforts on the aquatic toxicity of Ni.

14.4 IMPACT ASSESSMENT METHODOLOGY

14.4.1 LCA and Product Environmental Footprint

14.4.1.1 *Life-Cycle Assessment (LCA)*

LCA is a standardised decision-support tool which aims at assessing the potential impacts on ecosystems, human health, and natural resources of a product, technology, or service throughout its life-cycle stages (ISO 14040, 2006 and ISO 14044, 2006). It focuses on the overall pressure on the environment (multiple chemical risks, water use, energy consumption, land use, carbon footprint, and ozone depletion, etc.) and is used both in comparative assessments and in process evaluations. Four different phases are defined in an LCA study (Fig. 14.4): (i) goal and scope definition, (ii) the LCI analysis, (iii) performance of the life-cycle impact assessment (LCIA), and (iv) the interpretation of the assessment results.

In principle, LCA is a relative approach evaluating the complete life cycle of a product or system based on a defined function of the product/system. During the goal and scope phase the selection of the function and the associated functional unit used as reference is of high importance as all the input and output streams will be assigned to this unit. Also, the boundaries of the system under investigation are set in this phase. In the LCI phase, all data are collected of input (materials and energy) and output flows (products, emissions, and waste) that are relevant for the boundaries of the studied system. Typically, this is the most resource-intensive phase of an LCA study. A differentiation is made between foreground and background systems. All processes that are immediately related to the scope and boundary conditions of the system are part of the foreground system and are mostly concerned with the manufacturing of the product. Background systems are the processes that are not covered in the scope of the study but deliver essential inputs to the foreground system, for example, production of an auxiliary input or energy. Typically, an LCI database (e.g. the Ecoinvent database, http://www.ecoinvent.org) is used to retrieve the necessary LCI data for the background systems or to estimate when data are lacking on processes in the foreground system. After gathering all the LCI data, impacts are assigned to all the input and output flows in the LCIA phase. Different LCIA methods, for example, ReCiPe (Goedkoop et al., 2013), CML (Guinée, 2001), and Eco-Indicator99 (Goedkoop et al., 2000), have been established, with all having different impact methodologies and impact indicators or categories under study (Fig. 14.5).

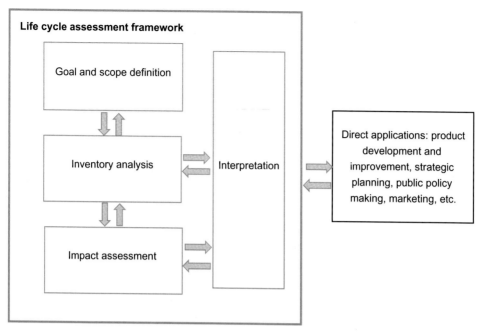

FIG. 14.4   The different stages in the life cycle assessment framework (ISO 14040, 2006).

The choice of the LCIA method will determine how the results of the LCI are translated into health and environmental impacts. Different LCAI methods use different sets of models that predict the relative contribution to the environmental impact of a substance, called the characterisation factor of a substance. Ultimately, the interest of an LCA study is to define the actual impacts of the product or system on human health, environment, and resources. This is because there is often a lack of robust evidence on the correlation between cause and effect due to the complexity of the environmental mechanisms in play. Therefore, impacts are mostly measured by midpoint indicators, measuring an environmental effect as an indicator for human health effects, environmental impacts, and resources.

Overall, the LCA will offer a broad perspective of the environmental impacts, enabling it to be an effective tool for general impact assessment between different products or services or of the different life-cycle phases of the product or system under investigation. As a downside, LCA will determine the average environmental impacts for steady-state conditions, thereby excluding the time–space distribution of stressors, the existence of toxicity thresholds, or varying acceptability of impacts (Harder et al., 2015). Furthermore, LCA is not (yet) regulated and not quite robust because results are highly sensitive to input assumptions.

14.4.1.2 *Product Environmental Footprint (PEF)*

Due to the increasing awareness of customers, of product environmental impacts, and the interest of companies for making 'green claims' on their products, there is a great demand for a standardised assessment of environmental impacts. Currently, there is a proliferation

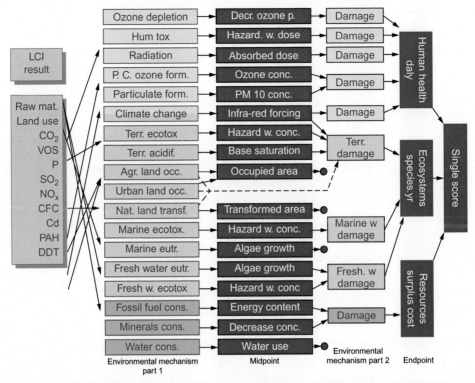

FIG. 14.5   Relationship between life-cycle inventory (LCI) parameters, midpoint indicators, and endpoint indicators in ReCiPe. *From Goedkoop M, Heijungs R, Huijbregts M, De Schryver A, Strujis J, van Zelm R: ReCiPe 2008: a life cycle impact assessment method which comprises harmonised category indicators at the midpoint and the endpoint level, ed 1 (version 1.08), The Netherlands, 2013, Ruimte en Milieu, Ministerie van Volkshuisvesting, Ruimtelijke Ordening en Milieubeheer.*

of ecolabels causing confusion among costumers and additional costs for manufacturers. To meet the demands for a standardised methodology, the European Commission has put forward the Product Environmental Footprint (PEF) and Organisation Environmental Footprint (OEF) as tools for measuring the environmental performance of a product or organisation. The basis for these tools is the LCA methodology adopted from the ILCD handbook (EC JRC, 2010), providing a guidance towards a robust and reliable LCA. In the PEF/OEF methodology, a default set of 14 obligatory impact categories with accompanying methods is provided, based on recommendations in the ILCD handbook (Table 14.3).

In contrast to a regular LCA, where there is a higher degree of freedom in impact categories/methodologies, the PEF/OEF methodology aims to increase the harmonisation of the impact assessment. Currently, the PEF/OEF methodology is being discussed among stakeholders and a pilot phase to test the methodology and draft PEF/OEF Category Rulings (PEFCR/OEFCR) for selected products and sectors has ended recently. The objective of the PEFCR/OEFCR is to identify the most relevant impact categories, life-cycle stages, processes, and elementary flows. This to achieve comparability, increased reproducibility, consistency, relevance, focus, and efficiency of PEF/OEF studies. From the metal sheet PEF assessment, the three toxicity and the resource depletion potential

TABLE 14.3 Default Environmental Footprint Impact Categories and Impact Assessment Models for Product Environmental Footprint Studies

Impact Category	Recommendation at Midpoint		
	Recommended Default LCIA Method	Indicator	Classification
Climate change	Baseline model of 100 years of the IPCC	Radiative forcing as Global Warming Potential (GWP100)	I
Ozone depletion	Steady-state ODPs 1999 as in WMO assessment	Ozone depletion potential (ODP)	I
Human toxicity, cancer effects	USEtox model (Rosenbaum et al., 2008)	Comparative toxic unit for humans (CTU_h)	II/III[a]
Human toxicity, noncancer effects	USEtox model (Rosenbaum et al., 2008)	Comparative toxic unit for humans (CTU_h)	II/III[a]
Particulate matter/ respiratory inorganics	RiskPoll model (Humbert, 2009)	Intake fraction for fine particles (kg PM2.5-eq/kg)	I
Ionising radiation, human health	Human health effect model as developed by Dreicer et al. (1995) (Frischknecht et al., 2000)	Human exposure efficiency relative to U^{235}	II
Photochemical ozone formation	LOTOS-EUROS (van Zelm et al., 2008) as applied in ReCiPe	Tropospheric ozone concentration increase	II
Acidification	Accumulated exceedance (Seppälä et al., 2006; Posch et al., 2008)	Accumulated exceedance (AE)	II
Eutrophication, terrestrial	Accumulated exceedance (Seppälä et al., 2006; Posch et al., 2008)	Accumulated exceedance (AE)	II
Eutrophication, aquatic	EUTREND model (Struijs et al., 2013) as implemented in ReCiPe	Fraction of nutrients reaching freshwater end compartment (P) or marine end compartment (N)	II
Ecotoxicity (freshwater)	USEtox model (Rosenbaum et al., 2008)	Comparative Toxic Unit for ecosystems (CTU_e)	II/III[a]
Land use	Model based on soil organic matter (SOM) (Milà et al., 2007)	Soil organic matter	III
Resource depletion, water	Model for water consumption as in Swiss Ecoscarcity (Frischknecht et al., 2008)	Water use related to local Scarcity of water	III
Resource depletion, mineral, fossil, and renewable	CML 2002 (Guinée, 2002)	Scarcity	II

For each of the recommended characterisation models a classification into three levels regarding their quality: *Level I*, recommended and satisfactory; *Level II*, recommended but in need of some improvements; and *Level III*, recommended, but to be applied with caution (EC JRC, 2011).

[a] *Metals are substances classified as Level III.*

indicators were considered as nonrobust. For the toxicity indicators, the USEtox model is considered unsuitable as some metal specificities (e.g. essentiality) are not considered and model parameters regarding the long-term behaviour are highly uncertain for metals (EC JRC, 2010). The abiotic depletion potential indicator is based on estimates of the total amount of reserves and this is not always in direct correlation with the environmental problem of resource depletion (van Oers et al., 2002). Typically, the life-cycle stages 'Mining and Concentration' and 'Smelting and Refining' contribute most to the overall environmental impact in the PEF assessment.

14.4.1.3 Metal-Specific Aspects in LCA and PEF

Recently, Santero and Hendry (2016) published a paper with guidelines to harmonise LCA methodologies for the metal and mining industry. Essential for the performance of an LCA for metals is the setting of the system boundaries. A cradle-to-grave assessment is preferred to capture all potential benefits of metals, for example, recycling, especially when comparing with other materials performing similar functions. When the scope is to evaluate the metal production process or the treatment of waste streams for the recovery of metals, a cradle-to-gate or a gate-to-gate approach is also often selected. The functional unit is typically based on the individual metals produced or on a specific raw input material that is treated to extract the metals. For example, the LCA assessment for the global production of primary zinc (Van Genderen et al., 2016) used the cradle-to-gate approach, studying the environmental impacts for the production of 1 MT of high-grade zinc from mining zinc ore. Alternatively, the LCA assessment evaluating the environmental impacts of treating electrical and electronic waste streams for the recovery of the metals used the electrical and electronical scrap input stream as bases for their functional unit (Iannicelli-Zubiani et al., 2017). The choice of functional unit will hence depend on the focus of the assessment (production of a specific metal or treatment of waste streams for the recovery of most of the valuable metals). Another important issue when dealing with metals in LCAs is the allocation of the impacts to the individual metals. Santero and Hendry (2016) make a distinction for the type of allocation between base and precious metals. For base metals the preferred approach would be to mass allocate the environmental impacts based on the metal content. When there is a large discrepancy between the market values of the different metals that are produced in a system, as is the case with precious metals, mass allocation will fail to capture the main purpose of the process. Therefore, economic allocation will be more appropriate, reflecting the main purpose of the metallurgical process. The type of allocation will have a huge impact on the overall assessment of the recovered copper and gold (Fig. 14.6).

As discussed elsewhere, an environmental and human health impact assessment of inorganic substances faces many specific issues, such as bioavailability, natural background, essentiality, data richness, etc. (see also Chapters 3, 5, 7, and 8). Although these aspects are often well considered in RA approaches, this is not (yet) the case for the LCIA. A first attempt is made in the USEtox model (Rosenbaum et al., 2008), which is proposed to assess the ecotoxicity for aquatic freshwater and human toxicity effects in the PEF framework (Table 14.3). USEtox is a scientific consensus model endorsed by the UNEP/SETAC life-cycle initiative for characterising human and ecotoxicological impacts of chemicals in LCIA and is based on several RA principles regarding ecotoxicity, fate, and exposure. However, for the purpose of comparing many substances and products in life-cycle analysis (LCA), several changes and

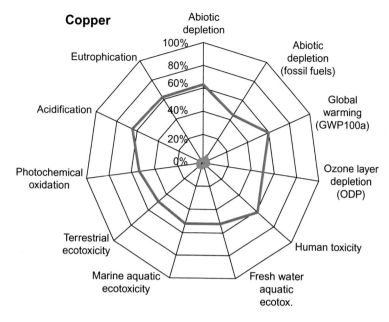

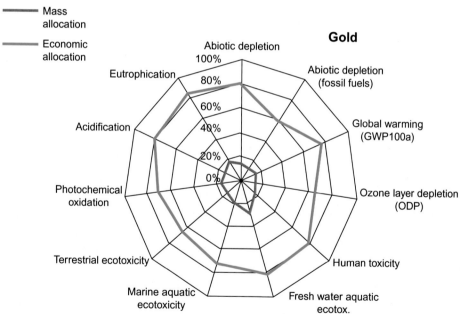

FIG. 14.6 Impact percentages due to the recovered copper and gold from *waste electrical and electronic equipment*, according to different allocation methods (mass and economic). *From Iannicelli-Zubiani EM, Giani MI, Recanati F, Dotelli G, Puricelli S, Cristiani C: Environmental impacts of a hydrometallurgical process for electronic waste treatment: a life cycle assessment case study, J Clean Prod 140:1204–1216, 2017.*

simplifications were made and USEtox has known limitations concerning its applicability to inorganic chemicals (e.g. Pizzol et al., 2011).

Despite metal-specific LCA developments published in the literature (e.g. Gandhi et al., 2010) and high-quality datasets generated under the REACH Regulation umbrella, USEtox version 1.0, and in general, hazard, LCA, or footprint-based environmental impact assessments, are insufficiently mature to be used for assessing the environmental impact of metals/inorganics in general because of the following key issues:

- (Eco)toxicity databases have often been insufficiently subject to quality control processes such as the use of nominal versus measured concentration in ecotoxicity testing, availability of test media composition impacting bioavailability, etc. Moreover, there is a lack of transparency in the selection of the ecotoxicity data.
- Metals are naturally occurring substances, some of which—such as copper—are essential to living organisms. Essentiality (and homeostatis) and natural background are not properly considered.
- Only a small fraction of the metals found in soils and in natural waters is bioavailable. Ignoring bioavailability intrinsically assumes a worst-case assumption of maximal bioavailability and ecotoxicity. In LCA philosophy, worst-case assumptions should not be made to avoid difficulties when comparing the environmental impact of substances/products with different levels of conservatism. Accounting for the bioavailability of metals, using techniques such as the BLM resolves many of these difficulties and is standard practice in a RA and environmental compliance context, but is only partly covered in the current available LCIA frameworks (e.g. USEtox) through complexation with organic matter because it does not include speciation with inorganic ligands as well as interactions with the biotic ligand. It has been demonstrated in the literature that the rank of metals, from best to worst environmental impact, changes significantly depending on the bioavailability conditions (Gandhi et al., 2010). A robust comparative analysis as in LCA or PEF should therefore consider variability in environmental impact due to the spatial (European) and temporal variability of the bioavailability conditions (Gandhi et al., 2011).
- The consideration of the relevant and metal-specific fate processes such as (irreversible) sulphide binding and precipitation processes to the sediment is lacking. An overview of all relevant processes can, for example, be found in the TICKET—Unit World Model (Farley et al., 2011, see also Chapter 9). Considering calculated solid water partition coefficients that account for these additional binding processes could temporarily circumvent this.

In summary, the current data quality underlying characterisation factors and the insufficient consideration of bioavailability result in an uncertain impact score hindering the detection of significant differences between products. The lack of consideration of natural background/essentiality and relevant inorganic longer term fate processes ('comparable' to 'degradation' processes for organics) result in a biased higher impact score for inorganics. It is recommended to carry out an uncertainty/sensitivity analysis to demonstrate the effect of natural background, bioavailability, and the quality of the (eco)toxicity data set on the robustness of the comparative assessment.

Santero and Hendry (2016) recommend only the following set of five impact categories for use in LCAs involving metals:

- Global warming potential
- Acidification potential
- Eutrophication potential
- Photochemical oxidant creation potential (e.g. smog potential)
- Ozone depletion potential

For these impact categories, the impact characterisation has been well-established and broadly accepted among LCA practitioners. As mentioned above, other categories, such as resource depletion, toxicity to humans and ecosystem, and land use change, rely on more controversial assumptions and methods and are thus currently less widely used and accepted in LCAs with metals. Therefore, reporting the impact from these methods is not recommended at present. Although these impacts are relevant environmental concerns, it is felt that the characterisation of these impacts from the inventory data does not yet adequately support decision-making. As the supporting science improves and the LCI data become more robust (e.g. higher spatial resolution), inclusion of these impact categories could be reconsidered in time (Santero and Hendry, 2016).

14.4.2 Water Footprint

Recently, the relevance of supply chain thinking for water resources management and freshwater allocation has been acknowledged, while the LCA community recognised the importance of water use and included the related impacts in LCA studies (Boulay et al., 2013). There are currently two main approaches enabling us to perform a comprehensive water footprint (WF) assessment of products, namely WF and LCA.

The WF concept was introduced by Hoekstra (2003), and subsequently elaborated as an indicator of human appropriation of freshwater resources that incorporates both direct and indirect water use of a consumer or producer. This method has a wide applicability and allows deriving the WF of an individual, a community, a business, or a nation. In the case of products, the WF is the total volume of freshwater used to produce the product, summed over the various steps of the production chain. The WF is a multidimensional indicator, showing water consumption volumes by source and polluted volumes by type of pollution while all components of a total WF are specified geographically and temporally (Hoekstra et al., 2009).

Simultaneously, the LCA community developed comprehensive methodologies to include environmental impacts related to water in LCA studies and started to frame the main concepts in the international standard on WF (ISO 14046, 2014). The LCA methodology aims at quantifying potential environmental impacts, generated by a human activity, on a wide range of environmental issues (climate change, human respiratory impacts, and land use, etc.). One of the potential causes of impact is water use. LCA therefore includes potential impacts from depriving human users and ecosystems of water resources, as well as specific potential impacts from the emitted contaminants affecting water, through different environmental impact pathways and indicators (mainly eutrophication, acidification, and toxicity to human and ecosystems). Quantitative impact indicators are at the core of the impact assessment phase (Boulay et al., 2013).

14.4.2.1 Assessment Methods and Tools

Next to stand-alone WF methods, several methods have been developed in an LCA context. An overview of the developed assessment methods is provided by Berger and Finkbeiner (2010) and Kounina et al. (2013). According to these authors, the LCI schemes developed by Vince (2007), Bayart et al. (2010), and Boulay et al. (2011) propose a detailed accounting of water use which considers volumetric, geographical, watercourse, and quality information to satisfy inventory requirements of modern impact assessment methods. Endpoint methods assess potential damages resulting from water use or consumption at the end of the cause–effect chain on human health, ecosystems, and resources.

In addition to WF methods as such, several databases have been identified which provide water use and consumption data for various products and materials. LCI databases containing elementary flows for freshwater withdrawal and turbined water are the GaBi (http://www.gabi-software.com/) and Ecoinvent (http://www.ecoinvent.org) databases. Distinct WF databases are, for example, the Quantis Water database (http://www.quantis-intl.com) or the WaterStat database (http://waterfootprint.org). Moreover, additional tools such as the WF Assessment Tool (http://waterfootprint.org), the Global Water Tool (http://old.wbcsd.org), the Local Water Tool (http://gemi.org/localwatertool/), or the Water Risk Filter (http://waterriskfilter.panda.org) could be useful as they facilitate the accounting of a company's water use and assess environmental, operational, legal, and reputational risks.

For additional information on water use assessments and WFs, taking the life-cycle perspective, it is recommended to visit the WULCA website (http://www.wulca-waterlca.org).

14.4.2.2 Application of WF Methods to Primary Metal Production Systems

The methods available to calculate the WF of a product or process have developed significantly over the past several years. Recent methods recognise that there are two main impacts associated with water use: consumption water use (CWU) and degradation water use (DWU), and these impacts can occur either directly at a production facility, or indirectly within a producer's supply chain (Northey et al., 2014). The authors provided examples showing the application of LCA-based water footprinting to mining, mineral processing, and metal production systems, with a focus on copper, gold, and nickel production.

CWU is defined as a reduction in the volume of water contained within a water store. To account for regional differences in water availability, CWU estimates are converted to a reference unit, H_2O-e (water equivalent), using the characterisation procedure shown in the equation here below. The unit H_2O-e represents the global average impact caused by the consumption of $1\,L$ of freshwater. This representation is similar to the approach taken with greenhouse gas emissions that are reported in terms of carbon dioxide equivalent (CO_2-e), where each individual substance that is emitted undergoes characterisation procedures to convert it to this common unit (Eq. 1).

$$CWU = \Sigma\left(CWU_i \times WSI_i / WSI_{global}\right) \quad \left(\text{in volume } H_2O\text{-e} / \text{time}\right) \tag{14.1}$$

where CWU_i is the change in volume water contained in a store or catchment, WSI_i is a site/region's water stress index (Pfister et al., 2009), and WSI_{global} is the global average water stress index of 0.602.

The general recommendation is to include both direct and indirect WFs, whereas direct WFs are the traditional focus of companies, the indirect WFs focus on the supply chain and are mainly related to material and energy inputs. The Water Stress Index (WSI) for an individual region is calculated based on the water withdrawals of different end users within the region, the available water in a region, and the seasonal variability of this availability through time (Pfister et al., 2009). The results of this are normalised between 0.01 and 1 using a logistic function to produce the final estimate of a region's WSI. A high WSI may indicate that water demand is exceeding sustainable supply for a region, whereas a low WSI indicates there is relatively little water demand in an area and/or that the sustainable supply capacity is quite large.

According to Northey et al. (2014), the CWU could be calculated from the following steps:

1. Compile data on the water inventory of the facility under study for direct CWU calculation (water withdrawal, water discharge). An example of the water flows for two simplified copper production processes is presented in Fig. 14.7;
2. Compile data on the inventory of the major material and energy inputs to the production process, that is, indirect CWU;
3. Estimate the catchment-/country-specific WSI as provided in Pfister et al. (2009);
4. Catchment-/country-specific WSI × consumptive water use (direct + indirect) normalised towards the global average WSI = CWU in (in volume H_2O-e/time).

DWU represents the impacts that occur as a result of changes in the quality of water that are attributed to the process or product. DWU, in terms of H_2O-e, is derived by comparing the impacts associated with a production system to the global average impact for 1 L of CWU. This is calculated using the ReCiPe 2008 impact assessment methodology that provides estimates of the potential impacts of a system on human health, ecosystems, and resources (Goedkoop et al., 2013).

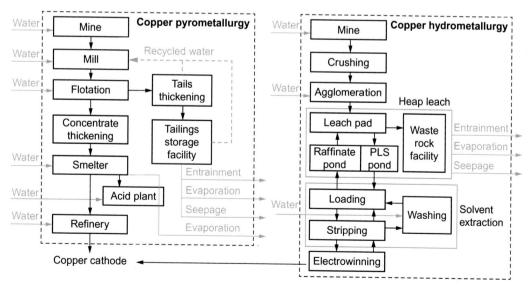

FIG. 14.7 Water flows for two simplified copper production processes. *From Northey S, Haque N, Lovel R, Cooksey MA: Evaluating the application of water footprint methods to primary metal production systems, Miner Eng 69:65–80, 2014.*

$$\text{DWU} = \text{ReCiPe points (emissions to water for product system)}/$$
$$\text{ReCiPe points (global average for 1 L of CWU)} \quad (\text{in volume } H_2O\text{-e})$$

where ReCiPe points (global average for 1 L of CWU) $= 1.86 \times 10^{-6}$ ReCiPe points (Ridoutt and Pfister, 2013).

DWU can also be calculated based upon separate global averages for impacts to human health (HH), ecosystems (ECO), and resources (RES), rather than the combination of all three. These are 7.05×10^{-8} points/m^3, 8.46×10^{-11} points/m^3, and 1.86×10^{-3} points/m^3, respectively (Ridoutt and Pfister, 2013).

To calculate the DWU for the primary metal production sites the following information and calculation steps are requested:

1. Compile data on emissions of chemicals/metals from facility to water (kg);
2. Compile a list of relevant ReCiPe characterisation factors (kg 1.4 dB eq/kg emission), usually freshwater, marine, terrestrial ecotoxicity, and human toxicity for metals;
3. Calculate the midpoint impacts (kg 1.4 dB eq);
4. Calculate the endpoint impacts (DALY);
5. Normalise and weigh the endpoint impacts (Points);
6. Sum of the normalised and weighed endpoint impacts (Points) × Ridoutt and Pfister factors (Points/m^3) = DWU in (in volume H_2O-e).

A single impact indicator for water use or a WF can be defined simply as the sum of CWU and DWU (Ridoutt and Pfister, 2013) (Eq. 2).

$$\text{Total water footprint} = \text{CWU} + \text{DWU} \quad (\text{in volume } H_2O\text{-e}/\text{time}) \tag{14.2}$$

The WF method as described by Northey et al. (2014) is very promising as it provides a broader description of the impacts associated with a production system and aims to aid decision-making by presenting information in a simple, directly comparable way that can account for both direct and indirect water use issues.

KEY MESSAGES

QICARs and AOP analysis are promising tools to predict toxic effects or mode of actions for less data-rich inorganic substances based on the physicochemical properties of the elements or a combination of the current knowledge from in vivo testing, biochemical and cellular assays, and computational predictive methods, respectively. Promising models are being developed for predicting the combined toxicity of inorganics. Such models will support decision-making for constituent-based RA of complex inorganic materials. Impact assessment methods, such as LCA, PEF, or WF are valuable tools to assess and compare impacts from substances or processes. However, these tools need further standardisation and implementation of specifics of inorganics to allow sound comparison of impacts from different types of (inorganic) materials. In conclusion, all these promising tools will complement the risk management of complex inorganic materials when further developed and accepted.

References

Bayart JB, Bulle C, Koehler A, et al: A framework for assessing off-stream freshwater use in LCA, *Int J Life Cycle Assess* 15(5):439–453, 2010.

Berger M, Finkbeiner M: Water footprinting—how to address water use in life cycle assessment? *Sustainability* 2(4):919–944, 2010.

Boulay AM, Bouchard C, Bulle C, Deschenes L, Margni M: Categorizing water for LCA inventory, *Int J Life Cycle Assess* 16(7):639–651, 2011.

Boulay AM, Hoekstra A, Vionnet S: Complementarities of water-focused life cycle assessment and water footprint assessment, *Environ Sci Technol* 47:11926–11927, 2013.

Brix KV, Schlekat CE, Garman ER: The mechanisms of nickel toxicity in aquatic environments: an adverse outcome pathway analysis, *Environ Toxicol Chem* 36:1128–1137, 2017.

Chen C, Mu Y, Wu F, Zhang R, Su H, Giesy JP: Derivation of marine water quality criteria for metals based on a novel QICAR-SSD model, *Environ Sci Pollut Res* 22:4297–4304, 2015.

Di Toro DM, Allen HE, Bergman HL, Meyer JS, Paquin PR, Santore RC: Biotic ligand model of the acute toxicity of metals. 1. Technical basis, *Environ Toxicol Chem* 20:2383–2396, 2001.

Dreicer M, Tort V, Manen P: ExternE, externalities of energy, vol. 5, nuclear, Centr d'étude sur l'Evaluation de la Protection dans le domaine nucléaire (CEPN), edited by the European Commission DGXII, Science, Research and development JOULE, Luxembourg, 1995.

EC JRC (European Commission-Joint Research Centre), Institute for Environment and Sustainability: *International reference life cycle data system (ILCD) handbook—recommendations for life cycle impact assessment in the European context*, ed 1, Luxemburg, 2011, Publications Office of the European Union. November 2011, EUR 24571 EN.

European Commission-Joint Research Centre, Institute for Environment and Sustainability: *International reference life cycle data system (ILCD) handbook—general guide for life cycle assessment—detailed guidance*, ed 1, Luxembourg, 2010, Publications Office of the European Union. March 2010, EUR 24708 EN.

Farley KJ, Carbonaro RF, Fanelli CJ, Costanzo R, Rader KJ, Di Toro DM: TICKET-UWM: a coupled kinetic, equilibrium, and transport screening model for metals in lakes, *Environ Toxicol Chem* 30:1278–1287, 2011.

Farley KJ, Meyer JS: Metal mixture modeling evaluation project: 3. Lessons learned and steps forward, *Environ Toxicol Chem* 34:821–832, 2015.

Farley KJ, Meyer JS, Balistrieri LS, et al: Metal mixture modeling evaluation project: 2. Comparison of four modeling approaches, *Environ Toxicol Chem* 34:741–753, 2015.

Frischknecht R, Braunschweig A, Hofstetter P, Suter P: Human health damages due to ionising radiation in life cycle impact assessment, *Environ Impact Assess Rev* 20:159–189, 2000.

Frischknecht R, Steiner R, Jungbluth N: *Methode der ökologischen Knappheit–Ökofaktoren 2006. Methode für die Wirkungsabschätzung in Ökobilanzen*, 2008, Umwelt-Wissen, 188.

Gandhi N, Diamond ML, Huijbregts MAJ, Guinée JB, Peijnenburg WJGM, van de Meent D: Implications of considering bio-availability in estimates of freshwater ecotoxicity: examination of two case studies, *Int J Life Cycle Assess* 16:774–787, 2011.

Gandhi N, Diamond ML, van de Meent D, Huijbregts MAJ, Peijnenburg WJGM, Guinée J: New method for calculating comparative toxicity potential of cationic metals in freshwater: application to copper, nickel and zinc, *Environ Sci Technol* 44:5195–5201, 2010.

Goedkoop M, Effting S, Collignon M: The Eco-indicator 99-A damage oriented method for life cycle impact assessment. In *Manual for designers*, 2 ed., Amersfoort, 2000, PRé Consultants B.V. 17-4-2000.

Goedkoop M, Heijungs R, Huijbregts M, De Schryver A, Struijs J, van Zelm R: *ReCiPe 2008: a life cycle impact assessment method which comprises harmonised category indicators at the midpoint and the endpoint level*, ed 1 (version 1.08), The Netherlands, 2013, Ruimte en Milieu, Ministerie van Volkshuisvesting, Ruimtelijke Ordening en Milieubeheer.

Guinée JB, editor: *Life cycle assessment: an operational guide to the ISO Standards; LCA in perspective; guide; operational annex to guide*, The Netherlands, 2001, Centre for Environmental Science, Leiden University.

Guinée, J.-B (editor), Gorrée M, Heijungs R, Huppes G, et al., Handbook on life cycle assessment: operational guide to the ISO standards. Series: Eco-efficiency in industry and science. Kluwer Academic Publishers. Dordrecht (Hardbound, ISBN 1-4020-0228-9; Paperback, ISBN 1-4020-0557-1), 2002.

Harder R, Holmquist H, Molander S, Svanström M, Peters GM: Review of environmental assessment case studies blending elements of risk assessment and life cycle assessment, *Environ Sci Technol* 49:13083–13093, 2015.

Hoekstra AY, editor: *Virtual water trade, Proceedings of the International Expert Meeting on Virtual Water Trade, Delft, the Netherlands, 12–13 December 2002, Value of Water Research Report Series,* vol. 12, Delft, 2003, UNESCO-IHE. Available from: www.waterfootprint.org/Reports/Report12.pdf.

Hoekstra AY, Chapagain AK, Aldaya MM, Mekonnen MM: *Water footprint manual: state of the art 2009,* Enschede, 2009, Water Footprint Network.

Humbert S: *Geographically differentiated life-cycle impact assessment of human health,* Doctoral dissertation, Berkeley, CA, 2009, University of California.

Iannicelli-Zubiani EM, Giani MI, Recanati F, Dotelli G, Puricelli S, Cristiani C: Environmental impacts of a hydro-metallurgical process for electronic waste treatment: a life cycle assessment case study, *J Clean Prod* 140:1204–1216, 2017.

ISO 14040: Environmental management—life cycle assessment—principles and framework. ISO, 2006.

ISO 14044: *Environmental management—life cycle assessment—requirements and guidelines,* 2006, ISO.

ISO 14046: *Environmental management—water footprint—principles, requirements and guidelines,* 2014, ISO.

Kaiser KLE: Correlation and prediction of metal toxicity to aquatic biota, *Can J Fish Aquat Sci* 37:211–218, 1980.

Kinraide TB: Improved scales for metal ion softness and toxicity, *Environ Toxicol Chem* 28:525–533, 2009.

Kounina A, Margni M, Bayart JB, et al: Review of methods addressing freshwater use in life cycle inventory and impact assessment, *Int J Life Cycle Assess* 18:707, 2013.

Le Faucheur S, Campbell PG, Fortin C: *Quantitative ion character activity relationships (QICARs),* Research report No R-1262, Final project report presented to Environment Canada, Science and Technology Branch, 2011, Ecological Assessment Division, Inorganics Unit.

Mendes LF, Zambotti-Villela L, Yokoya NS, Bastos EL, Stevani CV, Colepicolo P: Prediction of mono- bi-, and trivalent metal cation toxicity to the seaceed *Gracilaria domingensis* (Gracilariales, Rhodophyta) in synthetic seawater, *Environ Toxicol Chem* 32:2571–2575, 2013.

Meyer JS, Farley KJ, Garman ER: Metal mixture modeling evaluation: 1. Background, *Environ Toxicol Chem* 34:726–740, 2015.

Milà i, Canals L, Romanyà J, Cowell SJ: Method for assessing impacts on life support functions (LSF) related to the use of 'fertile land' in Life Cycle Assessment (LCA), *J Clean Prod* 15:1426–1440, 2007.

Mu Y, Wu F, Chen C, et al: Predicting criteria continuous concentrations of 34 metals or metalloids by use of quantitative ion character-activity relationships-species sensitivity distributions (QICAR-SSD) model, *Environ Pollut* 188:50–55, 2014.

Newman MC, McCloskey JT: Predicting relative toxicity and interactions of divalent metal ions: Microtox bioluminescence assay, *Environ Toxicol Chem* 15:275–281, 1996.

Newman MC, McCloskey JT, Tatara PT: Using metal-ligand binding characteristics to predict metal toxicity: quantitative ion character-activity relationships (QICARs), *Environ Heath Perspectives* 106:1419–1425, 1998.

Northey S, Haque N, Lovel R, Cooksey MA: Evaluating the application of water footprint methods to primary metal production systems, *Miner Eng* 69:65–80, 2014.

Norwood WP, Borgmann U, Dixon DG, Wallace A: Effects of metal mixtures on aquatic biota: a review of observations and methods, *Hum Ecol Risk Assess* 9:795–811, 2003.

OECD: *Guidance on the incorporation of bioavailability concepts for assessing the chemical ecological risk and/or environmental threshold values of metals and inorganic metal compounds,* Series on testing and assessment, No. 259, ENV/JM/MONO(2016)66, 2016.

OECD: *Revised guidance document on developing and assessing adverse outcome pathways,* Series on testing and assessment, No. 184, ENV/JM/MONO(2013)6, 2017.

van Oers L, de Koning A, Guinée JB, Huppes G: *Abiotic resource depletion in LCA,* The Netherlands, 2002, Road and Hydraulic Engineering Institute.

Ownby DR, Newman MC: Advances in quantitative ion character-activity relationships (QICARs): using metal-ligand binding characteristics to predict metal toxicity, *QSAR Comb Sci* 22:241–246, 2003.

Pearson RG: Hard and soft acids and bases, *J Am Chem Soc* 85:3533–3539, 1963.

Pfister S, Koehler A, Hellweg S: Assessing the environmental impacts of freshwater consumption in LCA, *Environ Sci Technol* 43(11):4098–4104, 2009.

Pizzol M, Christensen P, Schmidt J, Thomsen M: Eco-toxicological impact of 'metals' on the aquatic and terrestrial ecosystem: a comparison between eight different methodologies for Life Cycle Impact Assessment (LCIA), *J Clean Prod* 19:687–698, 2011.

Posch M, Seppälä J, Hettelingh JP, Johansson M, Margni M, Jolliet O: The role of atmospheric dispersion models and ecosystem sensitivity in the determination of characterisation factors for acidifying and eutrophying emissions in LCIA, *Int J Life Cycle Assess* 13:477–486, 2008.

Ridout BG, Pfister S: A new water footprint calculation method integrating consumptive and degradative water use into a single stand-alone weighted indicator, *Int J Life Cycle Assess* 18(1):204–207, 2013.

Rosenbaum RK, Bachmann TM, Gold LS, et al: USEtox—the UNEP-SETAC toxicity model: recommended characterisation factors for human toxicity and freshwater ecotoxicity in life cycle impact assessment, *Int J Life Cycle Assess* 13:532–546, 2008.

Santero N, Hendry J: Harmonization of LCA methodologies for the metal and mining industry, *Int J Life Cycle Assess* 21:1543–1553, 2016.

Seppälä J, Posch M, Johansson M, Hettelingh JP: Country-dependent characterisation factors for acidification and terrestrial eutrophication based on accumulated exceedance as an impact category indicator, *Int J Life Cycle Assess* 11:403–416, 2006.

Stockdale A, Tipping E, Lofts S, Ormerod SJ, Clements WH, Blust R: Toxicity of proton-metal mixtures in the field: linking stream macroinvertebrate species diversity to chemical speciation and bioavailability, *Aquat Toxicol* 100:112–119, 2010.

Struijs J, Beusen A, van Jaarsveld A, et al: Aquatic eutrophication. In *ReCiPe 2008—a life cycle impact assessment method which comprises harmonised category indicators at the midpoint and the endpoint level. Report I: characterisation*, 1 ed. (version 1.08), The Netherlands, 2013, Ruimte en Milieu, Ministerie van Volkshuisvesting, Ruimtelijke Ordening en Milieubeheer. (chapter 6).

Tatara CP, Newman MC, McCloskey JT, Williams PL: Use of ion characteristics to predict relative toxicity of mono-, di- and trivalent metal ions: *Caenorhabditis elegans* LC50, *Aquat Toxicol* 42:255–269, 1998.

Tipping E, Lofts S: Testing WHAM-FTOX with laboratory toxicity data for mixtures of metals (Cu, Zn, Cd, Ag, Pb), *Environ Toxicol Chem* 34:788–798, 2015.

Tollefsen EK, Scholz S, Cronin MT, et al: Applying adverse outcome pathways (AOPs) to support integrated approaches to testing and assessment (IATA), *Regul Toxicol Pharmacol* 70:629–640, 2014.

Van Genderen E, Adams W, Dwyer R, Garman E, Gorsuch J: Modeling and interpreting biological effects of mixtures in the environment: introduction to the metal mixture modeling evaluation project, *Environ Toxicol Chem* 34:721–725, 2015.

Van Genderen E, Wildnauer M, Santero N, Sidi N: A global life cycle assessment for primary zinc production, *Int J Life Cycle Assess* 1–14, 2016.

Vijver MG, Elliott EG, Peijnenburg WJGM, De Snoo GR: Response predictions for organisms water-exposed to metal mixtures: a meta-analysis, *Environ Toxicol Chem* 30:1482–1487, 2011.

Villeneuve DL, Crump D, Hecker M, et al: Adverse outcome pathway (AOP) development I: strategies and principles, *Toxicol Sci* 142:312–320, 2014.

Vince F: *UNEP/SETAC Life Cycle Initiative working group: assessment of water use and consumption within LCA*, Paris, 2007, Veolia Environnement.

Walker JP, Enache M, Dearden JC: Quantitative cationic-activity relationships for predicting toxicity of metals, *Environ Toxicol Chem* 22:1916–1935, 2003.

Wang Y, Wu F, Mu Y, et al: Directly predicting water quality criteria from physcochemical properties of transition metals, *Sci Rep* 6(22515), 2016.

Williams MW, Turner JE: Comments on softness parameters and metal ion toxicity, *J Inorg Nucl Chem* 43:1689–1691, 1981.

Wolterbeek HT, Verburg TG: Predicting metal toxicity revisited: general properties vs specific effects, *Sci Technol Environ* 279:87–115, 2001.

Wu F, Mu Y, Chang H, Zhao X, Giesy JP, KB W: Predicting water quality criteria for protecting aquatic life from physico-chemical properties of metals or metalloids, *Environ Sci Technol* 47:446–453, 2013.

van Zelm R, Huijbregts MAJ, den Hollander HA, et al: European characterization factors for human health damage of PM10 and ozone in life cycle impact assessment, *Atmos Environ* 42:441–453, 2008.

Zhou KM, Li LZ, Peijnenburg WJGM, et al: A QICAR approach for quantifying binding constants for metal-ligand complexes, *Ecotoxicol Environ Saf* 74:1036–1042, 2011.

Index

Note: Page numbers followed by *f* indicate figures, *b* indicate boxes and *t* indicate tables.

Printed in the United States
By Bookmasters